Digital Photogrammetry

Volume I

Digital Photogrammetry

Volume I

Background, Fundamentals,
Automatic Orientation Procedures

Toni Schenk

The Ohio State University

TerraScience

Cover photograph by Woolpert Consultants, Dayton, Ohio
Cover design by Beáta Csathó

Printed in the United States of America

ISBN 0-9677653-0-7 perfect bound
ISBN 0-9677653-1-5 case bound

Published by

TerraScience
16218 Bailor Rd., Laurelville, OH 43135
Tel. (740) 332-9043, Fax (614) 292-2957
e-mail TSchenk@TerraScience.net
http://www.TerraScience.net

To

Bea & Mia

Contents

Preface

Photogrammetry and cats share a common, most important trait: both have several lives. The end of photogrammetry has been predicted several times. For example, with the emergence of GPS in the eighties, some saw a rapid end of photogrammetry and surveying. Others are convinced that photogrammetry will submerge with computer vision or computer graphics. Pundits of platform orientation systems predict a partial end—aerial triangulation is a thing of the past. Despite these bleak prospects, I, for myself, think that photogrammetry is very much alive—digital photogrammetry excites researchers and practitioners alike. It offers a great potential to solve existing tasks more effectively. Equally important, new problems can be tackled.

Digital photogrammetry is a relatively young and fast-growing field. Many essential concepts and methods have been adopted from image processing and computer vision. Despite the strong influx of methods and algorithms originally developed in other fields, digital photogrammetry is a discipline in its own right. Consequently, I present the material in this book from a photogrammetry viewpoint. Although the book is written by a photogrammetrist for photogrammetrists, I hope that it may contain useful information for readers with a background in other fields. Over the past few years, numerous papers have been written about various aspects of digital photogrammetry, but a coherent treatment of the subject is missing. This book fills that gap.

The book is intended to be useful as a textbook or as a reference book. I have adopted a conceptual approach with emphasis on principles. At the same time I kept the presentation as elementary as possible by sacrificing mathematical rigor, when necessary, for an intuitive description. Implementation details and specific algorithmic aspects are not deeply covered—the book contributes only marginally at the tool level. I strongly believe that one should first understand and solve the basic problem before becoming too involved in implementation issues.

The first volume of this book developed from lecture notes for a graduate course in digital photogrammetry that I have taught at OSU for 12 years. To enhance the use of the book in the classroom, I have included a problem section for every chapter, except for the first part. The problems are mostly from homework assignments and quizzes, ranging from conceptual to computational questions.

The material covered in the first volume is now taught in two consecutive courses. The first volume is organized into three parts. Digital photogrammetry is very much interdisciplinary. Strong links exist to image processing, computer vision, pattern recognition, and computer graphics. Rather than treating these subjects in appendices, I have opted to start this book with a concise summary of the fields that distinctly influence digital photogrammetry. The "Background" part is organized into five chapters, beginning with signal and image processing, followed by computer vision and human vision. A better understanding of human vision may provide insights into the solution of so intricate a problem as determining surfaces, perceiving shapes and recognizing objects. The efforts of deriving useful information about objects from images benefit greatly from an understanding of how images are formed in the first place. Chapter 6 provides background information on radiometry and photometry to the extent necessary to understand the basic image formation process. Each chapter in the first part ends with a short section on suggested readings.

The second part covers fundamental aspects of digital photogrammetry with a natural progression from data acquisition (digital cameras and scanners) to digital photogrammtric workstations and on to image matching that is covered in two chapters. Digital images are either obtained directly by digital cameras (Chapter 7) or by scanning analog photographs (Chapter 8). Probably the single most significant product of digital photogrammetry is the digital photogrammetric workstation. Chapter 9 provides an account of these workstations, emphasizing their functionality and working principles rather than detailed system descriptions.

This book does not cover interactive digital photogrammetry (softcopy photogrammetry) well—the focus is on automating photogrammetric processes, treated in the third part of the book. Three chapters are devoted to the problem of performing the basic orientation procedures automatically. After briefly describing existing approaches, I take a fresh look at the problem of automating interior, exterior, and relative orientation and present solutions that have not yet been published elsewhere. The emphasis is on including linear and areal features into the orientation process in the quest to move from a point-based photogrammetry to a truly feature-based photogrammetry that is much better suited for automation. I hope that the third part will also be of interest to the "seasoned" digital photogrammetrist.

This book is an entirely digital production. I have generated nearly all the figures in PostScript. I use a rescaled version of the figures as overheads for teaching. More recently, I have converted some figures for online presentations with a considerable degree of animation to better explain key concepts. The book is directly printed from the PostScript file of the manuscript.

The second volume of this book will contain parts that are devoted to automatic aerial triangulation, reconstruction of surfaces from different sensors, including lidar, generation of orthophotos, and object recognition.

I greatly appreciate the patience of my students, who suffered through using the manuscript from embryonic stages to the present form. There is no better feedback about the quality of lecture notes and presentations as students giving you a strange look or realizing their consternation about questions whose subjects were not properly explained. I am grateful for the very valuable feedback I obtained from my students and colleagues when discussing aspects of digital photogrammetry.

Chapter 1

Introduction

This chapter provides a brief introduction to digital photogrammetry and a view of things to come. I will begin with some brief historical remarks followed by comments on terminology. After sketching a generic digital photogrammetry environment, I will introduce the concept of a digital image and contrast it to a photograph. Finally, digital photogrammetry is compared with related disciplines.

1.1 Background

Digital photogrammetry is rapidly emerging as a new subfield of photogrammetry. While digital photogrammetry has its roots in the 1950s, major research activities only began in the 1980s, sparked by significant advances in electronics and computers, such as digital cameras, parallel processing, and increased storage capacity. Considerable progress has been made during the last few years. Several digital photogrammetry products are now commercially available, such as softcopy workstations, automatic DEM software, and digital orthophotos—just to mention a few.

Fig. 1.1 on the following page depicts the four generations of photogrammetry and relates digital photogrammetry to the other subfields of photogrammetry.

first generation had its beginning with the invention of photography in 1839. It was very much a pioneering phase with remarkable achievements in terrestrial and balloon photogrammetry.

analog photogrammetry is characterized by the invention of stereophotogrammetry. Between the two world wars, the main foundations of

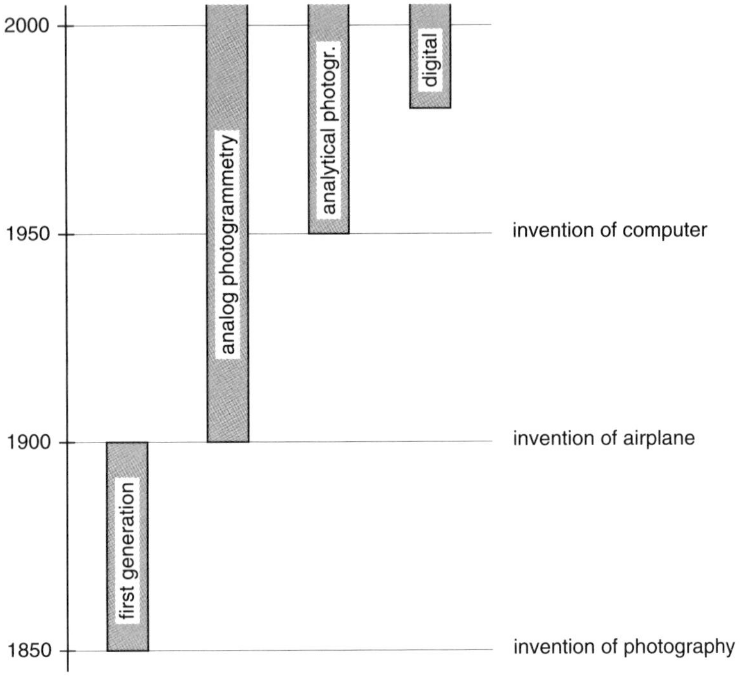

Figure 1.1: Photogrammetric generations.

aerial surveying techniques were built and they still stand today. Photogrammetry established itself as an efficient surveying and mapping method.

analytical photogrammetry had its beginning with the invention of the computer. The developments in computer technology very much determine the progress in photogrammetry. Table 1.1 summarizes the different computer generations and their impact on photogrammetry.

digital photogrammetry deals with digital imagery rather than (analog) photographs. Only with the fifth computer generation (see Table 1.1) has the necessary hardware and software become available to tackle the problems posed by storing, retrieving and manipulating large size images.

Table 1.1: Computer generations and photogrammetric disciplines.

<table>
<tr><th>Generation</th><th>Hardware</th><th>Software</th><th>Photogrammetric Disciplines</th></tr>
<tr><td>1st</td><td>- vacuum tubes</td><td>- machine code</td><td rowspan="3">analytical photogrammetry
- aerial triangulation
- correlation
- analytical plotter</td></tr>
<tr><td>2nd</td><td>- transistors
- magnetic core memory</td><td>- higher level languages
- (FORTRAN, COBOL)</td></tr>
<tr><td>3rd</td><td>- IC memory
- minicomputers
- mag. disk storage</td><td>- time sharing
- operating systems
- virtual memory</td></tr>
<tr><td>4th</td><td>- microprocessors, PC
- VLSI
- networking</td><td>- new languages
- (PASCAL, MODULA)
- IGS, DBMS</td><td>computer-assisted photogrammetry</td></tr>
<tr><td>5th</td><td>- parallel processing
- RISC architecture
- VHSIC
- optical disk storage</td><td>- knowledge based SW
- expert systems
- natural language processing</td><td>digital photogrammetry
- real-time photogrammetry</td></tr>
</table>

1.2 Terminology—Or the Lack Thereof

Whenever a new technology, such as digital photogrammetry evolves, there is no unified terminology, let alone accepted definitions. Even the term *digital photogrammetry* is disputed. Instead, the term *softcopy photogrammetry* is used, especially in the United States. Soft copy refers to the display of a digital image. Presumably those who coined this term wanted to contrast it with traditional photogrammetry that uses hard copies, such as films and maps. This is questionable since its counterpart, "hardcopy" photogrammetry, does not exist. This is why the term digital photogrammetry is preferred, because it expresses the distinction from the accepted terms of analog and analytical photogrammetry much better. Moreover, digital photogrammetry captures the notion of using digital imagery instead of photographs more distinctly. It is also a more comprehensive term—some processes in digital photogrammetry do not require a soft copy, but they always require a digital image.

As we shall see throughout this book, digital photogrammetry is closely related to the field of computer vision. Both disciplines use their own terminology and, inevitably, the same thing may be called quite differently. In this text, preference is given to photogrammetric terms and concepts.

The single most significant product of digital photogrammetry is the *softcopy workstation (WS)*, or *digital photogrammetric workstation (DPW*. One may argue over the names, but we use them interchangeably here. In this book, a *digital image* refers to a picture stored in a computer, for example on a softcopy workstation. The terms *photograph* or *diapositive* are used for a picture in analog form.

A final comment on the use of the term *automatic*. We often read about an automatic system or process. Take "automatic" generation of DEMs as an example. This means in practical terms that the software generates a DEM and human interaction is only occasionally necessary. The better the program, the less human intervention is required. In this context "automatic" obviously does not mean that a process is entirely automated. We read statements "... found 80 % of all points automatically". Such processes are usually implemented in an *interactive* environment. If a process is really "100 %" automatic and does not require human intervention—sort of a black box process—I prefer to call it *autonomous*.

1.3 Generic Digital Photogrammetry Environment

Fig. 1.2 on the next page depicts a schematic diagram of a digital photogrammetry environment. On the input side we have a digital camera or a scanner with which existing aerial photographs are digitized. At the heart of the processing side is a softcopy workstation (WS). The output side may comprise a film recorder to produce hard copies in raster format and a plotter for providing hard copies in vector format.

As indicated in Fig. 1.2, digital images are obtained directly by using digital cameras, or indirectly by scanning existing photographs. Most experts agree that it will take several years before digital data acquisition systems will eventually replace aerial cameras. Scanners, then, are an important component. They determine, together with the algorithms used on the softcopy WS, the accuracy of digital photogrammetry products.

The softcopy WS looks much like an ordinary graphics workstation with some additional features, such as stereo display, 3-D cursor, and, most likely, increased storage capacity to hold all the digital images of an entire project.

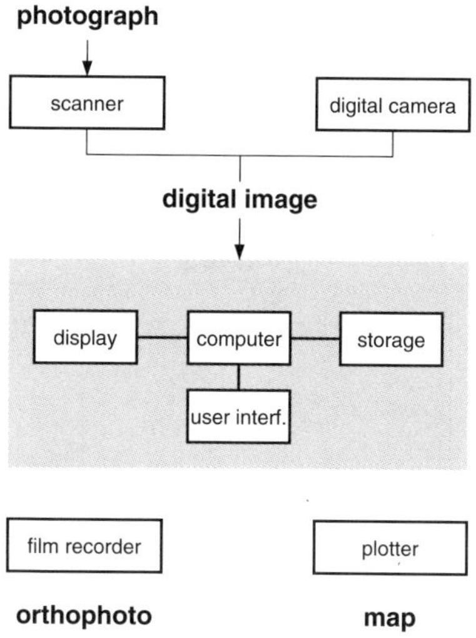

Figure 1.2: Schematic diagram of generic digital photogrammetry environment. The shaded area comprises the components of a classical digital photogrammetric workstation (DPW).

1.4 Properties of Digital Imagery

As the name indicates, digital photogrammetry uses digital imagery rather than (analog) photographs. This is the chief distinction from traditional photogrammetry, and almost all other differences are a consequence of this. Therefore, it is important to understand the properties of digital imagery.

1.4.1 Definition of Digital Image

Fig. 1.3 on the following page is a schematic diagram of a digital image as obtained, for example, by scanning an aerial photograph. The photograph can be described by a continuous function $F(x, y)$, called an *image*, where the coordinates x, y are spatial variables and the function value (amplitude) is the density. In reality, this continuous function is discretized, both in the spatial variables and the amplitude. The resulting discrete function $f(x, y)$ is called a *digital image*. The process of discretizing the spatial variables $\Delta x, \Delta y$ is known as *sampling*. Discretizing the amplitude g is referred to as *quantizing*. The discrete element $\Delta x, \Delta y$ is the *pixel* (derived from picture element) and Δg is a *gray level*, informally also called *brightness*. Then, a digital image can be expressed by $f(\Delta x \cdot i, \Delta y \cdot j, i = 0, \ldots, N-1;\ j = 0, \ldots, M-1)$ with i, j the pixel addresses, N the number of rows and M the number of columns. The image function is usually written as $f(x, y)$. Bear in mind that the spatial variables are discrete values, usually integer values.

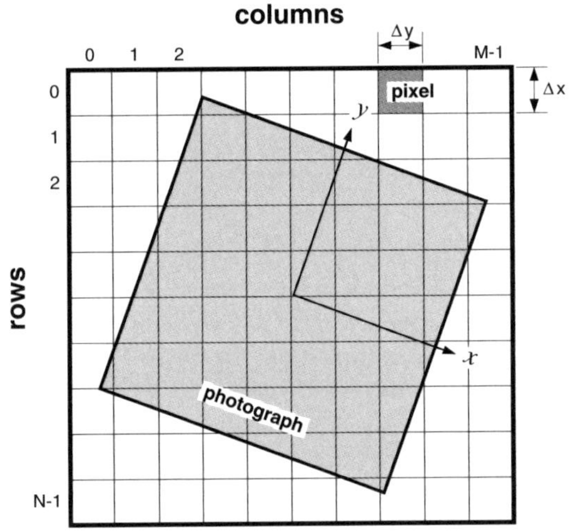

Figure 1.3: Image coordinate system of digital image (rows,columns). The shaded, rotated square indicates a digitized photograph with x, y the photo-coordinate axes.

1.4.2 *Spatial Resolution and Geometric Accuracy*

Obviously, the smaller the discrete variables $\Delta x, \Delta y$ ("pixel size") are chosen, the better they approximate the continuous function of the original diapositive. What values should be used to faithfully represent the diapositive obtained with a modern aerial camera? The theoretical limitations are given by the sampling theorem (see Chapter 2). It states that the smallest pixel size should be less than half of the highest frequency of the continuous function. If this condition is satisfied, then the continuous function can be recovered from the discrete function, assuming that the quantization effects can be neglected. That is, no information is lost during the discretization process.

An example may illustrate the case. Assume the resolution of a photograph is 70 lp/mm (line pairs per millimeter). In order to faithfully represent the photograph by the digital image, a pixel size of about 7 μm is required. As can be seen from Table 1.2, this would require almost 1 GB of storage for a typical $9'' \times 9''$ aerial film. Therefore, it may not be desirable to digitize images to their theoretical sampling rates.

Now the concerned photogrammetrist will inevitably ask the accuracy question. Do we sacrifice accuracy for the benefit of less storage and faster processing times? Not necessarily so. It is important to differentiate between *resolution* (pixel size) and *accuracy*. This can readily be appreciated when considering an algorithm that determines the location of a feature by computing its centroid, for example. Suppose 100 pixels cover the feature. Then, the accuracy of the location is $1/10^{th}$ of a pixel. *Subpixel accuracies* of $1/10^{th}$ or even better can be obtained for automatically determining conjugate points, for example. As a consequence, pixel sizes of 30 to 60 μm can render accuracies that are comparable to those obtained in analytical photogrammetry.

Table 1.2: Pixel size, number of pixels (resolution) and storage requirements for a digital image of an aerial photograph.

Pixel Size [micron]	*Number of Pixels*	*Storage Requirement (uncompressed)* [MB]
960	240×240	0.058
480	480×480	0.230
240	960×960	0.922
120	1920×1920	3.686
60	3840×3840	14.746
30	7680×7680	58.982
15	15360×15360	235.931
7.5	30720×30720	943.721

A closer examination of a digital image with a resolution of 30 to 60 μm reveals that small details are lost, however. If such details are important for interpretation purposes, then the image must have a smaller pixel size. By and large, the pixel size is driven by the interpretability of small features and to a lesser degree by accuracy considerations.

Table 1.2 summarizes the storage requirements for digital images of various pixel sizes with the number of pixels required to represent an aerial photograph. The examples are based on the assumption that the images are quantized to 256 different gray levels. That is, one pixel requires one byte.

The resolution as listed in Table 1.2 may change slightly, for example, when using the popular format of powers of two. Instead of a resolution of, say, 960×960, a resolution of 1024×1024[1] would be chosen. Note that the storage requirements can be substantially decreased by image compression techniques. The gray levels of images of natural scenes are not totally random. In fact, there is a considerable dependency of adjacent pixels. Imagine a road, for example, and a pixel size on the ground of one square foot. In that case, it is quite likely that neighboring pixels show a similar shade of gray. Image compression techniques exploit this dependency. For example, the JPEG[2] compression algorithm can reduce the storage requirement by one third without noticeable loss in image quality.

1.4.3 Radiometric Resolution

Quantizing an aerial photograph raises the question of how many gray levels should be used to represent the brightness. The solution is predominantly guided by practical considerations. A variable occupying one byte can assume 256 different values. This is more than sufficient to represent the gray shades of black and white photographs. Usually, the human operator cannot discriminate more than 50 gray shades. However, it does not pay to use only six bits to represent the gray levels. For all practical purposes, one byte is more than adequate to store the gray value of one pixel.

1.5 Classification of Processes and Tasks in Digital Photogrammetry

We may divide the different processes and tasks in digital photogrammetry into the four categories summarized in Table 1.3. The classification follows the computer vision paradigm (see Chapter 5). Computer vision is concerned with generating descriptions about scenes from images. The descriptions must be explicit and meaningful, depending on the application. There is a huge representational gap between the original image (iconic representation) and the symbolic scene description. This gap cannot be bridged in one giant step, rather the problem is approached in several steps. The major steps are usually associated with low-, middle- and high-level vision.

---

[1] which is equivalent to $2^{10} \times 2^{10}$, popularly called $1K \times 1K$.

[2] JPEG stands for Joint Photographic Experts Group.

1.5.1 System Level Tasks

Manipulating digital images is a fundamental task in digital photogrammetry. The size of digital images typically encountered in photogrammetry poses a real challenge to efficiently solving image manipulation tasks, such as storing, retrieving and displaying images. Suppose an aerial photograph is digitized to a resolution of 10 μm pixel size. This requires 450 MB (uncompressed) storage capacity. Photogrammetric projects involve dozens if not hundreds of photographs. It is desirable to have on-line access to all images of small or medium size projects. Another challenging system task is to display images, to zoom in and out, and to move within them without time delay. Yet another intriguing problem is stereo display—essential for photogrammetric applications.

Table 1.3: Classification of photogrammetric processes and tasks.

Category	*Processes, Algorithms*	*Tasks*
system level	store, access, display images	manipulate digital imagery
low level	process, match images extract features	image processing; orientations, digital orthophotos, DEM, AT
middle level	group, segment images	surface reconstruction feature reconstruction
high level	understand images	object recognition image interpretation

1.5.2 Low Level Tasks

The system level provides the necessary functionality to softcopy workstations. The basic functionality necessary for photogrammetric operations is provided by the next level, termed low level here, in analogy to low or early vision. Image processing tasks allow for enhancing the appearance of images, for example, histogram modifications such as contrast stretching. Typical low-level tasks are interior, relative and absolute orientation, usually followed by a geometric transformation to force epipolar lines to be parallel to

More advanced tasks on this level include the automatic generation of DEMs, computer supported aerial triangulation, and the production of digital orthophotos. Even though these tasks may be quite complex in terms of their implementation, their underlying algorithms, such as matching (finding conjugate points) and extracting features, are low vision problems.

1.5.3 Middle Level Tasks

As indicated in Table 1.3, typical middle-level tasks involve surface reconstruction and feature reconstruction. By surface reconstruction I mean a surface that is primarily reconstructed for the purpose of guiding subsequent vision processes, such as object recognition. Unlike the task of generating DEMs, the goal of surface reconstruction is not to represent the surface as accurately and densely as possible, but as explicitly as possible. Rather than dealing with a cloud of unrelated 3-D points, it is preferred to segment the surface into continuous patches, which are related to surfaces of objects.

Grouping and segmenting processes belong to the subject of perceptual organization—the ability of humans to detect relevant groupings and structures in images which are the input for object recognition and image understanding.

1.5.4 High Level Tasks

For a person to respond properly to the environment he must analyze, interpret, and understand visual stimuli. Ideally, the same feat must be accomplished by a robot or an autonomous vehicle. To pick a part from a bin or to navigate through a cluttered environment, the robot must understand its environment from sensor data and stored knowledge. The net result of image understanding is a fully interpreted scene. Image understanding or image interpretation is application dependent.

Object recognition is an important subtopic of image understanding. While image understanding is essential for many machine vision applications, its role in digital photogrammetry is less important. Here, the emphasis is on recognizing and locating objects. Subsequent tasks, such as analyzing objects and their interrelationships, are typically performed by a geographic information system (GIS).

1.6 Relationship of Digital Photogrammetry To Other Disciplines

Having digital images stored suggests processing them further by the computer. For example, it may be desirable to change the brightness of an image, or to enhance its contrast. In traditional photogrammetry, such processes are performed in the photo lab. It is certainly a major advantage of digital photogrammetry to have these processes available at the fingertips of an operator. *Digital image processing* is concerned with improving pictorial information for human interpretation, and processing of images for autonomous machine perception. As indicated in Fig. 1.4, image processing takes an image as input, modifies it, and produces a new image.

In addition to digital image processing, we may also want to use the computer to perform other tasks, such as finding features. It is quite conceivable to automatically locate the fiducial marks, for example, which would lend itself to an autonomous interior orientation. A more challenging problem is to find control points, particularly if no approximate locations are known. These two examples are *pattern recognition* problems. The goal of pattern recognition is to find a description in an environment, in our case, in an image. If the description is geometric or iconic it is called a pattern. The recognition task of symbolic descriptions is more challenging.

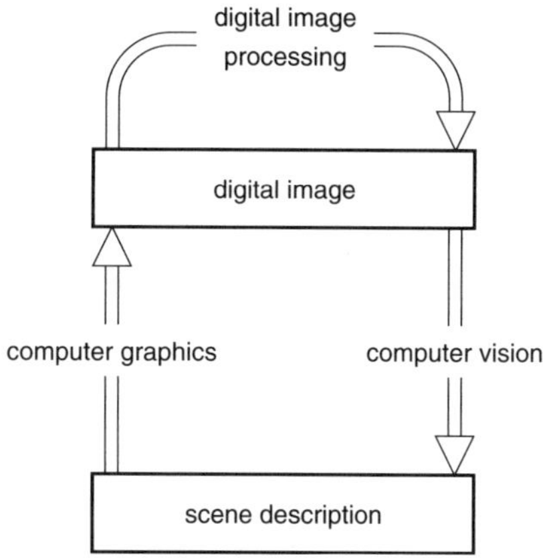

Figure 1.4: The relationship of digital photogrammetry to other disciplines.

Can we automatically locate buildings, fences, or roads in a digital image with pattern recognition techniques? The methods and tools provided by pattern recognition are not powerful enough to solve this difficult problem. *Computer vision* or *machine vision* attempts to solve these sorts of problems. Computer vision is one of the most intensively studied subfields of artificial intelligence. It is a relatively new and rapidly changing field. Many of the essential concepts have only evolved during the last ten years. By and large, computer vision pursues the same goals as human vision: generate descriptions about a scene from images. The descriptions ought to be explicit and meaningful to allow other system components to carry out a task. In that sense, computer vision is part of an entire system, such as a robot.

We conclude this subsection by referring again to Fig. 1.4 on the preceding page where the field of *computer graphics* is shown as the opposite process of computer vision. That is, computer graphics begins with a description of the scene and produces an image. This process is known in CAD (computer aided design) systems as rendering.

1.7 A View of Things to Come

This introductory chapter may be regarded as an "executive summary" of digital photogrammetry. The following chapters will fill in the necessary details. The book is organized in two volumes. This volume contains three parts: background, fundamentals of digital photogrammetry, and automatic orientation procedures; the second volume addresses issues related to applications of digital photogrammetry, including aerial triangulation, DEM and orthophoto generation, and object recognition.

Part I provides essential background information. For a good understanding of the working principles of digital photogrammetry, one needs a good grasp of image processing (Chapter 3) and computer vision (Chapter 5). These disciplines, in turn, draw on signal processing. Chapter 2 provides a short summary of digital signal processing, including the most important properties of linear systems and digital filtering. One important aspect of digital photogrammetry is the automation of photogrammetric tasks that human operators perform with great ease, thereby making use of our most impressive sense, namely "seeing". A better understanding of human vision (Chapter 4) may provide insights into the solution of so intricate a problem as determining surfaces, perceiving shapes and recognizing objects. Chapter 6 concludes Part I by providing background information on radiometry and photometry. This serves as preparation for the subject of data acquisition.

Part II is concerned with the fundamentals of digital photogrammetry. Digital images are at the heart of digital photogrammetry, and a solid understanding of their properties is essential. Images are obtained by digital cameras and scanners. In the quest of understanding the image formation process, Chapter 7 is devoted to digital cameras and provides a short summary about CCD sensors. Chapter 8 deals with scanners that play an important role in digital photogrammetry as long as aerial frame cameras are being used for acquiring photographs. Probably the single most significant product of digital photogrammetry is the digital photogrammetric workstation—or softcopy workstation if you prefer. Chapter 9 provides an account of these workstations, emphasizing the system functionality and working principles rather than detailed system descriptions.

Two chapters are devoted to the most fundamental task in digital photogrammetry: image matching. Chapter 10 presents an analysis of the major problems and possible solutions, followed by the most popular image matching methods. More advanced matching methods are treated in Chapter 11, including feature-based and relational matching. The last chapter of digital photogrammetry fundamentals is devoted to the computation of normalized images, a process usually referred to as registering images in epipolar geometry.

Part III contains three chapters devoted to the problem of performing the basic orientation procedures automatically. Interior orientation is the subject of Chapter 13. After describing known solutions, a new method of determining the fiducial marks is presented. Chapter 14 deals with the interesting problem of incorporating linear features into the computation of the relative orientation. How to deal effectively with features is also the general theme of automatic exterior and absolute orientation.

Part I

Background

Digital photogrammetry is very much interdisciplinary. The cross-disciplinary link to computer vision is so vigorous that some see digital photogrammetry as a subfield of computer vision. Other links exist to image processing, pattern recognition, and computer graphics. Hence, it makes perfect sense to start this book with a concise summary of the fields that distinctly influence photogrammetry. However, the discussion is restricted to the most relevant principles.

Part I is organized into five chapters. Although computer vision has the most profound impact on digital photogrammetry, we begin with signal processing, followed by image processing. Digital image processing starts with an image and produces another image as the output. Image processing techniques are used to improve the visual appearance of images to human observers. Moreover, images are processed and suitably represented for automatic analysis by a machine.

The goal of computer vision is to obtain information about objects from imagery, such as their physical, geometric, and topological properties. Such information is used in robotics, for example, to endow machines with vision.

We see and analyze our environment continuously, nearly in real time. Just because we do this without conscious effort does not imply that we know how we analyze and understand scenes, however. In fact, the lack of a detailed understanding of vision is the reason why it is so difficult to program a computer to analyze and understand images. It seems only natural, then, that someone who attempts to solve part of the puzzle should have a basic understanding of human vision.

The efforts in deriving useful information about objects from images benefit greatly from an understanding of how images are formed in the first place. In fact, you may consider the former task as the inversion of image formation. Chapter 6 concludes Part I by providing background information on radiometry and photometry to the extent necessary to understand the basic image formation process. Moreover, understanding camera and scanner specifications requires familiarity with radiometric and photometric concepts and quantities.

Chapter 2

Digital Signal Processing

Digital photogrammetry uses methods from various disciplines, such as artificial intelligence, computer vision, and digital image processing. Digital signal processing provides a common base to these fields. Thus, some knowledge of signal processing is helpful for a better understanding of the fundamental processes in digital photogrammetry. Examples are abundant, ranging from image formation (sampling theorem, aliasing) to object recognition.

We begin with a brief discussion of linear, shift-invariant systems in the time domain, followed by a representation of signals in the frequency domain. This paves the way for highlighting some key issues of digital filtering.

2.1 Signals and Systems

Signals occur in almost every engineering discipline. With the electronic revolution, the processing of signals was no longer restricted to analog solutions; digital signal processing established itself as a new subfield. Advances in digital signal processing brought profound changes and future developments are likely to promote advances in other fields.

Signals are represented mathematically as functions of one or more variables. Take an image, for example. Here, the brightness (gray level) is represented as a function of two spatial variables. Usually, the independent variable is referred to as time, although it may not represent time at all, as in the case of a digital image.

The independent variable is either continuous or discrete. Consequently, we differentiate between *continuous-time* and *discrete-time* signals. If the representation of the dependent variable (amplitude) is also discrete, then we call it a *discrete signal.* An *analog signal,* in contrast, is continuously represented along both axes. Discrete signals may be obtained from analog signals by *sampling* the independent variables and *quantizing* the amplitude. An example of this process, also referred to as *A/D* conversion[1], is the scanning of a photograph (diapositive). The photograph may be considered an analog signal while the result of the scanning process—the digital image—is a digital signal, where the spatial variables and the amplitude are discrete.

We are only concerned with digital signals here. To further simplify matters, we assume that the signals are *deterministic.* Therefore, digital signal processing is concerned with the representation of discrete signals by sequences of numbers, and by their transformation.

2.1.1 Sequences

Discrete-time signals are represented by sequences. A *sequence* of numbers x is formally written as:

$$x = \{x(n)\} \qquad \text{where} \qquad -\infty < n < \infty \tag{2.1}$$

with $x(n)$ being the nth number (sample) of sequence x. Usually, the sequence is written as $x(n)$, or simply x instead of Eq. 2.1. Note that $x(n)$ is only defined for integers n. It is not defined elsewhere.

Operations on Sequences

A sequence, x, can be scaled and shifted as shown below

$$\begin{aligned} s \cdot x &= s \cdot x(n) \\ x'(n) &= x(n - \text{shift}) \end{aligned}$$

Two sequences, x and y, can be added, subtracted or multiplied. We have

$$\begin{aligned} x \pm y &= x(n) \pm y(n) \\ x \cdot y &= x(n) \cdot y(n) \end{aligned}$$

[1]Analog to digital conversion.

Examples of Sequences

The *unit-sample sequence* is defined as:

$$\delta(n) = \begin{cases} 0 & n \neq 0 \\ 1 & n = 0 \end{cases} \tag{2.2}$$

and typically represents a discrete-time impulse, simply called *impulse.* A *unit-step sequence* is defined as:

$$u(n) = \begin{cases} 1 & n \geq 0 \\ 0 & n < 0 \end{cases} \tag{2.3}$$

A unit-step sequence can be expressed as the sum of unit-samples

$$u(n) = \sum_{k=-\infty}^{n} \delta(k)$$

Likewise, a unit-sample sequence is the difference of unit-step sequences

$$\delta(n) = u(n) - u(n-1)$$

Periodic sequences are defined as:

$$x(n) = x(n+T) \qquad \text{with} \qquad T = period \tag{2.4}$$

An arbitrary sequence, $x(n)$, can be expressed as the sum of scaled, shifted unit samples:

$$\boxed{x(n) = \sum_{k=-\infty}^{\infty} x(k)\delta(n-k)} \tag{2.5}$$

Fig. 2.1(d) depicts an arbitrary sequence. The unit-sample is shown in (a) and the unit-step sequence is sketched in (b). Fig. 2.1(c) illustrates the case of a scaled and shifted (delayed) impulse.

2.1.2 Linear Shift-Invariant Systems

A *system* generates, processes, or measures signals. The complexity of a system varies from such mundane things as a toggle switch to the highly

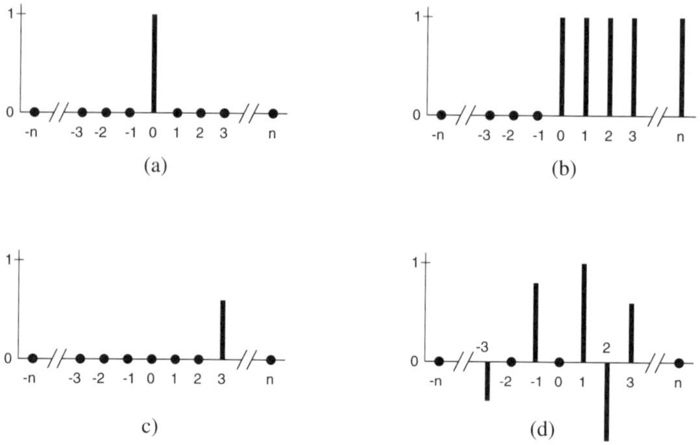

Figure 2.1: Examples of sequences: (a) unit-sample sequence (impulse), (b) unit-step sequence, (c) scaled and shifted impulse, (d) arbitrary sequence.

developed stage of living organisms. Systems can be analog, digital, deterministic, or random. Moreover, there is a wide range of system implementations: electrical, chemical, biological, mechanical, or economic systems.

A system is mathematically defined as a unique transformation (operator) that maps an input sequence $x(n)$ into the output sequence $y(n)$:

$$y(n) = S\,[x(n)] \tag{2.6}$$

Of all the possible properties of systems, the following are the most important: linearity, shift-invariance, stability, causality, and invertibility.

Linearity

The linearity property allows multiple effects to be superimposed. The sum of two effects is equal to the effect of their sums:

$$S\,[a \cdot x_1(n) + b \cdot x_2(n)] = a \cdot S\,[x_1(n)] + b \cdot S\,[x_2(n)] \tag{2.7}$$

The operator S operates on x_1 and x_2 while a and b are scalar numbers. A familiar example of linearity is integration:

$$\int [a \cdot f(x) + b \cdot g(x)]\, dx = a \int f(x)dx + b \int g(x)dx$$

There are two aspects to linearity, namely superposition and proportionality. The right side of Eq. 2.7 contains contributions from x_1 and x_2 exactly in proportion to their content in the input. If there is only one input and that input is increased by a certain factor, the output will increase by exactly the same factor. Most real systems are non-linear but they may be approximated by linear systems, at least in certain parts.

Let $x(n)$ in Eq. 2.6 be an arbitrary sequence, substituted by Eq. 2.5:

$$y(n) = S\left[\sum_{k=-\infty}^{\infty} x(k)\delta(n-k)\right]$$

Now, the principle of superpositioning leads to:

$$y(n) = S\left[\sum x(k)\right] S\left[\delta(n-k)\right]$$

Since $x(k)$ does not depend on n, we can further simplify:

$$y(n) = \sum x(k) S\left[\delta(n-k)\right] \tag{2.8}$$

The output of the system to an impulse input is called the *impulse response.* It is customary to denote the impulse response by h. With $S[\delta(n-k)] = h_{n,k}$ we have

$$y(n) = \sum_{k=-\infty}^{\infty} x(k) h_{n,k} \tag{2.9}$$

Note that h still depends on the two variables n and k. From a computational point of view we have not gained anything by the principle of linearity.

Shift-Invariance

The principle of shift-invariance requires that an impulse applied to the input of a system will produce the same output, independent of where the input occurs, except for a shift. If $y(n)$ is the response to $x(n)$, then the response to $x(n-k)$ will be $y(n-k)$. In a similar manner, if $h(n)$ is the response to $\delta(n)$, the response to $\delta(n-k)$ will be $h(n-k)$. Thus, Eq. 2.9 becomes:

$$\boxed{y(n) = \sum_{k=-\infty}^{\infty} x(k)h(n-k)} \tag{2.10}$$

Eq. 2.10 is commonly called the *convolution sum.* If a sequence $y(n)$ is related to the values of two sequences $h(n)$ and $x(n)$ as in Eq. 2.10, we say that $y(n)$ is the convolution of $h(n)$ with $x(n)$. Convolution is independent of the order in which two sequences are convolved:

$$y(n) = x(n) * h(n) = h(n) * x(n) \tag{2.11}$$

Note the popular notation $*$ (convolution operator) that is used to simplify the writing of the convolution sum.

In conclusion, linear shift-invariant (LSI) systems are characterized by their unit-sample response $h(n)$. Fig. 2.2 is a graphical demonstration of this important property.

Let $x(n) = -0.5 \cdot \delta_0 + 0.5 \cdot \delta_1 + 0.7 \cdot \delta_2$ be an input (see Fig. 2.2(e), top) and let $h(k) = 0.5 \cdot \delta_0 - 0.5 \cdot \delta_1 + 1.0 \cdot \delta_2 + 0.8 \cdot \delta_3$ be the impulse response (Fig. 2.2(a), bottom). Linearity implies that we can add up the system responses to every individual scaled impulse. Moreover, shift-invariance states that the response does not change. Fig. 2.2(b) shows the first scaled impulse of the input $x(n)$ to enter the LSI system. In a similar fashion, the other scaled and shifted input impulses are entering the system, causing appropriate responses, depicted in Fig. 2.2(c,d). Note that the response is always identical, except for the shift and scale. The response of the LSI system to the entire input sequence is obtained by adding the individual responses in Fig. 2.2(b)–(d). The result is shown in Fig. 2.2(e).

Now we emulate the LSI system numerically by $y = S \cdot x$ where y is the system response, S a Toeplitz matrix with the impulse response as column vectors, and x the input sequence.

$$\begin{bmatrix} y_0 \\ y_1 \\ y_2 \\ y_3 \\ y_4 \end{bmatrix} = \begin{bmatrix} 0.5 & 0.0 & 0.0 \\ -0.5 & 0.5 & 0.0 \\ 1.0 & -0.5 & 0.5 \\ 0.8 & 1.0 & -0.5 \\ 0.0 & 0.8 & 1.0 \\ 0.0 & 0.0 & 0.8 \end{bmatrix} \begin{bmatrix} -0.5 \\ 0.5 \\ 0.7 \end{bmatrix} \tag{2.12}$$

We obtain $y = [-0.25, 0.5, -0.4, -0.25, 1.10, 0.56]^T$, confirming the graphical result.

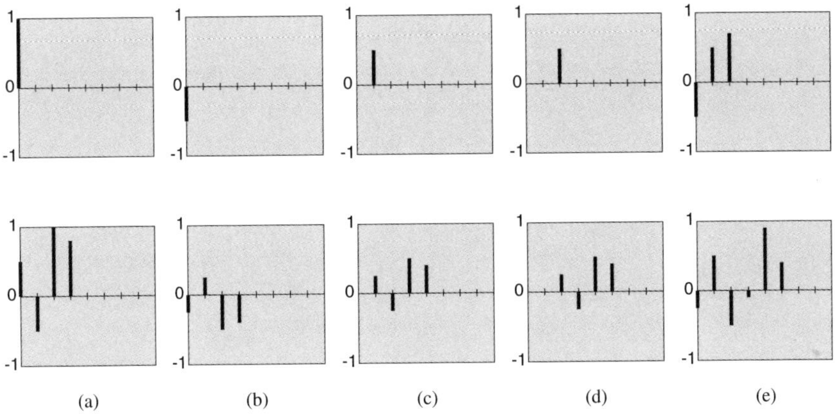

Figure 2.2: Illustration of the convolution process. The top row represents input and the bottom row is the response. In (a), the impulse response is shown. An arbitrary sequence (input) and its system response is depicted in (e). This result is obtained by superimposing the impulse responses of the input. The top row of (b)–(d) shows the decomposition of the signal in (e) as scaled and shifted impulses. The bottom row of (b)–(d) shows the respective responses. The system response of the signal in (e), then, is obtained by adding the individual responses (b)–(d).

Stability

A digital system is considered stable if and only if its impulse response is absolutely summable:

$$S = \sum_{k=-\infty}^{\infty} |h(k)| < \infty \tag{2.13}$$

Eq. 2.13 allows for infinitely long impulse responses. Such systems are called *IIR* systems. Obviously, all finite impulse response systems are stable.

Causality

Causality holds in systems for which the output for any $n = n_0$ depends on the input for $n \leq n_0$ only. In a time series, the output does not depend on future events. These systems have no energy source and are called passive systems. A linear shift-invariant system is *causal* if and only if the unit-sample response is zero for $n < 0$:

$$h(n) = 0 \qquad \text{for} \qquad n < 0$$

A simple moving average is acausal because it depends on future values.

Invertibility

A linear shift-invariant system (LSI) is invertible if Eq. 2.11 can be solved for x:

$$y = h * x \quad \Rightarrow \quad x = h^{-1} * y$$

Inverse operators are important in deconvolution problems.

2.2 Frequency Representation

In this section we represent sequences and signals by sinusoids or complex exponentials. As will be shown, the representation in the frequency domain makes some of the operations on sequences more transparent. Moreover, the use of complex exponentials usually leads to more compact formulas.

As far as the term *frequency* is concerned, there are two concepts in use. The *rotational frequency* refers to the number of periods, T, a sequence repeats per second. We use the symbol f in this text to denote rotational frequencies. The rotational frequency is the reciprocal of the period and is expressed in units of Hertz ([Hz]).

The *angular frequency* ω expresses the same concept, except in terms of revolutions. That is, a period of 1 second corresponds to a full revolution 2π. We have

$$\begin{aligned} f &= \tfrac{1}{T} && \text{[Hz]} \\ \omega &= 2\pi f && \text{[rad]} \end{aligned} \tag{2.14}$$

We now approximate a function by complex exponentials. Let $x(n)$ be an arbitrary sequence and let it be represented by a complex exponential using angular frequency notation:

$$x(n) = e^{i\omega n} \tag{2.15}$$

On the left-hand side of Eq. 2.15 is the original representation of the sequence, usually referred to as the *time domain*, while the right-hand side shows the representation in the *frequency domain*. Since we are concerned with images in digital photogrammetry, the independent variables are coordinates.

Let us now shift sequence $x(n)$ by s and study the effect in the frequency domain

$$
\begin{aligned}
x(n+s) &= e^{i\omega(n+s)} = e^{i\omega s}e^{i\omega n} \\
x(n+s) &= \lambda(\omega)e^{i\omega n}
\end{aligned}
\tag{2.16}
$$

We realize that the shift operation did not change the original frequency representation of $x(n)$, except for a multiplication by $\lambda(\omega)$. Since the same function emerges after the shift, it is called an *eigenfunction*. The complex exponentials are the eigenfunctions of the shift operation. In other words, a shift operation in the spatial domain amounts to a scalar multiplication in the frequency domain. The scalar factors, $\lambda(\omega)$, are the *eigenvalues*.

This encouraging outcome leads us to apply the same principle to an LSI system. In the spatial domain, the LSI system is characterized by Eq. 2.10, the convolution sum. We are now interested to see the properties of linearity and shift-invariance in the frequency domain. We proceed as before by representing the convolution sum by complex exponentials.

$$
\begin{aligned}
\sum_{k=-\infty}^{\infty} h_k x(n-k) &= \sum_{k=-\infty}^{\infty} h_k e^{i\omega(n-k)} = \sum_{k=-\infty}^{\infty} e^{i\omega n} h_k e^{-i\omega k} \\
\sum_{k=-\infty}^{\infty} h_k x(n-k) &= e^{i\omega n} \sum_{k=-\infty}^{\infty} h_k e^{-i\omega k}
\end{aligned}
\tag{2.17}
$$

To show the analogy to the previous process of a shift operation (or translation for that matter), let

$$
\lambda(\omega) = \sum_{k=-\infty}^{\infty} h_k e^{-i\omega k} \tag{2.18}
$$

Then Eq. 2.17 becomes

$$
y(n) = \lambda(\omega)e^{i\omega n} \tag{2.19}
$$

exhibiting the same remarkable characteristic: the complex exponentials are the eigenfunctions of LSI systems. They emerge unchanged except for the multiplication with the eigenvalues.

It is more customary to use the symbol H instead of λ, however. Thus, we have

$$
\begin{aligned}
y(n) &= H(\omega)e^{i\omega n} \\
H(\omega) &= \sum_{k=-\infty}^{\infty} h_k e^{-i\omega k}
\end{aligned}
\tag{2.20}
$$

The quantity $H(\omega)$ is called the *frequency* or *spectral response* (or simply *spectrum*) of a system. H(ω) is a continuous, periodic function of the frequency variable ω in the case that $y(n)$ is non-periodic and discrete. It is completely determined by the system's unit-sample response h_k. Because the eigenvalue $H(\omega)$ describes the change between input and output, it is also called the *transfer function.*

Now the question arises whether the impulse response h_k can be obtained directly from the frequency response. This is indeed possible because $H(\omega)$ in Eq. 2.20 is invertible. The relationship between spatial and frequency domain is defined by the transform $h_k, H(\omega)$. It is one kind of Fourier transform. Before discussing the subject of Fourier transforms as a means to express the relationship between spatial and frequency domain, we shall conclude this section by adding that complex exponentials are also the eigenfunctions of differentiation and integration as shown below

$$
\begin{aligned}
\frac{d}{dn}e^{i\omega n} &= i\omega e^{i\omega n} \\
\int e^{i\omega n}dn &= \frac{e^{i\omega n}}{i\omega}
\end{aligned}
$$

2.3 Fourier Transforms

In the previous section we have established a relationship between a discrete spatial domain and a continuous frequency domain. We now complete the picture. Table 2.1 shows the four possible combinations between spatial and frequency domain. The third column indicates what Fourier transform applies.

Since the purpose of this chapter is to provide a summary of those aspects of digital signal processing that are relevant to digital photogrammetry, we do not derive the different Fourier transforms here but merely point out the consequences and issues involved. The traditional approach would now be to begin with Fourier Series for sampling the continuous spatial or

Table 2.1: Combinations of different representations of spatial and frequency domains.

Representation *spatial*	*frequency*	*Transform*
continuous	continuous	Fourier integral
continuous	discrete	Fourier series
discrete	continuous	Fourier transform
discrete	discrete	Discrete Fourier transform

frequency domains, followed by the Fourier Integrals to establish a relationship between two continuous domains. This sequence of events makes eminent sense because the derivation and discussion of Fourier Integrals requires a different level of mathematical skills than, say, the discussion about series of trigonometric functions. As stated before, we summarize here only some of the more important properties. Therefore, we can afford to reverse the traditional approach and begin with the mathematically most involved case of continuous domains.

2.3.1 Fourier Integral

The Fourier Integral Transform provides a transformation between a continuous-time and continuous-frequency domain. The transform pair, simply called the *Fourier transform* is given below:

$$\boxed{\begin{aligned} F(\omega) &= \int_{-\infty}^{\infty} f(n)e^{-i\omega n}dn \\ f(n) &= \int_{-\infty}^{\infty} F(\omega)e^{i\omega n}d\omega \end{aligned}} \tag{2.21}$$

The duality of the transform pair is manifested by interchanging the spatial and frequency variables. The mathematical challenge arises from existence and convergence considerations of the integrals that involve all variables from minus infinity to plus infinity. One implication is that in order to compute the frequency response, for example, an infinite number of data points must be available—obviously a condition that does not exist in physical systems. Another major obstacle to overcome is the fact that periodic functions, such as sinusoids and step functions, are not absolutely integrable.

Theorems

We present three important theorems here.

a) The Convolution Theorem states that the Fourier transform of a convolution of two functions is the product of their Fourier transform.

Let $f(n)$ and $g(n)$ be two functions in the spatial domain. The convolution is defined as follows:

$$h(n) = \int f(n)g(n-k)dk \tag{2.22}$$

The spatial variable n is fixed and the integration variable is k.
Let $H(\omega)$ be the Fourier transform of the convolution $h(n)$. We have

$$\begin{aligned} H(\omega) &= \int_{-\infty}^{\infty} h(n)e^{-i\omega n}dn \\ &= \int_{-\infty}^{\infty} f(n)e^{-i\omega n}\left(\int_{-\infty}^{\infty} g(n-k)e^{-i\omega n}dn\right)dk \\ &= F(\omega)G(\omega) \end{aligned}$$

b) The Time-Limited Band-Limited Theorem says that no signal can be both limited in the time and in the frequency domain. A time-limited signal is characterized by

$$E_t = \int_{-a}^{a} |f(n)|^2 \, dn$$

that is, its total energy is confined to the interval $-a$ to a.

A similar definition holds for a band-limited signal

$$E_s = \int_{-b}^{b} |F(\omega)|^2 \, d\omega$$

where the total energy of the spectrum is contained in the interval $-b$ to b.

The consequence of this theorem is that a signal must be infinitely long in either the time domain or in the frequency domain. This requirement has a profound impact on digital signal processing. In practice, signals are likely to be band-limited. Moreover, they are only observed over a limited period of time. As a consequence, the time-limited band-limited theorem is hardly ever satisfied for signals related to physical quantities. Take an image, for example. It has a finite size, therefore it is limited in the spatial domain. Is its spectrum unlimited? Hardly, because the crystal size of the emulsion sets a limit to the resolution, or to the spectrum for that matter.

c) The Sampling Theorem states that it is possible to reconstruct a band-limited function from its equally spaced samples if the spacing is such that at least two samples occur in the highest frequency. We will discuss sampling in the next section.

Some Transform Pairs

For a better understanding of the relationship between the spatial domain and the frequency domain, we present a few transform pairs here. Fig. 2.3 is a graphical representation of the signal in both domains.

We begin with the *delta function*, also called impulse for short. Evaluating $\int \delta(t)dt$ reveals that the Fourier transform is constant at unit height (see also Fig. 2.3(a)). Likewise, a constant time signal at 2π has a spike as the Fourier transform. Note that in order to get a spike in the spectrum (*line spectrum*), the signal must be infinitely long.
The *rect function*, also called *boxcar function*, is defined by

$$f(t) = \begin{cases} 1 & \text{for } -T < t < T \\ 0 & \text{otherwise} \end{cases}$$

Its Fourier transform is the *sinc* function

$$\boxed{F(\omega) = 2T\frac{\sin(\omega T)}{\omega T}} \tag{2.23}$$

The sinc function oscillates at the constant rate of π/T and the amplitude decays by $1/t$. The main lobe extends from $-\pi/T$ to π/T. Note that the wider the rect function is, the narrower the central lobe and the oscillations. The "inverse" pair, sinc function in time domain and rect function in spectral domain also exists. An important application of the rect function is truncating a signal. Applied to the time domain, it cuts out a finite signal. In the frequency domain, it can be used as an ideal bandpass filter as it truncates all frequencies outside the interval $-\omega_c$ to ω_c. The rect/sinc pair plays an important role in optics, where the rect function refers to an

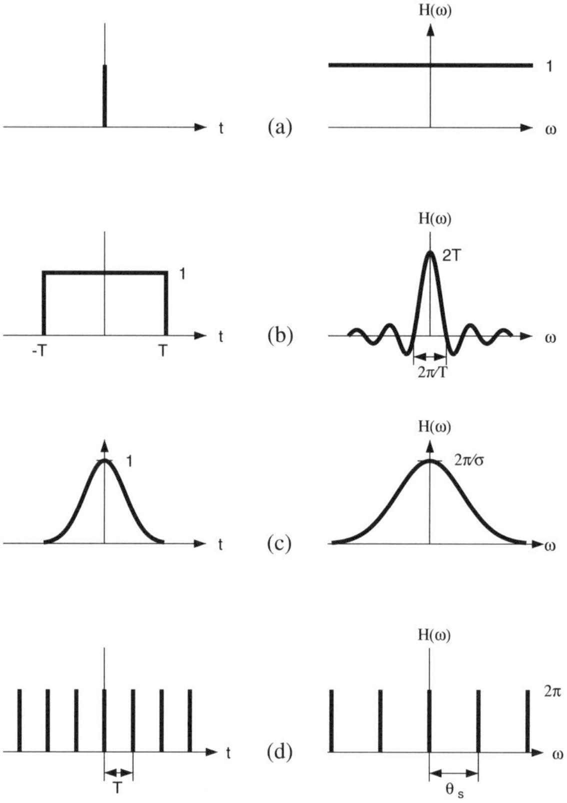

Figure 2.3: Examples of Fourier transform pairs: (a) a unit-sample/constant pair, (b) rect/sinc pair, (c) Gaussian/Gaussian pair, (d) sampling/sampling pair.

aperture of rectangular shape and the sinc function (its Fourier transform), to the diffraction pattern in the image plane.

The next two pairs are unique in that the Fourier transform does not change their functional form. The *Gaussian function* and its transform are

$$\begin{aligned} f(t) &= e^{-\sigma^2 t^2} \\ F(\omega) &= \frac{\pi}{\sqrt{\sigma}} e^{\frac{-\omega^2}{4\sigma^2}} \end{aligned} \tag{2.24}$$

An interesting property of this transform pair is that the product of the widths in the time and frequency domain is $w_t \cdot w_f \geq 1/2$. We observe again the characteristic spreading of transform pairs: the wider the Gaussian in the time domain the narrower the spectrum and vice versa. Gaussian functions are important in smoothing because they retain their functional form.

The *sampling function* also preserves the functional dependence in both domains. The transform pair is given by

$$\begin{aligned} f(t) &= \sum_{n=-\infty}^{\infty} \delta(t - nT) \\ F(\omega) &= \frac{2\pi}{T} \sum_{n=-\infty}^{\infty} \delta(\omega - n\omega_s) \end{aligned} \tag{2.25}$$

The sampling interval in the time domain, T, and in the frequency domain, w_s, are related by $T\omega_s = 2\pi$. This relationship shows again the reciprocity of the spacing of samples in both domains. Sampling functions are important for digitizing (sampling) continuous signals. This is true for both domains. For example, multiplying a continuous time signal with the sampling function $f(t)$ produces equally spaced samples.

Discrete Fourier Transform

So far the frequency response $H(\omega)$ is a continuous, periodic function. That is, the Fourier transform results in closed-form expressions of which we have presented several examples in the previous section. However, for computational evaluations, such as the determination of operator coefficients, we need a discrete representation of the spectrum. This amounts to sampling the frequency response $H(\omega)$.

We sample $H(\omega)$ at equally spaced points. For simplicity reasons we make the number of samples equal to the number of data points (samples in the time domain). We repeat Eq. 2.20 on page 26 and substitute the continuous frequency variable by $\omega = 2\pi\nu/N$, $\nu = 0, 1, 2, \ldots, N-1$.

$$\begin{aligned} H(\omega) &= \sum_{k=0}^{N-1} h_k e^{-i\omega k} \\ H(\nu) &= \sum_{k=0}^{N-1} h_k e^{-i2\pi\nu k/N} \end{aligned} \tag{2.26}$$

Now, we wish to express h_k as a function of H_ν. This is accomplished by multiplying both sides of Eq. 2.26 with $e^{i2\pi\nu/N}$ and by exploiting the orthogonality relation of exponential functions. This leads to

$$\boxed{\begin{aligned} h_k &= \frac{1}{N}\sum_{\nu=0}^{N-1} H_\nu e^{i2\pi\nu k/N} \\ H(\nu) &= \sum_{k=0}^{N-1} h_k e^{-i2\pi\nu k/N} \end{aligned}} \tag{2.27}$$

Eq. 2.27 constitutes the discrete Fourier transform pair (DFT). This transform is computationally implemented as the fast Fourier transform. Note that due to the periodicity of $H(\omega)$, the sampling is limited to the interval 2π.

As a result of sampling the frequency domain, the time domain also becomes periodic. However, signals in the time domain are hardly ever periodic. For instance, an image is certainly not periodic.

Evaluating the DFT at the zero positions in both domains ($h_k = 0,\ \nu = 0$) results in the following two equations:

$$\begin{aligned} H_0 &= \sum_{k=0}^{N-1} h_k \\ h_0 &= 1/N \sum_{\nu=0}^{N-1} H_\nu \end{aligned}$$

The zero frequency is the sum of the impulse responses while the first value of the spatial sequence is the average of the frequency response.

2.3.2 *Summary*

We summarize some of the results presented in this section by following the procedure of sampling a signal in both domains. We will describe the steps involved and the effects in both domains.

Fig. 2.4(a) shows a continuous, unlimited signal in the spatial domain. The Fourier transform provides a relationship between the spatial and the frequency domain. Both representations are continuous. According to the time-limited, band-limited theorem at least one function must be unlimited. Since we are considering a band-limited signal, it follows that it should be unlimited in the spatial domain—a practical impossibility.

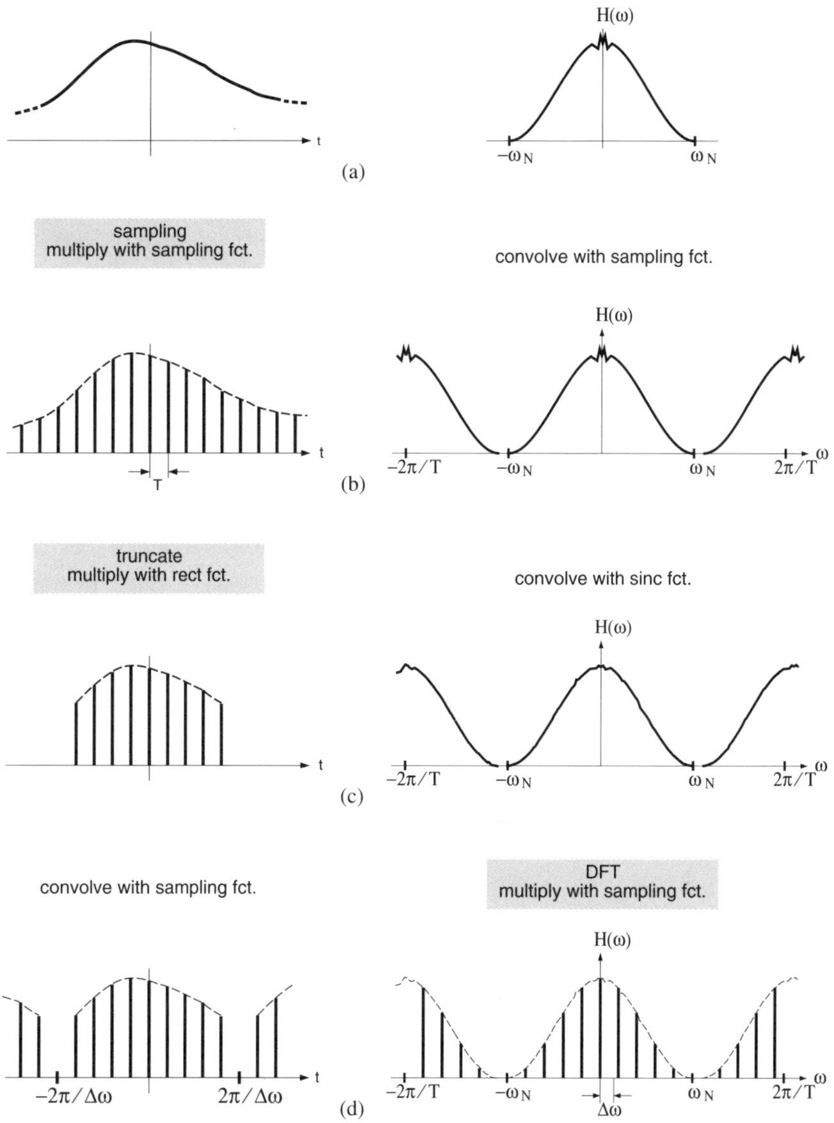

Figure 2.4: Continuous and unlimited signal in the spatial and frequency domain (a); sampled spatial domain and corresponding frequency domain (b); truncated spatial domain and its corresponding frequency domain (c); sampled frequency domain by DSP and corresponding spatial domain (d).

Now we digitize the continuous signal in the spatial domain by multiplying it with the sampling function defined by Eq. 2.25 on page 31. Fig. 2.3(d) shows the sampling function in both domains. Remember that the sampling function does not change its functional characteristic. Therefore, spikes in the spectrum occur with a spacing of $2\pi/T$.

The multiplication of the original signal with the sampling function renders the discrete function $f(n)$ that is only defined at the positions $k \cdot T, k = 0, 1, 2, \ldots, \infty$. From the convolution theorem we know that multiplication in one domain requires convolution in the other domain. As a consequence, the spectrum must be convolved with the Fourier transform of the sampling function. The result is a replication of the spectrum at the positions $2\pi/T \cdot k, k = 1, 2, \ldots, \infty$. This step is depicted in Fig. 2.4(b). Sampling of the spatial domain induces periodicity of the spectrum. If the spacing $2\pi/T$ is smaller than the band-limited spectrum, then the replicated spectra will overlap and the frequencies in the overlapping portion are lost. Fig. 2.5 depicts this situation known as *aliasing*.

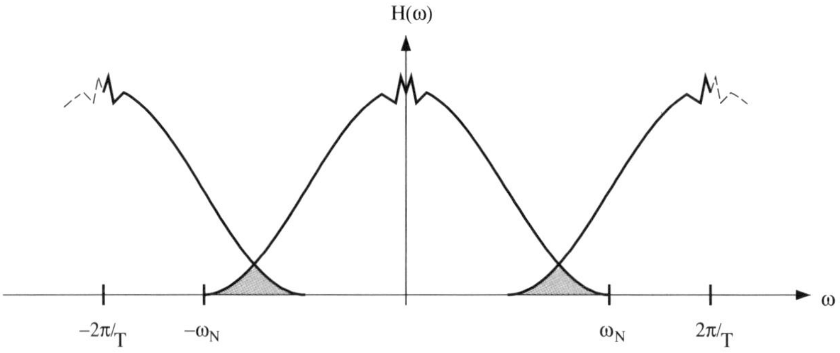

Figure 2.5: Sampling the spatial domain too coarsely causes overlap *aliasing* of higher frequencies in the periodic spectrum.

In order to obtain a discrete signal of finite spatial extent we multiply it by a rect function, shown in Fig. 2.3(b). The Fourier transform is a sinc function. This step has the desired effect in the spatial domain: a discrete, finite signal. The effect in the frequency domain is less optimal because the convolution of the spectrum with the sinc function results in a loss of spectral resolution. This is no surprise since the time-limited band-limited theorem is now violated. The smaller the width of the rect function the more spread out the sinc function and the higher the loss in spectral resolution. Fig. 2.4(c) illustrates this step.

The last step is sampling the frequency domain by DFT. The sampling interval of $\Delta\omega$ corresponds to a spacing of $2\pi/\Delta\omega$ in the spatial domain.

The result of this process, discussed in the section about the discrete Fourier transform, is graphically shown in Fig. 2.4(d). We note that sampling the spectrum induces periodicity in the spatial domain. The discrete signal is repeated at positions spaced by $4\pi/\Delta\omega$.

2.4 Digital Filtering

The concept of filtering originates from electrical engineering where it is used to manipulate electrical signals. In general, any linear shift-invariant system may be considered a filter. In a more restrictive sense a filter is used to filter out part of the spectrum. A digital filter operates on a digital signal.

Fig. 2.6 depicts a few filters that are characterized by the way the spectrum is affected. A *lowpass filter* lets lower frequencies pass through while blocking (stopping) higher frequencies. The contrary holds for a *high-pass filter* where high frequencies pass and low frequencies are stopped. Fig. 2.6(c) shows an example of a *bandpass filter* and in Fig. 2.6(d) we have a *bandstop filter.*

Filters can also be characterized by the way they change the data domain. This is particularly true in disciplines other than electrical engineering. The most popular example is the moving average filter that is used for smoothing data. Examples from numerical analysis include integrators and difference operators.

2.4.1 Moving Average Filter

We take the moving average filter to emphasize the importance of analyzing the effects of filters in both domains. When confronted with the problem of smoothing a noisy data stream presumably most everybody would proceed as follows: replace the value (amplitude) of a data point by the average amplitude of immediate neighbors (including the point under consideration); repeat this process for all data points. The neighbors to the left and right of the data point form a window. This simple procedure is the result of fitting a straight line through the points within the window by least-squares.

We formulate the process described as a convolution, defined by Eq. 2.10 on page 22 and note that the impulse response h_k is

$$h_k = \frac{1}{2N+1}(1, 1, 1, \ldots, 1) \tag{2.28}$$

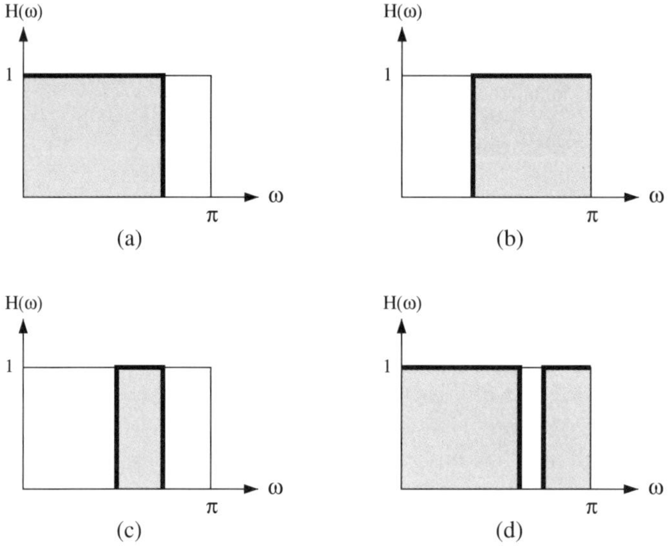

Figure 2.6: Four filters that filter out portions of the spectrum: (a) lowpass filter; (b) highpass filter; (c) bandpass filter; (d) bandstop filter.

Let us now determine the frequency response, $H(\omega)$, according to Eq. 2.20 on page 26.

$$\begin{aligned} H(\omega) &= \frac{1}{2N+1}\left(e^{-i\omega N}, \ldots, e^{0}, \ldots, e^{i\omega N}\right) \\ &= \frac{1}{2N+1}\left(1 + 2\sum_{k=1}^{N}\cos(k\omega)\right) \end{aligned} \tag{2.29}$$

In the spatial domain the moving average filter is a rect function. We recall that its Fourier transform is a sinc function. Eq. 2.29 can also be written as follows:

$$H(\omega) = \frac{\sin\left((N+0.5)\omega\right)}{(2N+1)\sin(\omega/2)} \tag{2.30}$$

This makes the connection to the sinc function more transparent.

Fig. 2.7 shows the frequency response of a three-point and a nine-point moving average filter with equal weights. We recognize the desired smooth-

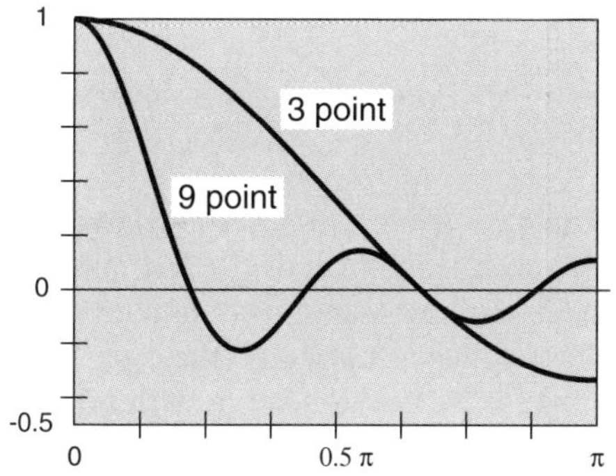

Figure 2.7: Frequency response of a three-point and nine-point moving average filter.

ing effect in that low frequencies are passed through while higher frequencies are stopped. Comparing the two responses also confirms the obvious, namely that the smoothing effect is higher with the nine point average filter (a larger portion of the higher spectrum is stopped). The figure also demonstrates clearly how far the moving average filter is from an ideal lowpass filter. For example, the three-point filter has a wide transition zone. Moreover, both filters have spectral ranges with negative values. We conclude by adding that this rather unfavorable spectral response does not change significantly if the smoothing is performed with a polynomial fit to the data points.

2.4.2 Ideal Lowpass Filter

As shown in the previous section, the moving average filter has an unsatisfactory behavior in the frequency domain. We now take a different approach and design the filter in the frequency domain. Let us try to solve a common problem—noise removal. Under the assumption that noise is a high frequency phenomenon we can solve this problem easily—at least in theory—by a lowpass filter. The ideal lowpass filter is sketched in Fig. 2.6(a). The stopband begins at ω_c, the cut-off frequency. It is chosen at that point in the spectrum where we expect that frequencies are not caused by the signal but by noise. Obviously, this requires good knowledge about the signal.

We can define the frequency response of an ideal lowpass filter by

$$H(\omega) = \begin{cases} 1 & -\omega_c \leq \omega \leq 0 \\ 0 & \text{otherwise} \end{cases} \tag{2.31}$$

For computing the filter coefficients in the spatial domain we convolve the signal with the Fourier transform of the lowpass filter—the rect function. We recall that the Fourier transform of the rect function is the sinc function (see Fig. 2.3 on page 30). Now, the sinc function is of infinite length and we truncate it by multiplication with a rect function. The effect of this operation in the spectrum is a convolution of the ideal lowpass filter with a sinc function with the result that there is no longer a sharp transition between passband and stopband. Moreover, the filter has wiggles caused by the side lobes of the sinc function. These undesirable effects can be reduced by a larger spatial extent of the filter, but with a finite size we never can completely eliminate the problem. Consequently, an ideal lowpass filter cannot be realized.

Incidentally, the same problems occur if the smoothing is performed in the frequency domain. The frequency response $H(\omega)$ is a continuous, periodic function that can be expanded into a Fourier series, for example. In a practical implementation, the Fourier series must be finite—truncating it introduces approximation errors that are larger near discontinuities (Gibb's phenomena). The ideal lowpass filter has a discontinuity between passband and stopband. So, regardless of the approach, the fundamental problem persists.

One approach in designing digital filters is known as *windowing*. Here, filter coefficients are determined such that they have as few side effects in the spectrum as possible. This rather broad objective can only be transformed into firm specifications by considering the application: what may be perfectly acceptable in one application may be a problem in another one.

2.5 To Probe Further

There are numerous books about digital signal processing and related subjects, such as: digital filtering, time series, and Fourier transforms. Most of these texts are written for electrical engineers, mathematicians or physicists and are not easy reading material for non-specialists. Several books on the subject of digital image processing (see next chapter) contain introductory chapters about signal processing.

J. H. Karl's *An Introduction to Digital Signal Processing* (Academic Press, Inc.) is a very readable book and holds what the author promises: to provide a fast track to digital signal processing that does not require previous knowledge of the subject nor sophisticated mathematical skills. A more comprehensive treatment of the subject is *Digital Signal Processing* by A. Oppenheim and R. Schafer (Prentice-Hall, Inc., Englewood Cliffs, New Jersey). An excellent introduction to the subject of digital filtering is the book by R. W. Hamming *Digital Filters* (Prentice-Hall, Inc., Englewood Cliffs, New Jersey).

Chapter 3

Digital Image Processing

Digital image processing is concerned with acquiring, transmitting, processing, and representing images. Image processing techniques are used to improve the visual appearance of images to a human observer and to process pictorial information suitable for machine perception. Digital image processing can be broadly divided into the following areas:

Acquisition deals with ways of obtaining images; for example, by using digital cameras or by digitizing analog images (photographs).

Storage and Compression are concerned with suitable techniques to store digital images that require huge storage capacities. For example, an aerial photograph, digitized to 15 μm pixel size, requires 232 MB. Compressing, storing and retrieving images effectively is particularly important in a production environment.

Enhancement and Restoration techniques are used to improve the visual appearance of images or to recover degraded images.

Segmentation deals with partitioning the image into meaningful regions.

Visualization techniques are used for presenting images on a variety of media, including monitors, printers, and film recorders.

Image processing has many facets and is related to several disciplines such as signal processing, computer science, pattern recognition, artificial intelligence, and computer vision. A clear division is often not possible.

Obviously, digital image processing plays an important role in digital photogrammetry. One is tempted to draw the analogy to the photo lab and its role in photogrammetry. However, this analogy does not take into account the full potential of image processing that is certainly not limited to the traditional photolab processes.

The purpose of this chapter is to present relevant aspects that are not covered elsewhere in this book. After a brief discussion of image models and image properties, the main focus is on enhancement and restoration, followed by a short introduction to segmentation. Acquiring images is the subject of Chapter 7. Storage and compression, as well as visualization, will be discussed in the chapter about digital photogrammetric workstations. Edge detection, feature extraction, grouping, and object recognition—subjects usually also covered in image processing textbooks—will be discussed in computer vision, Chapter 5.

3.1 *Image Models*

This section summarizes mathematical models that are useful for representing images. An image can be thought of as a continuous function $g(x, y)$ where x, y are two spatial variables. The dependent variable is the gray level (brightness). Images of natural scenes, such as aerial photographs, cannot be expressed analytically, that is, no closed-form expression exists. The only way to deal with images computationally is to seek a discrete function. The process of converting the continuous representation into a discrete one is called *digitizing.* The discrete function is a *digital image.* It is obtained by *sampling* the spatial variables and by *quantizing* the gray levels (amplitude). From the sampling theorem discussed in the previous chapter we remember that a continuous function can be sampled at regular intervals if certain conditions are met. One such condition is that the function must be well-behaved and band-limited. This is the case for images. For example, the recorded intensity (energy) is positive and finite, and the size of an image is limited. Fig. 3.1 shows an image with different sampling sizes.

We use the following matrix representation for a digital image:

$$g(x, y) = \begin{vmatrix} g(0,0) & g(0,1) & \cdots & g(0, C-1) \\ g(1,0) & g(1,1) & \cdots & g(1, C-1) \\ \vdots & \vdots & \vdots & \vdots \\ g(R-1,0) & g(R-1,1) & \cdots & g(R-1, C-1) \end{vmatrix} \tag{3.1}$$

where:	$x = 0, 1, \ldots, C-1$	row
	$y = 0, 1, \ldots, R-1$	column
	R	maximum number of rows
	C	maximum number of columns
	$g(x, y) = \{0, 1, \ldots, \max\}$	gray level

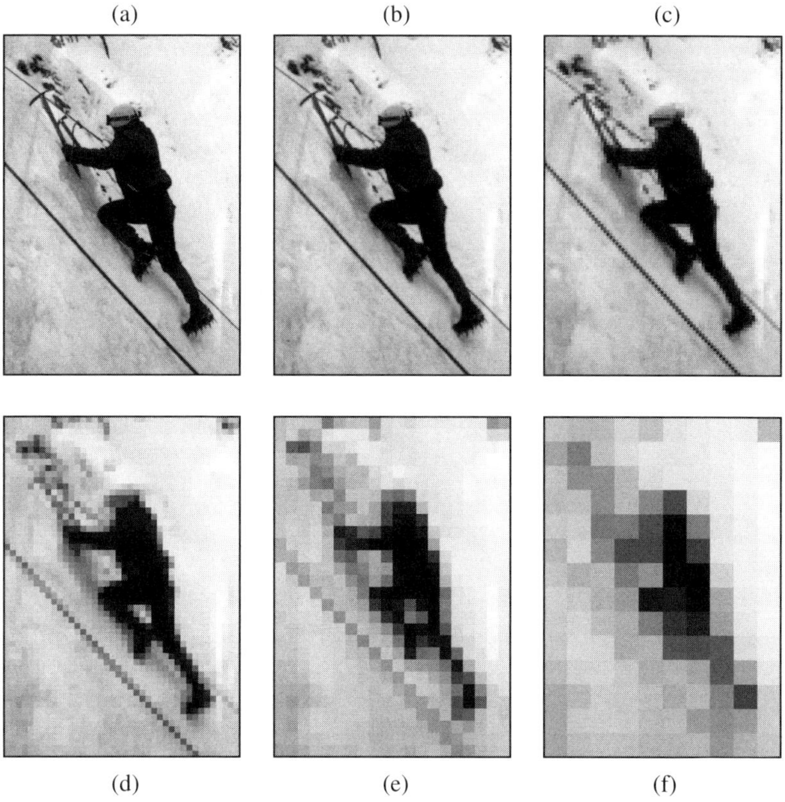

Figure 3.1: Image represented at different sampling sizes: (a) 448×320, (b) 224×160, (c) 112×80, (d) 56×40, (e) 28×20, (f) 14×10.

With the notation of Eq. 3.1, a color image with three channels can be represented as

$$G_c = \{g_1(x, y), g_2(x, y), g_3(x, y)\} \tag{3.2}$$

If the set of gray levels comprises only one value, say g_h, then we have a *homogeneous image*. A *binary image* contains two gray values, usually $\{0, 1\}$, but not necessarily so. Binary images are obtained, for example, by separating gray level images into foreground and background where the foreground contains objects. A *black-and-white image* has one channel, that is, one gray level function $g(x, y)$. In contrast, a *multi-channel image* has n image functions $g_i(x, y), i = 0, 1, \ldots, n-1$. In a similar fashion, time-sequence images, such as *video sequences*, can be expressed by $g_t(x, y), t = 0, 1, \ldots, T_e$. Image sequences play an important role in mobile mapping systems.

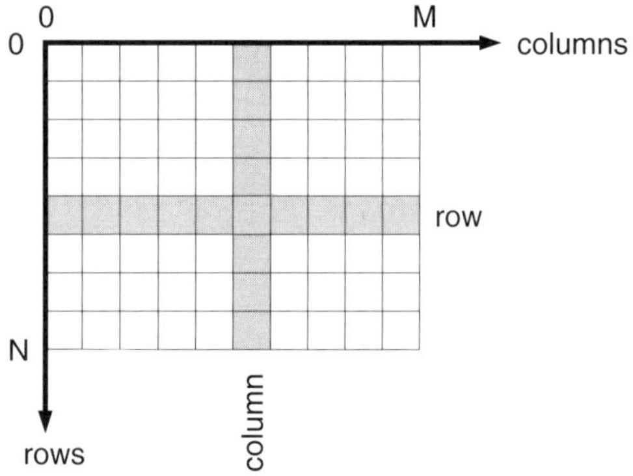

Figure 3.2: Discrete representation of an image as a matrix.

Now, the dependent variable may be considered deterministic or stochastic. In the latter case, the gray level is the result of a random process. The stochastic representation is useful for solving problems related to transmitting or compressing images, or for statistical analysis.

3.2 *Image Characteristics*

There are several ways to characterize images. We mention here a few useful statistical properties.

3.2.1 *Average and Standard Deviation*

The average gray level, g_a, of an image and its standard deviation, σ, can be easily determined by

$$g_a = \frac{1}{R \cdot C} \sum_{x=0}^{R-1} \sum_{y=0}^{C-1} g(x,y) \tag{3.3}$$

$$\sigma = \frac{1}{R \cdot C} \sum_{x=0}^{R-1} \sum_{y=0}^{C-1} (g(x,y) - g_a)^2 \tag{3.4}$$

The average g_a reflects the overall brightness of an image while the standard deviation is a measure for the contrast. A small number refers to a "flat" image with low contrast, that is, an image with low information content. Another quantity to express the information content is the entropy.

3.2.2 Entropy

Entropy measures the randomness of a set of random variables. Applied to a digital image, the entropy expresses the uncertainty of a gray value. Let $p_i, i = 0, 1, \ldots, \max$ be the probabilities of the gray values of a digital image. Then, the entropy H is defined as

$$H = -\sum_{i=0}^{\max} p_i \log_2 p_i \tag{3.5}$$

in bits. A binary image has two gray levels. Suppose the probability for either value is 0.5. Then, the entropy will be one. This makes perfect sense as we need just one bit per pixel to represent a binary image. For a typical image with 256 gray levels the entropy is eight, assuming the same probability of $1/256$ for all gray levels.

3.2.3 Histogram

The histogram of an image is important for several operations that are performed on images. Let $p_i(z)$ denote the relative frequency with which the gray level z occurs in an image $f_i(x,y)$ and let $[z_l, \ldots, z_u]$ be the range of gray levels. Assuming that the gray levels are normalized in the interval $[0,1]$ it follows that the histogram is the probability density function $p_i(z)$ of image i. For quantized gray levels, we have the cumulative density function:

$$h_i(z) = \sum_{z=l}^{u} p_i(z) \tag{3.6}$$

The histogram can be represented in tabular form. For analysis by a human, a graphical representation is more suitable (see Fig. 3.3).

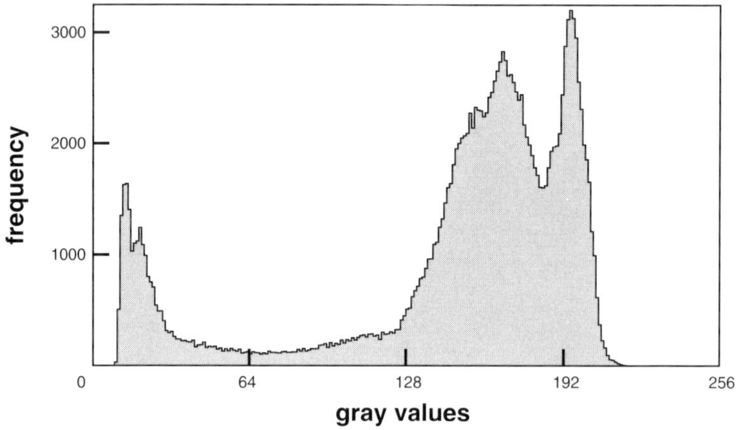

Figure 3.3: Histogram of image shown in Fig. 3.1(a). The area under the curve (gray shaded) is identical to the total number of pixels (448 × 320). The area from gray level 8 to 40 is about 15% of the total area. This range of gray levels refers to the climber, rope, ice ax, and a few rocks, all of which appear fairly dark.

3.2.4 *Moments*

One of the most fundamental processes in photogrammetry is finding conjugate points in two or more overlapping images. Humans solve this problem effectively by identifying regions, called *image patches*, covering the same area in the object space. For automating this process it would be ideal if the image patches had a small set of characteristic properties that change monotonically with changes of the location of image patches. Moreover, such properties should be invariant to variations in translations, rotations, and scale. Under certain circumstances, moments satisfy these requirements.

The moment M_{pq} is defined by

$$m_{pq} = \int_{-\infty}^{\infty} \int_{-\infty}^{\infty} x^p y^q f(x, y) dx dy \tag{3.7}$$

where $p, q = 0, 1, 2 \ldots$ define the order of the moment. If $f(x, y)$ is piecewise linear and has no infinite zero values, then the sequence of moments m_{pq} exists. Moreover, the inverse process exists, that is, $f(x, y)$ can be determined from the moments.

The central moments for a binary image are expressed by

$$\mu_{pq} = \sum_x \sum_y (x - x_o)^p (y - y_o)^q f(x, y) \tag{3.8}$$

where x_o, y_o is the centroid of $f(x, y)$, or the first moment.

Central-moment based features convey descriptive information about an image patch. In fact, there are seven invariant moments that can be derived from the second and third moments.

3.3 Image Enhancement and Restoration

The purpose of image enhancement techniques is to improve the appearance of an image for a human observer. This is a rather subjective process, typically performed interactively. The selection of the proper methods and the choice of the right parameters depends on the quality of the original image and on the application. Image processing enjoyed early success in the 1960's when images from the moon were processed and enhanced.

There is a multitude of operations that can be performed on images for the purpose of enhancement or restoration. There are several ways to categorize these operations. Frequently, they are divided into the following four categories:

Point operations: The result of a point operation depends only on the input gray level at a single point. Typical point operations include stretching and thresholding of gray levels.

Local operations: Here, several input pixels in the neighborhood contribute to the resulting output pixel. Many operations are local, e.g., smoothing (averaging), feature extraction, and edge enhancement.

Global operations: The entire input image contributes to the output. Global operations are often performed in the frequency domain.

Geometric operations: The result of a geometric transformation depends on gray levels from different positions in the input image. Scaling, rotating, translating and rectifying are typical examples.

Another useful classification is guided by applications. Here, the desired effect of an operation serves as a criterion. This suggests a grouping into the following three broad categories:

Smoothing: Includes techniques that render an image of smoother appearance, e.g. by removing noise or suppressing details.

Sharpening: The purpose is to make certain features of an image more pronounced, e.g., edges.

Correcting defects: These operations pursue the goal of correcting image defects, e.g., remove blunders in the gray levels.

The domain in which an operation is performed is yet a third possibility for classifying operators. The previous chapter demonstrated the usefulness of representing signals and performing operations on them in the original domain (e.g. spatial domain) and in the frequency domain. The advantages of analyzing operations in both domains are profound—transform techniques are an indispensable tool in image processing. Fig. 3.4 depicts schematically the duality of the two representations. An image $f(x,y)$ is convolved with the unit-sample response $h(x,y)$ in the spatial domain. Performed in the frequency domain, the same operation amounts to multiplying the frequency representation of the image, $F(\mu,\nu)$, by the transfer function $H(\mu,\nu)$. The unit-sample response and the transfer function are related by the Fourier transform and its inverse, respectively.

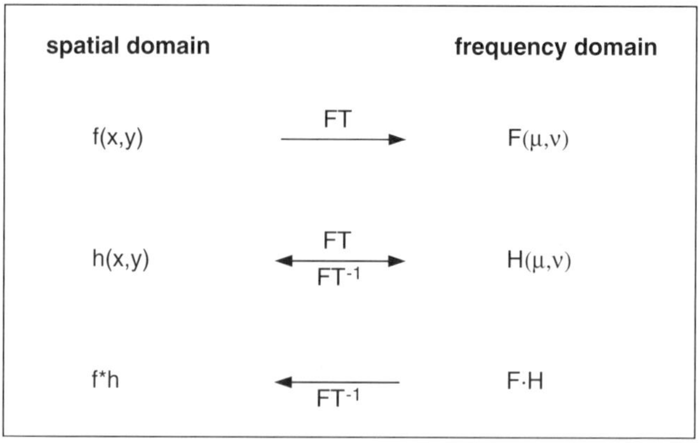

Figure 3.4: Duality of operations performed in the spatial domain and frequency domain.

The Fourier transform provides the connection between the two domains. To cope with images, the relevant equations derived in the previous chapter ought to be extended to two-dimensional, discrete functions.

Convolution of image $f(x,y)$ with unit-sample response $h(i,j)$

$$\boxed{f(x,y) * h(x,y) = \sum_{i=0}^{M-1}\sum_{j=0}^{N-1} f(x-i, y-j)h(i,j)} \tag{3.9}$$

Discrete Fourier Transform

$$\boxed{\begin{aligned} F(\mu,\nu) &= \frac{1}{M \cdot N}\sum_{x=0}^{M-1}\sum_{y=0}^{N-1} f(x,y)e^{-i2\pi(\mu x/M+\nu y/N)} \\ f(x,y) &= \sum_{\mu=0}^{M-1}\sum_{\nu=0}^{N-1} F(\mu,\nu)e^{i2\pi(\mu x/M+\nu y/N)} \end{aligned}} \tag{3.10}$$

The forward Fourier transform of Eq. 3.10 transforms the spatial domain to the frequency domain and the inverse Fourier transform provides the frequency to spatial domain relation.

The following sections are concerned with image enhancement and restoration techniques. Specifically, point operations, smoothing, sharpening, and correcting operations on images are presented. The operations are performed in the spatial domain or, if more appropriate, in the frequency domain.

3.3.1 *Histogram Modifications*

Suppose $g_1(x,y)$ is an image whose gray levels we want to modify by a transformation T to obtain a new image $g_2(x,y)$. We have

$$g_2(x,y) = T\left[g_1(x,y)\right] \tag{3.11}$$

If we want to preserve the order of the gray scale, e.g., from black to white, then T must be monotonic. Another reasonable constraint on T is that the inverse operation should exist.

Contrast Stretching

In contrast stretching, the gray levels are modified so that they occupy a wider range of permissible values. A linear stretching, for example, is accomplished by

$$\boxed{g_2(x,y) = (g_1(x,y) + t_1)\,t_2} \quad (3.12)$$

The transformation parameters t_1, t_2 can be selected so that the gray levels in g_2 are confined to the interval $< g_2^{\text{min}}, g_2^{\text{max}} >$. Let g_1^{min} be the minimum gray level and g_1^{max} be the maximum gray level in image g_1 (e.g. determined from its histogram). Then we have for the transformation coefficients

$$t_1 = g_2^{\text{min}} - g_1^{\text{min}} \quad (3.13)$$

$$t_2 = \frac{g_2^{\text{max}} - g_2^{\text{min}}}{g_1^{\text{max}} - g_1^{\text{min}}} \quad (3.14)$$

Example 3.1:

Suppose the minimum gray level in the original image is 89 *and the maximum gray level is* 176*. We want to stretch the gray levels to an interval* $[0, 255]$ *(full gray scale of an 8-bit image). Then*

$$t_1 = 0 - 89 = -89$$

$$t_2 = \frac{256}{87} \approx 2.94$$

A gray level of 100 *in the original image transforms with Eq. 3.12 to* 32*, for example.*

The stretching function does not need to be linear. Sometimes, a logarithmic contrast stretching is more pleasing for a human observer. The human visual system perceives brightness differences on a logarithmic scale.

Histogram Equalization

Histogram equalization defines a transformation from gray levels g_1 to gray levels g_2 in such a way that the distribution of g_2 becomes even, that is, the histogram of g_2 is uniform. Fig. 3.5 illustrates the procedure.

Suppose for a moment that the gray level distribution is continuous. Fig. 3.5 shows the histogram of g_1 in its normal position. The histogram of g_2, $h(g_2)$, in the left half of the figure is rotated by 90 degrees. We are seeking a transformation function such that $h(g_2) = (R \cdot C)/n$, where R, C are the number of rows and columns, and n is the number of gray levels. To achieve this we divide the area under the histogram curve into $n/(R \cdot C)$

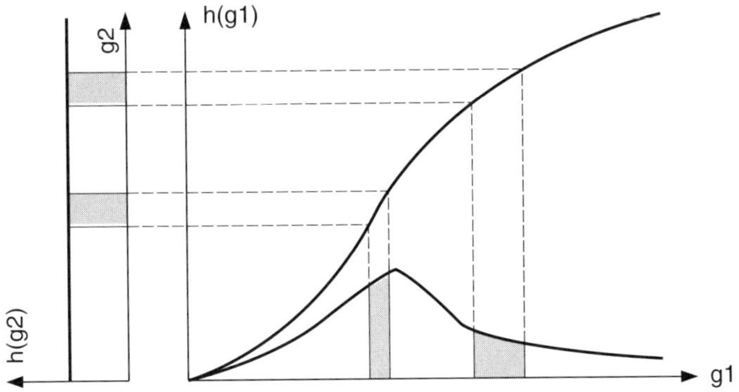

Figure 3.5: Histogram equalization.

equal "parts". Then, the following equation expresses the condition for histogram equalization:

$$h(g_2) = \frac{n}{R \cdot C} \int_0^1 h(r) dr \tag{3.15}$$

with r the dummy variable of integration. The integral is the cumulative distribution function. It follows that the transformation function needed to obtain a constant histogram is the cumulative distribution function.

For digital images, the gray levels are discrete. The discrete form of the cumulative distribution function is obtained simply by adding up all gray levels (see Eq. 3.6 on page 45 and Fig. 3.3 on page 46).

Fig. 3.6 demonstrates the effects of histogram equalization. The original image of Fig. 3.6(a) has a poor contrast, also clearly visible from the histogram in Fig. 3.6(c). The equalized histogram (Fig. 3.6(d)) provides a much improved rendition of the image, as shown in Fig. 3.6(b).

3.3.2 Smoothing Operators

Smoothing operators are primarily used for

- removing or reducing noise. If the noise in an image is a high-frequency phenomenon then it can be reduced by lowpass filters;
- reducing resolution. Imagine an aerial image with 10^9 pixels. For viewing purposes, for example, the full resolution is not required.

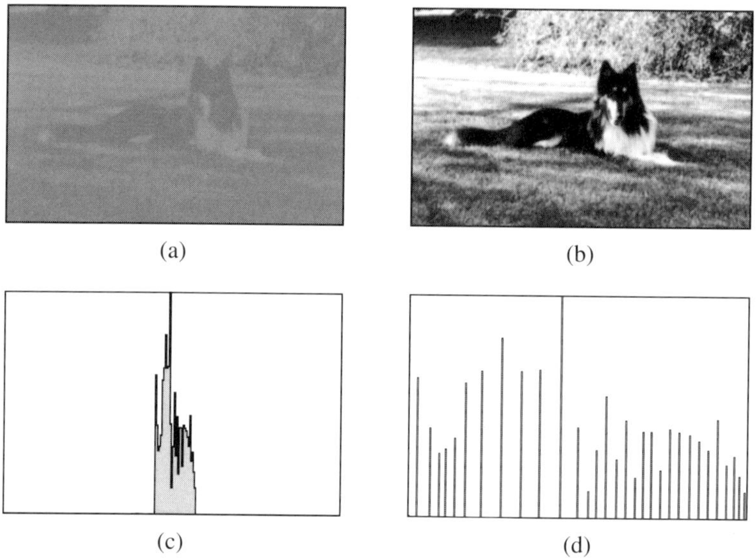

Figure 3.6: Example of histogram equalization. The original image (a) has a poor contrast, also evident in its histogram (c). The equalized histogram in (d) leads to a new image (b) with much better rendition.

The same holds for certain automatic processes that perform better with lower resolutions. Reducing the resolution in predefined steps lends itself to the generation of image pyramids.

Averaging Operator

We discussed this operator in the previous chapter where it is called moving average filter. In image processing literature it is usually called *neighborhood averaging*. Since the operator is linear and shift-invariant, the procedure in the spatial domain is governed by convolution. We expand the concept discussed in the previous chapter to two spatial variables. The equivalent to the one dimensional impulse response of Eq. 2.28 on page 35 reads

$$h = \frac{1}{(2N+1)(2M+1)} \begin{vmatrix} 1 & 1 & \dots & 1 \\ 1 & 1 & \dots & 1 \\ \vdots & \vdots & \vdots & \vdots \\ 1 & 1 & \dots & 1 \end{vmatrix} \tag{3.16}$$

This two-dimensional impulse (unit sample) response is sometimes called a *window* or *mask*. The numbers in the window are the filter coefficients.

The smoothed image $g_s(x, y)$ is obtained by convolving the original image, $g_o(x, y)$ with the averaging operator defined in Eq. 3.16. This operation is expressed by

$$\boxed{g_s(x,y) = \sum_{i=-N}^{N-1} \sum_{j=-M}^{M-1} g_o(x-i, y-j) \cdot h(i,j)} \tag{3.17}$$

where $(2N + 1) \times (2M + 1)$ is the size of the window. Normally, the window is square, that is, $N = M$. Neighborhood averaging is popular, easy to implement and for many applications quite effective. However, as discussed in Section 2.4, the behavior in the frequency domain is less than perfect because the sinc function—the Fourier transform of the averaging operator—has a rather wide transition band and oscillates about the frequency axis (see Fig. 2.7 on page 37).

In order to reduce this undesirable effect, the filter coefficients are sometimes changed ("tampered"). An example is replacing the coefficients in the first and last rows and columns by 0.5 Then, Eq. 3.16 reads

$$h = \frac{1}{(2N+1)(2M+1)} \begin{vmatrix} 0.5 & 0.5 & \dots & 0.5 & 0.5 \\ 0.5 & 1 & \dots & 1 & 0.5 \\ \vdots & \vdots & \vdots & \vdots & \vdots \\ 0.5 & 1 & \dots & 1 & 0.5 \\ 0.5 & 0.5 & \dots & 0.5 & 0.5 \end{vmatrix} \tag{3.18}$$

Gaussian Filter

The Gaussian function has the unique property that its Fourier transform is also a Gaussian. That is, the bell-shaped form of the Gaussian is preserved. This is of considerable interest because it alleviates one of the fundamental problems in filtering: finite window size in the spatial domain results in an infinite transfer function and vice versa. The impulse is an extreme example—its Fourier transform is a constant.

The two-dimensional Gaussian transform is defined by

$$
\begin{aligned}
g(x,y) &= \frac{1}{\sqrt{2\pi\sigma}} e^{-(x^2+y^2)/2\sigma^2} \\
G(\mu,\nu) &= e^{-(\mu^2+\nu^2)/2\sigma^2}
\end{aligned}
\tag{3.19}
$$

The uniqueness of the Gaussian makes it the operator of choice for generating image pyramids and representing images in scale space.

Fig. 3.7 shows the effect of smoothing the original image depicted in Fig. 3.1 on page 43. The images on the top row are the result of convolving the original image by a moving average filter. A closer inspection reveals the ringing effect of this filter, particularly for large operators. The images on the bottom row are the result of using Gaussian filters of different sizes.

3.3.3 Sharpening Operators

For certain applications it is desirable to enhance small details—high frequency components—of an image. A typical example is to emphasize edges. Sharpening can be performed in the spatial or frequency domain.

3.3.4 Difference Operators

The goal of difference operators is to detect changes in the gray levels that occur within small regions. The method is analogous to differentiating continuous functions. Instead of taking the derivatives of a function we take differences of the discrete image function. The first derivatives are approximated by

$$
\begin{aligned}
\frac{\Delta f(x,y)}{\Delta x} &= \frac{f(x+1,y)-f(x,y)}{(x+1)-x} = f(x+1,y)-f(x,y) \\
\frac{\Delta f(x,y)}{\Delta y} &= \frac{f(x,y+1)-f(x,y)}{(y+1)-y} = f(x,y+1)-f(x,y)
\end{aligned}
\tag{3.20}
$$

This leads to the concept of the image gradient.

Image Gradient

The gradient of a continuous function $f(x,y)$ is a vector and its magnitude expresses the rate of change per unit distance in the direction of the vector. The gradient and the magnitude are expressed by

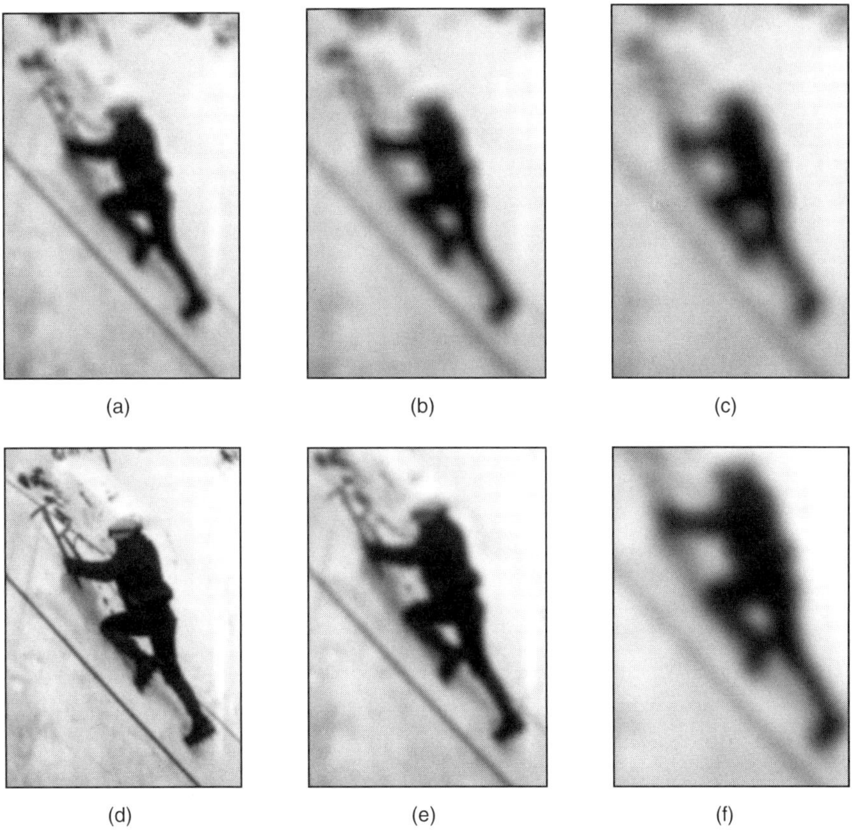

Figure 3.7: Examples of smoothing an image. The original image of Fig. 3.1(a) is smoothed by a moving average operator. The mask in (a) is 15×15, in (b) it is 25×25 and in (c) 35×35 pixels. The bottom row represents images smoothed by a Gaussian with increasing standard deviations. Sigma varies from 2 (d) to 4 (e) and to 8 (f).

$$
\begin{aligned}
\mathrm{grad}[f(x,y)] &= [\vartheta f/\vartheta x, \vartheta f/\vartheta y]^T \\
|\,\mathrm{grad}[f(x,y)]\,| &= \left[\left(\frac{\vartheta f}{\vartheta x}\right)^2 + \left(\frac{\vartheta f}{\vartheta y}\right)^2\right]^{1/2}
\end{aligned}
$$

Substituting the differences given in Eq. 3.20 for the derivatives leads to the image gradient

$$\text{grad}[f(x,y)] = [(f(x+1,y)-f(x,y))^2 + (f(x,y+1)-f(x,y))^2]^{1/2} \quad (3.21)$$

Sometimes this is further approximated by

$$\text{grad}[f(x,y)] \cong |\, f(x+1,y)-f(x,y)\,| + |\, f(x,y+1)-f(x,y)\,| \quad (3.22)$$

Note that the image gradient, or gradient for short, is a scalar—the magnitude of the gradient vector.

From the above discussion, it follows that the gradient operator windows are defined by

$$G_x = \begin{vmatrix} 0 & -1 & 0 \\ 0 & 1 & 0 \\ 0 & 0 & 0 \end{vmatrix} \qquad G_y = \begin{vmatrix} 0 & 0 & 0 \\ -1 & 1 & 0 \\ 0 & 0 & 0 \end{vmatrix} \quad (3.23)$$

Operations based on taking differences between immediate neighbors are quite sensitive to noise. To reduce this effect, different windows have been proposed. We present two popular operators.

a) Sobel Operator: performs better for noisy images because it takes differences between every other row or column. Moreover, the direct neighbors have a higher weight.

$$S_x = \begin{vmatrix} 1 & 2 & 1 \\ 0 & 0 & 0 \\ -1 & -2 & -1 \end{vmatrix} \qquad S_y = \begin{vmatrix} 1 & 0 & -1 \\ 2 & 0 & -2 \\ 1 & 0 & -1 \end{vmatrix} \quad (3.24)$$

b) Roberts Operator: uses a different approximation of the derivatives—or the gradient for that matter. The gradient is approximated by taking the differences along diagonals:

$$\text{grad}[f(x,y)] = [(f(x+1,y+1)-f(x,y))^2 + (f(x,y+1)-f(x+1,y))^2]^{1/2} \quad (3.25)$$

Gradient Image Representation

There are numerous ways of representing the results of gradient operations. A gradient image is obtained by entering the gradient as a new gray level at the location of the central pixel. Smooth regions render small gradient values. As a consequence, the gradient image appears mostly dark because in natural scenes smooth regions dominate. Therefore, a better way to represent a gradient image is to threshold it in such a fashion that only large gradients are entered; small gradient values leave the original image unchanged. Such a thresholding operation may be formulated as follows:

$$g(x,y) = \begin{cases} \mathrm{grad}[f(x,y)] & \text{if grad} \geq T \\ f(x,y) & \text{otherwise} \end{cases} \tag{3.26}$$

where $f(x,y)$ is the original image, $\mathrm{grad}[f(x,y)]$ the response of gradient operation, and $g(x,y)$ the gradient image used for representation.

Fig. 3.8 on the following page demonstrates the result of a Sobel operation. The original image in Fig. 3.8(a) is convolved with the two masks of Eq. 3.24. The resulting gradient images in Fig. 3.8(c) and (d) illustrate that in natural images the gradients do not change much. The histogram of the gradient image (d) is shown in Fig. 3.8(b). It confirms the fact that the gradient images appear gray, without much contrast. A better appearance is obtained by binarizing the gradient images and by superimposing the result on the original image. Based on a histogram analysis, all gray levels in the gradient images smaller than 80 have been set to zero. On the upper end, all gray levels larger than 180 were set to 255. The resulting binary images are then superimposed on the original image. The comparison between Fig. 3.8(c) and (e) and Fig. 3.8(d) and (f) shows the improved visualization effect.

Laplace Operator

The differencing operators discussed so far are first derivative (gradient) operators. They are all direction dependent. For detecting discontinuities in all directions, different masks must be used. For example, the gradient operator has two masks and detects discontinuities in rows or columns. For certain applications it may be desirable to use an operator that is orientation independent. A simple operator with this property is the Laplace operator. For a continuous function $f(x,y)$ it is defined by

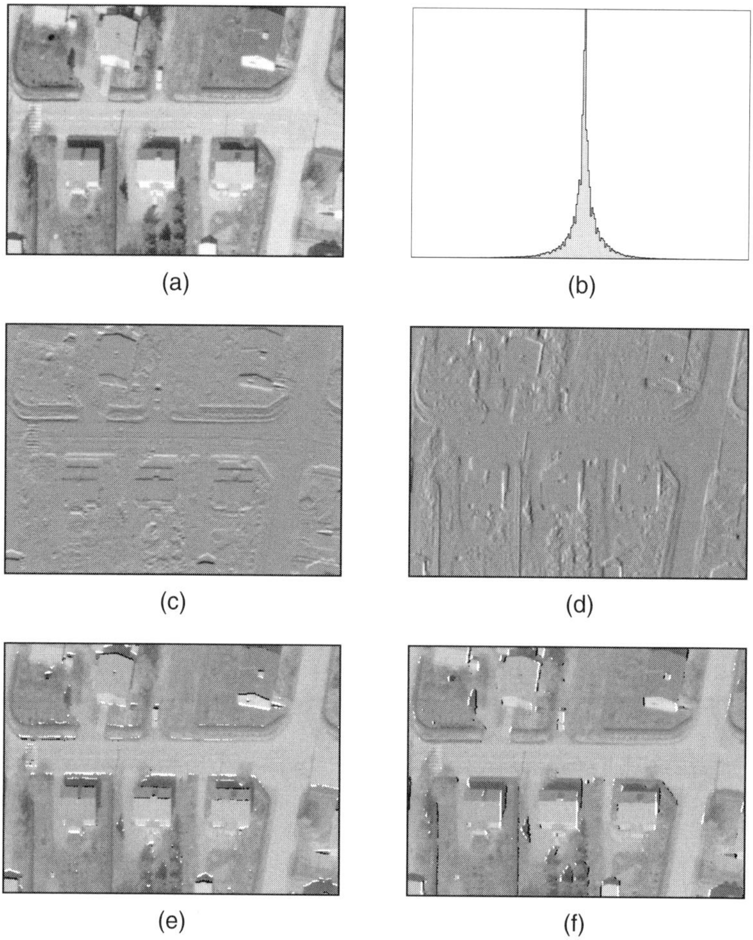

Figure 3.8: Result of convolving an image with the Sobel operator. In (a) the original image is shown, while (c) and (d) are the images after convolving the original image with the horizontal and vertical Sobel masks. The histogram in (b) shows that the gray levels of the Sobel images are concentrated about the mean. The gray levels at the tails of the histogram are set to zero and 255, respectively, and superimposed on the original image. The result is shown in (e) and (f). Compare text for details.

$$\nabla^2 = \frac{\vartheta^2 f(x,y)}{\vartheta x^2} + \frac{\vartheta^2 f(x,y)}{\vartheta y^2} \tag{3.27}$$

Its discrete realization can be derived from taking the second differences in rows and columns. That is, we take the differences of the gradients as shown below

$$\begin{aligned}\nabla_x^2 &= [f(x+1,y) - f(x,y)] - [f(x,y) - f(x-1,y)] \\ &= f(x+1,y) - 2f(x,y) + f(x-1,y) \quad (3.28)\\ \nabla_y^2 &= [f(x,y+1) - f(x,y)] - [f(x,y) - f(x,y-1)] \\ &= f(x,y+1) - 2f(x,y) + f(x,y-1) \quad (3.29)\end{aligned}$$

Adding the two component equations together results in the Laplace mask

$$L = \begin{vmatrix} 0 & 1 & 0 \\ 1 & -4 & 1 \\ 0 & 1 & 0 \end{vmatrix} \quad (3.30)$$

Since the Laplacian is a second derivative operator, it is even more sensitive to noise than the gradient operators discussed earlier. Sometimes, the Laplacian is modified as follows:

$$L = \begin{vmatrix} 1 & 1 & 1 \\ 1 & -8 & 1 \\ 1 & 1 & 1 \end{vmatrix} \quad (3.31)$$

A possibility to represent the result of convolving an image by the Laplacian is to subtract it from the original image. Let I_o be the original image, I_c the convolved image, and I_r the image to be represented. The two-step operation of $I_c = I_o \star L$ and $I_r = I_o - I_c$ can be combined as follows: $I_r = I_o \star (Z - L)$ where Z is an operator that leaves the original image unchanged. The new mask $S = Z - L$ is called the *sharpening operator*. It is obtained by

$$S = \begin{vmatrix} 0 & 0 & 0 \\ 0 & 1 & 0 \\ 0 & 0 & 0 \end{vmatrix} - \begin{vmatrix} 1 & 1 & 1 \\ 1 & -8 & 1 \\ 1 & 1 & 1 \end{vmatrix} = \begin{vmatrix} -1 & -1 & -1 \\ -1 & 9 & -1 \\ -1 & -1 & -1 \end{vmatrix} \quad (3.32)$$

3.3.5 *Image Correction*

Image correction pursues the goal of removing spurious image defects. This may include individual pixels with gray levels radically different from their immediate neighbors (blunders), or groups of pixels, usually arranged along straight lines. Such image defects are likely to be caused by improperly working cameras and scanners. Median filters and frequency domain filters are often used to eliminate image defects.

Median Filter

Unlike the other operators presented so far, the median filter is not linear. The median filter is based on neighborhood ranking. The pixels in the neighborhood are ranked according to brightness. Then, the median value in this ordered list is used as a representative gray level for the central pixel. Fig. 3.9 demonstrates the principles of median filtering.

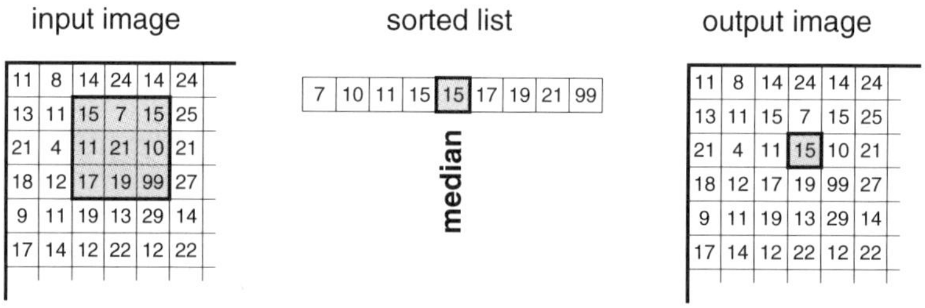

Figure 3.9: Example of the median operator.

As the example in Fig. 3.9 demonstrates, the median filter is well suited to eliminate "point errors"—individual pixels that are corrupted or missing. This is not an uncommon problem, for example, individual photo receptors in digital cameras or scanning systems may be defective.

Eliminating Erroneous Pixels

The median filter eliminates point errors but at the cost of smoothing the image. If smoothing is not desired then the following approach is better suited to eliminate single point errors.

Suppose we compare a pixel with the average of its eight neighbors. If the difference exceeds a threshold then it is considered wrong and the gray level is replaced by the average. We modify the averaging mask of Eq. 3.16 on page 53 by changing the center value and obtain the convolution mask as follows:

$$h = \frac{1}{8}\begin{vmatrix} 1 & 1 & 1 \\ 1 & 0 & 1 \\ 1 & 1 & 1 \end{vmatrix} \tag{3.33}$$

A pixel is considered wrong if $\|h * f - f\| > T$. Here, f is the image function and T is a threshold value.

Eliminating Erroneous Rows or Columns

A real digital image is obtained by using a camera or a scanner. These devices may occasionally malfunction, corrupting entire rows or columns. One possibility to detect corrupted rows or columns is to compare them with neighboring rows or columns. The detection of errors is based on the assumption that neighboring rows have similar gray level distributions—a reasonable assumption for natural scenes. A suitable measure for the similarity of the gray levels is obtained by correlation.

The covariance $r(i, i+1)$ between rows i and $i+1$ is defined by

$$r(i, i+1) = \frac{1}{C}\sum_{j=0}^{C-1}[f(i,j) - f_a(i)][f(i+1,j) - f_a(i+1)] \tag{3.34}$$

where C is the number of columns, $f_a(i), f_a(i+1)$ the average gray levels of rows i and $i+1$, respectively.

The correlation factor is obtained by normalizing the covariances. We have

$$\rho(i, i+1) = \frac{r(i, i+1)}{[r(i,i) \cdot r(i+1, i+1)]^{1/2}} \tag{3.35}$$

If the correlation factor $\rho(i, i+1)$ is low then one of the two rows is most likely wrong. If this procedure is repeated for all pairs $i, i+1, i = 0, 1, \ldots, C-2$, a suitable threshold value can be determined by analyzing the correlation coefficients. Fig. 3.10 demonstrates the procedure. The test image in Fig. 3.10(a) has two bad rows, one with gray levels that are nearly 255 and the other one with very small gray levels. All neighboring rows are correlated by the procedure described above. The resulting correlation coefficients are plotted in Fig. 3.10(c). This graph shows a high correlation for neighboring rows, except for the two bad rows where the correlation coefficient is zero. Analyzing the correlation coefficients leads to the detection of the bad scan lines. They are then corrected by averaging the neighboring pixels.

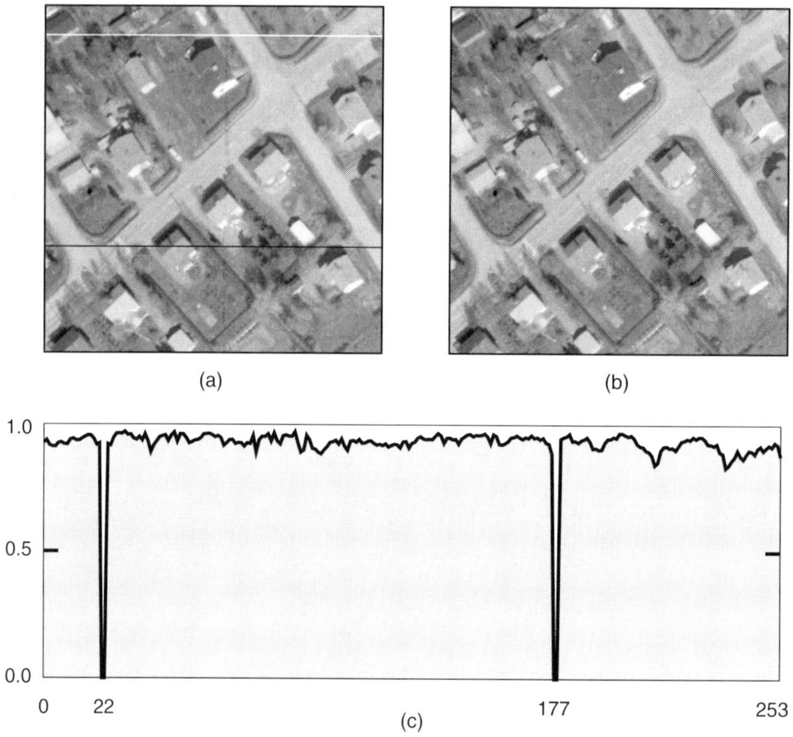

Figure 3.10: In (a) an image with two bad rows is shown. These two rows are detected by correlating all neighboring rows. The correlation coefficients for the 254 row pairs are shown in (c). The bad rows have no correlation to their immediate neighbors. The corrected image in (b) is obtained by replacing the bad rows with the average of the two immediate neighbors. Note the high correlation between neighboring rows.

3.4 Geometric Transformations

Unlike the operations discussed so far, geometric transformations change the tesselation. The resulting image differs in size and perhaps in shape. Geometric transformations are quite common in digital photogrammetry. We mention a few examples:

- after the interior orientation it may be desirable to perform a transformation such that the image coordinate system becomes parallel to the photo coordinate system and is of the same scale;

- after the relative or absolute orientation aerial image pairs are usually transformed to their normalized position (see Chapter 12). In this position, rows are epipolar lines and conjugate points are in the same rows;
- rectification is an important photogrammetric procedure, ranging from simple plane rectification to the more complex process of generating digital orthophotos;
- draping is a popular way to display perspective views of surfaces (DEMs) that include gray shades or colors;
- image registration, also called rubber-sheeting, is a polynomial transformation of an image, to a set of control points. This procedure is typically performed to register satellite imagery to maps.

Various mathematical models exist to formulate the relationship between two images, or between an image and the object space. A generic class of transformations can be expressed by

$$\mathbf{x}' = \mathbf{S} \cdot \mathbf{R} \cdot \mathbf{x} + \mathbf{t} \tag{3.36}$$

where $\mathbf{x}$ is a point vector in the input image, $\mathbf{x}'$ the corresponding point vector in the transformed image, $\mathbf{S}$ a scale matrix, $\mathbf{R}$ a rotation matrix (not necessarily orthogonal), and $\mathbf{t}$ the translation vector.

We do not elaborate further on the mathematical models here but focus on the consequences of changing the tesselation. Let us use the rotation of an image as an example to demonstrate the problems of determining the size of the transformed image and its the gray levels.

3.4.1 Size of Transformed Image

Fig. 3.11 shows the effect of rotating an image $f(x, y)$, size $R \times C$. Let $f'(x, y)$ be the rotated image, size $R' \times C'$.

We have for the size of the rotated image

$$R' = R\cos\alpha + C\sin\alpha \tag{3.37}$$

$$C' = C\cos\alpha + R\sin\alpha \tag{3.38}$$

Note that the transformed image contains regions with no information. Sometimes it may be desirable to cut out a square-shaped image from

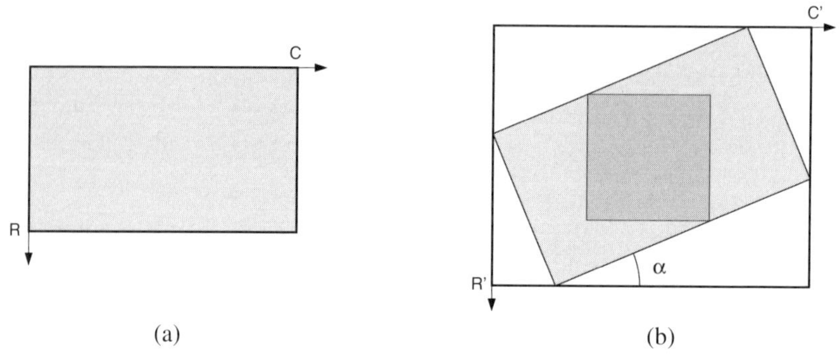

Figure 3.11: Example of image rotation. The original image in (a) is rotated about α. The rotated image in (b) is larger and has parts without information.

the rotated image that has no regions uncovered by the original image. Fig. 3.11(b) shows this image square.

The rotated image has larger dimensions. For representing the entire rotated image we must either increase the pixel size or allocate more pixels.

3.4.2 Resampling

Most geometric transformations lead to pixels that do not coincide with the original image. Fig. 3.12 shows the tesselation of the original image and—superimposed—that of the transformed image. Let the left upper corner of a pixel be its origin. We now have the problem of assigning a proper gray level to the transformed pixels. This can be accomplished by assigning an existing gray level of the original image or by interpolating one from neighboring pixels. The first approach amounts to truncating the real value of the transformed pixel to its integer value (Fig. 3.12(a)), or to rounding it (Fig. 3.12(b)). The latter approach is known as *nearest neighbor* resampling. The advantage of these two approaches is that the gray levels of the original image are preserved. As a consequence, the geometry of the transformed image is distorted: the pixel location is moved to locations of the original image.

The second approach to assigning gray levels in the transformed image is based on interpolation. Several methods can be used to interpolate the gray levels from neighboring pixels in the original image.

Fig. 3.13(a) illustrates the concept of *bilinear* interpolation. As the name suggests, it is a linear interpolation in the direction of rows and columns.

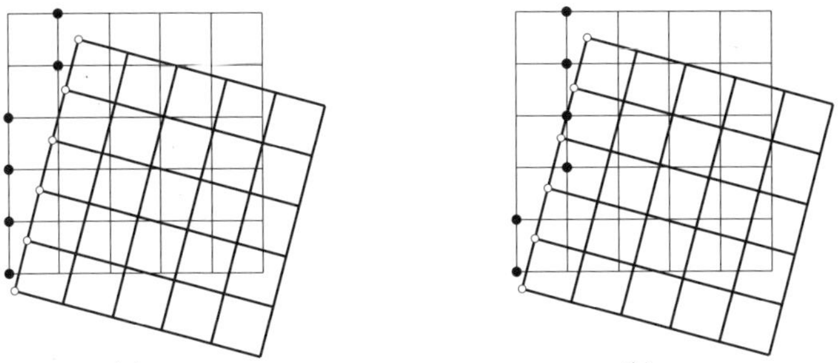

(a) (b)

Figure 3.12: Tesselation of the original image. Superimposed is the tesselation of the transformed image. The grid posts are obtained by the inverse transformation. In (a) gray levels are assigned to the transformed image from the original image at the truncated positions of the real values. In (b) the real-valued pixel locations are rounded to the nearest integer values (nearest neighbor method). Note that these two methods preserve the original gray levels but introduce geometric distortions due to truncating or rounding the transformed, real-valued pixels.

First, intermediate gray levels are computed at the positions R_1 and R_2, followed by an interpolation between these two points. Let x, y be the fractional parts of the transformed row and column, respectively. Then, we have for the bilinear interpolation:

$$\begin{aligned} R_1 &= x \cdot g_{i,j} + (1-x) \cdot g_{i+1,j} \\ R_2 &= x \cdot g_{i,j+1} + (1-x) \cdot g_{i+1,j+1} \\ P &= y \cdot R_1 + (1-y) \cdot R_2 \end{aligned}$$

Another approach is to interpolate the gray level as a weighted average of the four neighbors. Fig. 3.13(b) depicts the case of using the reciprocal of distances as weights.

Fig. 3.13(c) depicts the special case where the transformed pixels have irregular shapes. This situation is typical for orthophotos. Here, the regular tesselation of a DEM is transformed to the image. Depending on the topography, the pixels may be quite distorted and the correct way to interpolate a gray level is to consider all pixels in the image that are covered by the transformed pixel. The weights, then, would be chosen proportional to the contributing areas.

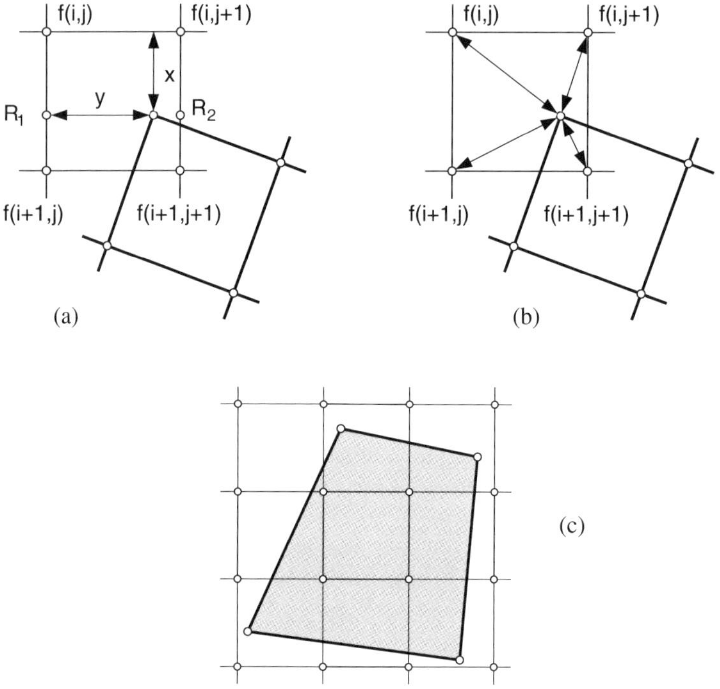

Figure 3.13: Resampling by interpolation. Bilinear interpolation is shown in (a). Two methods of weighted averaging are shown; in (b) with distances as reciprocal weights and in (c) where weights are proportional to the contributing areas of pixels.

Regardless of the approach, interpolating gray levels has the disadvantage that the original gray levels are distorted. Moreover, interpolation has a smoothing effect. This may render a more pleasing appearance of the image. On the other hand, smoothing may have an unwanted effect on edge detection (dislocalization).

3.5 Image Segmentation

The goal of image segmentation is to partition an image into regions that correspond to something "meaningful" in the scene, say an object such as a building or areas that share similar characteristics. From this rather vague description it is obvious that segmentation is application dependent. What may be an interesting region for one application may be totally irrelevant in another setting. Once an image is segmented it is then further analyzed, either automatically or interactively. Thus, image segmentation is a preprocess of image understanding (scene analysis).

Segmentation methods can be divided into two categories: point methods and region dependent methods. Region-based segmentation methods assign pixels to a certain region by taking local image properties into account. That is, the neighborhood of a pixel is analyzed for the occurrence of properties that are characteristic for a region. Finding regions of similar image properties can be approached in two ways: pixel classification or boundary determination. The first approach can be thought of as a region growing process: select a seed pixel in a region of interest and expand it in all directions until the image properties change. The boundary of the region is found implicitly in this case. The goal of the second approach is to explicitly locate the boundary. The region, then, is implicitly determined by its boundary. Again, the decision of which approach is better suited depends on the application. Sometimes, a combined approach is most adequate.

The next section describes histogram thresholding as an example of point-based segmentation. Two region-based methods—the co-occurrence matrix and texture segmentation by Gabor filters—will be briefly introduced.

3.5.1 Histogram Thresholding

A simple approach for segmenting an image is to define ranges of brightness values that hopefully correspond to meaningful regions in the scene. This is accomplished by choosing suitable threshold values from the gray level histogram, followed by dividing the image into regions. Fig. 3.14 exemplifies the procedure. The histogram of this figure shows three distinct bands of gray levels. Setting thresholds is straightforward and the segmented image is easily obtained (Fig. 3.14(c)).

The success of histogram thresholding depends entirely on the separability of gray level bands in the histogram and of the spatial occurrence of the gray levels. Consider an image where the gray levels occur with the same frequency as is the case in Fig. 3.15. Clearly, the histogram cannot be divided and segmentation based on histogram thresholding fails even though the original has separable regions.

The next example (Fig. 3.16) shows an image that has the same histogram as Fig. 3.14(a). Although this histogram can easily be divided into three regions, the segmentation is meaningless because the selected gray level bands do not correspond to spatially coherent regions.

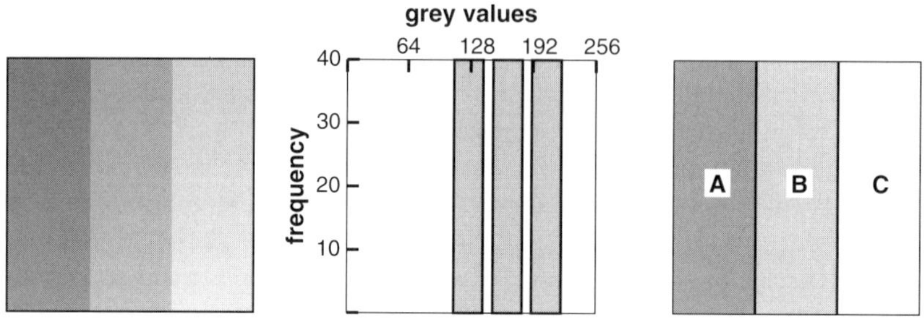

(a) (b) (c)

Figure 3.14: The synthetic image in (a) contains three distinctly different regions (gray wedges). The histogram in (b) shows three clusters and suggests to set threshold values at 148 and 190, respectively. Pixels with gray values from the first cluster are shown darkest in (c). Likewise, pixels belonging to the second cluster are shown lighter. The result is the three regions A, B, and C.

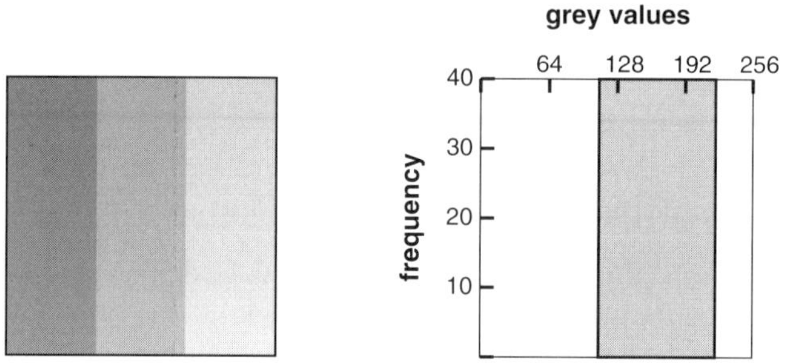

(a) (b)

Figure 3.15: Although the synthetic image in (a) contains three distinctly different regions, the histogram (b) cannot be thresholded because all gray levels occur with the same frequency.

The usefulness of histogram thresholding is further demonstrated in Fig. 3.17. Figure (a) shows a corner fiducial mark of an aerial photograph. For the purpose of interior orientation it is desirable to detect the fiducial marks automatically. First, we try to detect all pixels that make up the fiducial mark. The histogram in Fig. 3.17(b) shows three clusters: the dark pixels of the fiducial background; the bright pixels of the fiducial mark

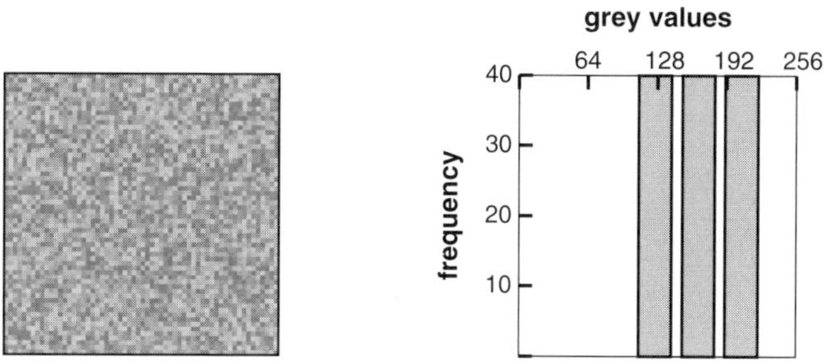

(a) (b)

Figure 3.16: The synthetic image in (a) contains the same gray values as Fig. 3.14(a) but the pixels are now randomly distributed. Although the histogram in (b) shows three clusters that can be easily thresholded, the segmentation is meaningless because the histogram clusters do not correspond to spatially coherent regions.

and bright image content; and the gray pixels from the surrounding image. Thresholding the histogram at about 128 would include the fiducial mark as well as bright pixels from the image (ice field). The dark pixels corresponding to the fiducial mark background can be easily thresholded, however. The segmented image in (c) is obtained in two steps. First, the dark background is separated from the rest of the image, followed by a second segmentation, but now only within the dark region of the background. The result, shown in Fig. 3.17(c), contains three regions. Gray corresponds to the image content, dark to the fiducial background, and white to the fiducial mark itself.

The previous example with the fiducial mark utilizes domain knowledge. It is reasonable to assume that the camera type is known and with it, the implementation of the fiducial marks, for example, a white circle and spikes against a dark background.

3.5.2 Co-Occurrence Matrix

The co-occurrence matrix is a region-based segmentation. We introduce the method by way of an example. Fig. 3.18(a) shows an aerial photograph of the Jakobshavn Glacier in Greenland. Suppose a glaciologist is studying ice flow patterns in order to determine ice velocities. This, in turn, may require segmenting the image into regions of similar surface texture. Fig. 3.18(b)

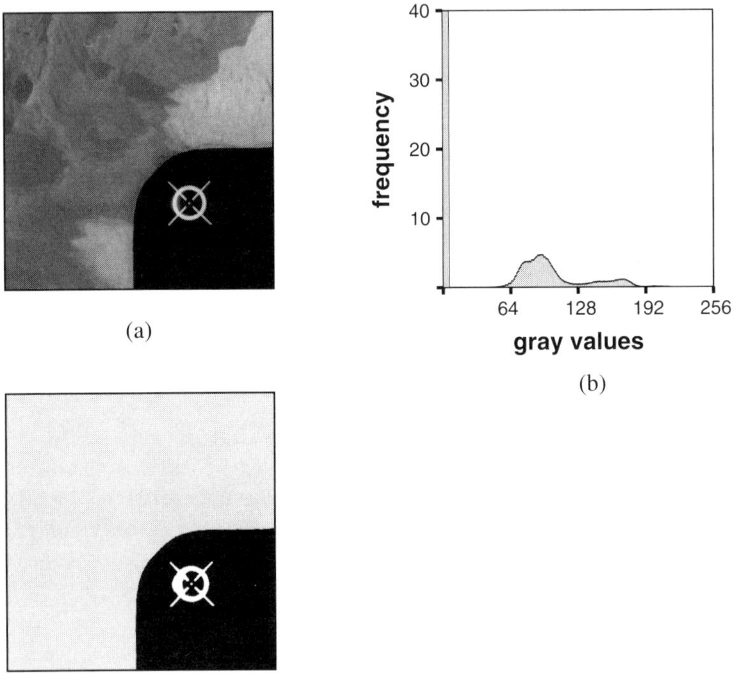

(c)

Figure 3.17: The image in (a) shows a corner of an aerial image containing a fiducial mark. The histogram in (b) shows three clusters. The segmented image in (c) is obtained by separating the dark fiducial background, followed by a segmentation confined to the background region.

depicts a small sub-area of the glacier that contains three slightly different texture patterns. As the histogram in Fig. 3.18(c) clearly indicates, there is no way that it can be thresholded for the purpose of segmenting the original image.

Texture is a variation of brightness in a small region. The variation is somewhat regular and repetitive. The variance of the gray levels of a region is a simple measure for the variation. However, the variance would not be suitable to detect orientation dependent structures such as stripes.

Another measure for texture is the co-occurrence matrix. Here, a matrix is constructed that contains the frequency of gray level pairs occurring in predefined spatial patterns. Consider Fig. 3.19 on page 72, for example. The synthetic image, size 8×8 pixels, contains four different gray

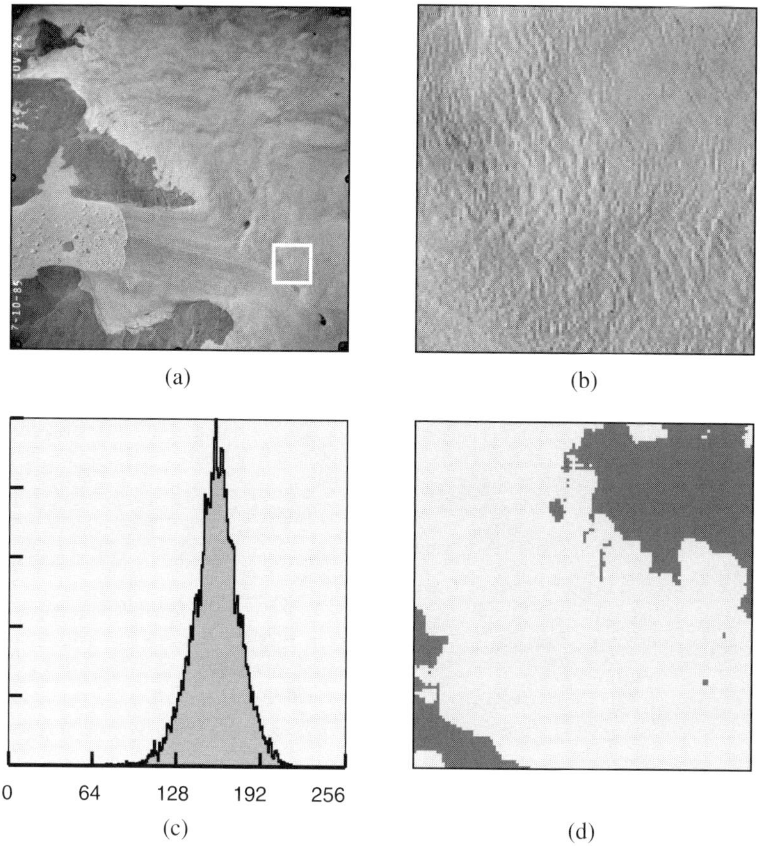

Figure 3.18: In (a) the Jakobshavn Glacier in Greenland at the calving front is shown. The highlighted area shown in (b) is approximately 2.8 km square. The histogram of this sub-image in (c) does not offer any clue about how to segment the sub-image. A segmentation based on the co-occurrence matrix is shown in (d) (see text for a more detailed explanation).

levels. The dimension of the co-occurrence matrix is equal to the number of gray levels (4 in our example). The spatial pattern is characterized by the position of a pixel with respect to the center pixel. Following the chain-code pattern, the direct neighbor to the right would have the number 0, the diagonal neighbors up and down the numbers 1 and 7, respectively. The determination of the co-occurrence matrix involves counting the pixel combination $g_n(i,j); g_m(i',j'), n = 0, \ldots, g_{max}, m = 0, \ldots, g_{max}$ where i, j is the location of the pixel under consideration and i', j' the pixel location determined by the spatial relationship. For example, with the relationship "0" we have $i' = i, j' = j + 1$.

columns

rows

3	3	3	3	3	3	3	3	3	3
1	3	3	3	3	3	3	3	3	2
1	1	3	3	3	3	3	3	2	2
1	1	1	3	3	3	3	2	2	2
1	1	1	1	3	3	2	2	2	2
1	1	1	1	0	0	2	2	2	2
1	1	1	0	0	0	0	2	2	2
1	1	0	0	0	0	0	0	2	2
1	0	0	0	0	0	0	0	0	2
0	0	0	0	0	0	0	0	0	0

(a)

	0	1	2	3
0	16	0	3	0
1	3	6	0	3
2	0	0	6	0
3	0	0	3	16

(b)

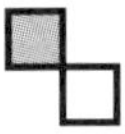

	0	1	2	3
0	10	0	0	0
1	5	8	0	0
2	0	0	6	0
3	1	0	4	15

(c)

Figure 3.19: Demonstration of co-occurrence matrix. The synthetic image in (a) contains four different gray levels. The co-occurrence matrix in (b) shows the occurrence (frequency) of gray level combinations between a pixel and its direct right neighbor (chaincode = 0). In (c) the neighboring pixels are related to chaincode 7. The co-occurrence matrices have dimensions of $n \times n$ with n the number of gray levels in the original image.

3.5.3 *Texture Segmentation by Gabor Filters*

Texture[1] provides useful information about shape, orientation, and depth of objects. The human visual system employs texture as a visual cue for object recognition and image interpretation. There is general agreement that texture is—together with size, tone, shape and pattern—a basic element for manual photo interpretation.

There is no universally accepted formal definition of texture. It can be described by its coarseness, contrast, density, orientation, frequency, repetitiveness of spatial patterns, for example. These texture primitives

[1]The text in this section, including the figures, is based on D.C. Lee's PhD dissertation *An adaptive texture segmentation approach for applications in digital photogrammetry*, Department of Geodetic Science and Surveying, Report No. 444, The Ohio State University (1997).

(elements), also called texels, are to be estimated in texture analysis. As you would expect from the lack of a rigorous definition, many approaches exist to extract texture primitives and to group them into meaningful regions.

Gabor Functions

Two-dimensional Gabor functions are attractive in texture analysis because they perform in a fashion that is remarkably similar to the human visual system. In fact, Gabor functions can be conceived as hypothetical structures of receptive fields in the visual cortex.

The two-dimensional Gabor function is defined by a complex sinusoidal, modulated by a Gaussian,

$$g(x,y) = \frac{1}{2\pi\sigma_x\sigma_y} e^{-\left(\frac{x^2}{2\sigma x}+\frac{y^2}{2\sigma y}\right)} e^{(2\pi i(Ux+Vy))} \tag{3.39}$$

with σ_x and σ_x the standard deviations with respect to the x- and y-axis, and $U = F\cos\theta$, $V = F\sin\theta$, where $\theta = \arctan(V/U)$.

Since the retinal receptive fields (see Sec. 4.2.3) in human vision have unity aspect ratio, that is, $\sigma_x = \sigma_y = \sigma$, Eq. 3.39 simplifies to

$$g(x,y) = e^{-\left(\frac{x^2+y^2}{2\sigma^2}\right)} \sin\left(\omega(x\cos\theta - y\sin\theta) + \varphi\right) \tag{3.40}$$

where the angular frequency is $\omega = 2\pi F$ and φ denotes the phase shift. Fig. 3.20 shows a perspective view of Gabor filters in the spatial domain. Parameters in Gabor filters are

- σ the standard deviation of the Gaussian envelope
- ω spatial frequency
- θ orientation
- φ phase

These four parameters characterize texture patterns. Thus, Gabor filters allow for the extraction of different texture primitives, depending on the choice of the parameters. This is an important issue: texture analysis as a means to extract useful information from images requires a variety of parameters.

Selection of Filter Parameters

Probably the most crucial step in texture segmentation by Gabor filters is the selection of appropriate filter parameters. As a first step, the image

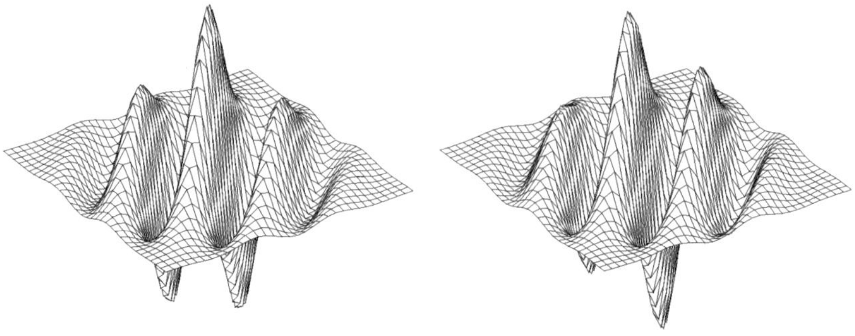

Figure 3.20: Perspective view of Gabor filters. The real component (left) is even-symmetric while the imaginary component (right) is odd-symmetric.

is subdivided into regions that serve as an initial segmentation. This may be accomplished by determining the entropy (see Sec. 3.2.2), for example. Entropy measures the degree of randomness of the gray levels in an image. The Gabor filter parameters are now independently determined in each subdivision.

a) Orientation: The underlying assumption is that texture is a dominant local orientation in each subdivision. A straightforward approach used to determine the dominant orientation, θ, is to compute the weighted average of the gradient vectors at every pixel within the region.

$$\theta_{ij} = \arctan\left(\frac{G_y(i,j)}{G_x(i,j)}\right) \tag{3.41}$$

$$l_{ij} = \sqrt{G_x^2(i,j) + G_y^2(i,j)}$$

$$\theta = \frac{\sum_{i=0}^{m-1}\sum_{j=0}^{n-1}\theta_{ij}l_{ij}}{\sum_{i=0}^{m-1}\sum_{i=0}^{m-1}l_{ij}} \tag{3.42}$$

where $G_x(i,j), G_y(i,j)$ are the x- and y-component of the gradient vector at pixel (i,j), and $m \times n$ is the window size.

The standard deviation of the estimated orientation in Eq. 3.42 serves as an indication for the significance of the local orientation. In aerial imagery depicting natural scenes, there is quite often no significant local orientation, and one has to resort to discretizing the orientations, for example in increments of 45°.

b) Spatial Frequency: The local spatial frequency is defined as $\omega = 2\pi/T$ with T the wavelength that is considered an average local dimension of the texture elements within the subdivision. Since the spatial frequency must be determined with respect to the dominant orientation it stands to reason to compute the orientation first.

c) Filter Size: The standard deviation of the Gaussian envelope controls the Gabor filter size. The selection of an appropriate size depends on the image content, scale, and resolution. We will see in Chapter 4 that the human visual system employs several different sizes of the retinal-ganglion-cell receptive fields. This is the motivation for selecting filter sizes that differ by an octave. Again, the entropy determines a proper filter size for every subdivision.

Fig. 3.21 shows the result of processing an aerial image with pairs of Gabor filters (0°- and 90°-phases). The convolution is performed region by region adaptively, starting with the initial subdivision explained earlier.

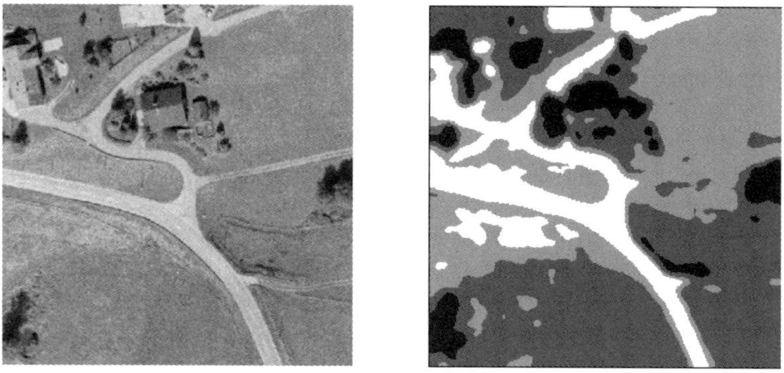

Figure 3.21: Aerial view of an urban scene (left), size 512×512 pixels. The texture segmentation by Gabor filters is shown in the right figure. Four filter sizes with $\sigma = 3, 6, 9, 12$ have been used.

3.6 To Probe Further

There are numerous books on the subject of digital image processing. The classic reference is Pratt's book *Digital Image Processing* (Wiley, 1978). An excellent reference is *Handbook of Image Processing* (CRC Press, 1996) by J. Ross. This book is easier to read and offers a wealth of examples. The two books *Digital Picture Processing* (Academic Press, 1982) by Rosenfeld & Kak, and *Digital Image Processing* by Gonzalez & Wintz (Addison-Wesley, 1987) also provide good introductions to the subject. A more recent addition is *Digital Image Processing* (Prentice-Hall, 1996) by K. Castleman. Jähne's *Practical Handbook on Image Processing for Scientific Applications* (CRC Press, 1997) merges basic research in digital image processing with key applications in various disciplines.

Chapter 4

Human Vision

For an animal or person to respond properly to a changing environment it must detect objects, events, and structures. This ability, called perception, requires that a living organism be sensitive to different stimuli, which carry important information about the environment. Most animals have some visual perception abilities. For people, vision is the most important sense. It is also by far the most impressive and complicated sense.

We see and analyze our environment continuously, nearly in real time. Doing this without conscious effort does not imply that we know *how* we analyze and understand scenes, however. In fact, the lack of a detailed understanding of vision is the reason why it is so difficult to program a computer to analyze and understand images. It seems only natural, then, that someone who attempts to solve part of this task should have a basic understanding of human vision. Consider this chapter as an exciting journey through the fascinating world of vision.

4.1 Overview of Human Visual System

Imagine you are about to cross a busy street. Reflected light from the street scene is incident on the 120 million or so rods and cones on the retina which transform the light intensities into nerve signals that are transmitted back to the visual cortex and higher brain centers. Somewhere there, the scene is interpreted and represented in such a fashion that you can make decisions, plan and execute proper actions.

Visual information is processed in various stages at centers of specialized nerve cells, from the retina via the lateral geniculate body to the primary visual cortex and on to higher cortical brain areas. The processing centers are connected by the visual pathway, depicted in Fig. 4.1, and can be thought of as a serial link—impulses traveling along axons, passing information to dendrites.

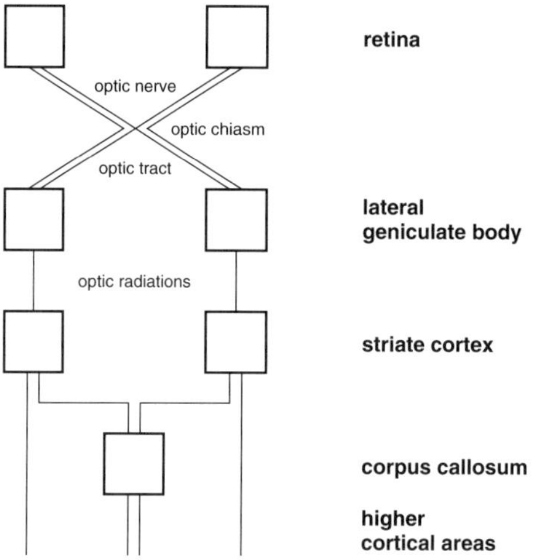

Figure 4.1: Schematic diagram of the visual pathway. Each structure consists of millions of cells. Information is sent to one or several higher order structures.

As illustrated in Fig. 4.2, a nerve cell, or neuron, consists of a globular shaped cell body (nucleus), an axon and dendrites. The axon transmits signals to other neurons. The nucleus and the tapering fibers are enclosed by the membrane. The signal traveling along the axon is called an impulse. Nerve cells contain salt water and they are immersed in water with different salts, such as potassium, sodium and calcium chloride.

The membrane has tiny holes (pores, pumps and channels) through which ions of the salt water can enter or leave the axon. The pores are specialized; they either let ions of one kind pass the membrane or eject them from the inside. The opening and closing of pores is controlled by the potential across the membrane or by chemicals inside or outside of the membrane.

During the resting state, potassium pores are open and eject K^+ ions from the axon to the outside. As a result, a negative potential across the membrane of 70 mV is built up. If an impulse occurs, the sodium pores open and Na^+ ions flow inside the axon, hence, reversing the negative resting potential to minus 40 mV. This process is called *depolarization* and it takes place over a relatively small range of the axon. On both ends of this active state, the membrane is in its resting state. The positive potential triggers the next section of the membrane to depolarize and the entire process of ions flowing in and out of the axon repeats further and further away from the cell body; the impulse travels to the branching ends of the axon. There, neurotransmitters are released through the tiny gap between axons and dendrites, termed synaptic claft.

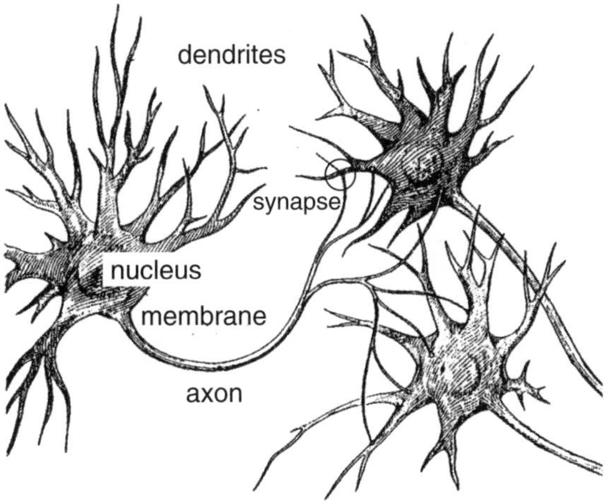

Figure 4.2: Nerve cells consist of the cell body, the axon, and dendrites, all enclosed by the membrane. Axons pass impulses to dendrites at synapses. (From *Eye, Brain, and Vision* by Hubel ©1988 by Scientific American Library. Used with permission by W.H. Freeman and Company).

The next sections provide a concise summary of human vision from a neurophysiological perspective. Neurophysiology is concerned with the processes that are performed by specialized tissues and cells of the nervous system. A detailed knowledge about cell structures and their interrelationships does not yet explain the way the world looks to us. This is why we will turn to the perception perspective of vision in the remaining part of this chapter.

An analogy may serve as an example to motivate the two approaches to vision. To understand a computer system requires analyzing the software (perception) and the hardware (neurophysiology, anatomy).

4.2 The Eye

Vision begins with an eye. Light reflected from objects in the scene is focused by the lens onto the retina where it is transformed by the photoreceptors into electrical signals that are immediately processed by the retina. The output is transmitted back to the visual cortex by the optic nerve.

This section describes the optical properties, such as image formation and resolution, as well as the photometric properties of the eye. It is tempting to compare the eye with a camera. The analogy must be met with caution, however. First, the quality of the retinal image is far inferior to that of any disposable camera. Aberrations of lens and cornea cause considerable distortions. The curvature of the retina makes straight lines in object space appear curved, disturbing the metrical relationship between image and object space. Moreover, the constant movements of the eye result in a blurred image. While the purpose of the camera is to render a static snapshot of the world, the eye's and brain's purpose is to extract useful information to guide a person's response to an ever changing environment.

The camera model does not explain why things appear to us the way they do. Extracting useful information for scene interpretation from the photoreceptor signals begins on the retina. This section concludes with describing receptive fields and how ganglion cells respond to incident light.

4.2.1 The Eyeball

The eye is nearly spherical in form with an average diameter of 20 mm. As illustrated in Fig 4.3, the eye is enclosed by three membranes. The innermost membrane is the retina which carries the photoreceptors. The next layer is the choroid followed by the sclera. At the anterior extreme, the sclera becomes the cornea, a tough, transparent tissue that contributes most to the refractive power of the eye.

The ciliary body and the iris are the anterior continuation of the choroid. The choroid contains blood vessels, which supply the eye with nutrition. Reflected light bouncing around in the eyeball is absorbed by black pigments submersed in the choroid.

The lens consists of concentric layers of cells. Fibers attach it to the ciliary muscles, which, in turn, control the curvature of the lens to change the focal length so that objects at different distances are sharply imaged on the retina.

Light is focused on the retina to form an image. Some 125 million light sensitive *photoreceptors* (rods and cones) are unevenly distributed over the entire posterior portion of the eyeball. The retina consists of three layers: photoreceptors, middle layer, and ganglion cells whose axons are bundled

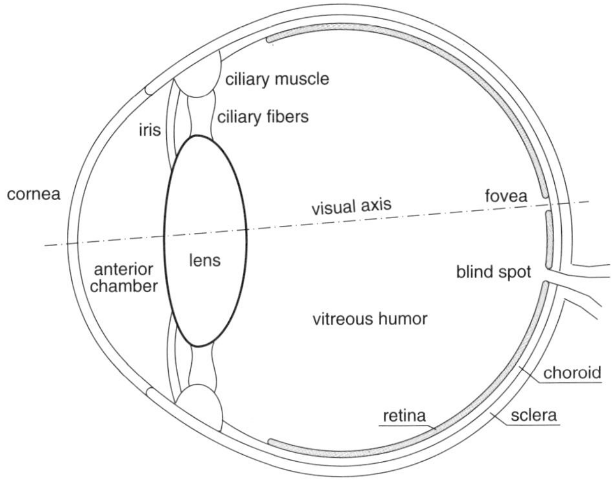

Figure 4.3: Simplified cross section of human eye.

together to form the optic nerve. Fig. 4.4 shows a cross section through the retina. Note that light passes through the ganglion cells and the middle layer before it reaches the photoreceptors! The small pit at the intersection of the visual axis and retina is the *fovea,* the site of acute vision. Here, at the very center, the other cell layers are pushed aside and only cones are present.

The optic nerve leaves the eyeball through a hole in the retina: the optic disc, also known as the *blind spot.* The latter name expresses vividly the fact that the physical image at the site of the optic nerve has a 'hole' since no photoreceptors are present. Referring again to the camera model implies that the photograph obtained from a scene would have a black disc somewhere near its center. Also, through the optic disc, arteries enter and veins leave.

4.2.2 Optical Properties of the Eye

Image Formation

Elements which contribute to the formation of an image include the cornea, aqueous humor, lens, and vitreous humor. The refractive power stems from the difference of refractive indices at interfaces. The most prominent difference occurs at the air/cornea interface ($n \approx 1.38$). As a consequence, the two principal planes are close to the cornea, 1.3 mm and 1.6 mm behind its vertex.

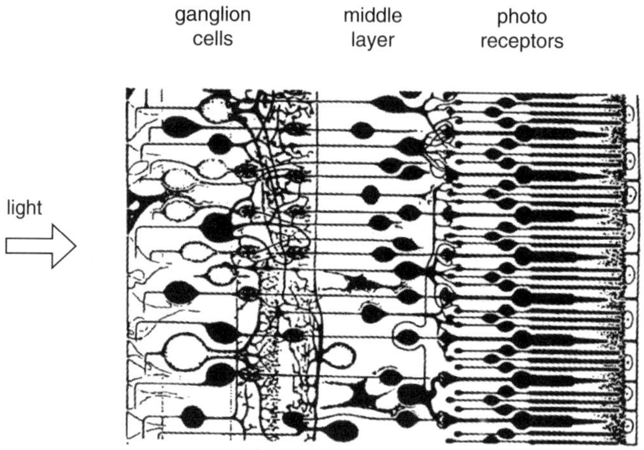

Figure 4.4: Cross section of retina. Light passes several layers of neurons before it reaches the photoreceptors. (From Levine, *Vision in Man and Machine*, 1985, McGraw-Hill, reproduced with permission of McGraw-Hill Companies).

The lens is embedded in media of similar refractive indices, hence, it contributes much less to the overall refractive power of the eye. The lens' purpose is to correct the fixed constant focal length of the cornea so that the image is always formed on the retina, regardless of the eye to object distance—a capability known as *accommodation.*

The ciliary muscles control the shape of the lens. With decreasing radius of curvature, the refractive power increases and objects closer to the eye are properly imaged on the retina. The distance between the retina and rear nodal point changes between 17 mm and 14 mm.

The iris controls the amount of light entering the eye, functioning as the eye's aperture stop. Its diameter can change from 2 mm to 8 mm. Other, more effective ways of controlling the light intensity are related to changing the sensitivity of the photoreceptors (brightness adaptation).

Resolution

The eye's ability to recognize fine details is usually called *visual acuity.* In its most simple form, acuity is expressed as a measure of resolution. Unfortunately, resolution is an ambiguous concept. For example, it can be defined by optical line pairs, modulation transfer functions, contrast ratios, or spot sizes.

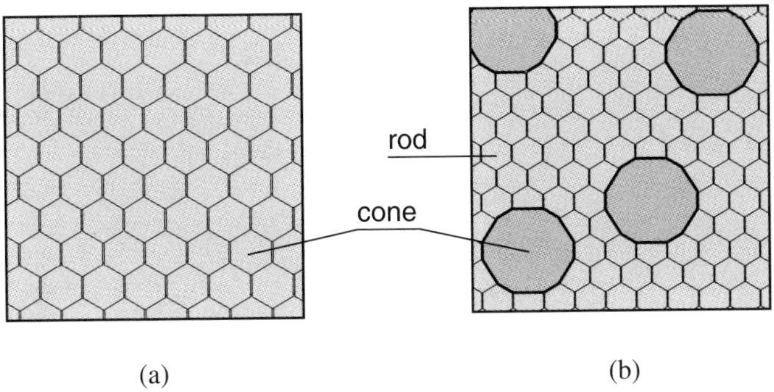

Figure 4.5: Schematic diagram of the spatial distribution of rods and cones. In the center of the fovea, only cones are present (a) with an approximate spacing of 2 to 3 microns. The hexagonal shape of the cones optimizes packing density. Further out in the visual field, cones are wider apart and the space is filled with rods (b).

Regardless of the method chosen for expressing the resolution, the distribution (spacing, mosaic) of the photoreceptors is an important factor. Fig. 4.5(a) depicts the arrangement of photoreceptors in the central fovea. First, we note that only cones are present. To provide the highest packing density (minimize loss of space), the cones have a hexagonal shape. The average distance between neighboring cones varies from 2 μm to 3 μm in the very center. This distance increases sharply away from the center. As a consequence, the visual acuity drops (see Fig. 4.6).

The spacing of 2.5 μm and the distance of 17 mm from the fovea to the nodal point subtend an angle of 0.5′ (half a minute). It stands to reason to consider this as a limiting factor of acute vision. Let us translate this number into a more familiar measure of resolution, such as line pairs per millimeter (lpm). At a viewing distance of, say, 30 cm, half a minute corresponds to 11 lpm. As a guiding number, 8 lpm is usually assumed.

What is the theoretical limit of the resolution of a diffraction limited system with the same optical parameters as the eye? If we take the size of the Airy disc as the limiting factor for discriminating two point sources that are close together, then the resolution becomes $\delta = 1.22\frac{\lambda}{D}$ with D the apparent diameter of the aperture and λ the wavelength of light. Using an average wavelength $\lambda = 550$ nm and a pupil diameter of 4 mm[1], the

[1]The iris can be open as much as 8 mm but then the increasing influence of aberrations reduces the resolution. The highest resolution is achieved with a pupil diameter of 4 mm.

maximum resolution computes to $\delta \approx 0.57'$. We conclude that the cones in the retina are packed just densely enough to provide maximum resolution of a diffraction limited system! Fig. 4.6 shows the visual acuity as a function of the cone spacing.

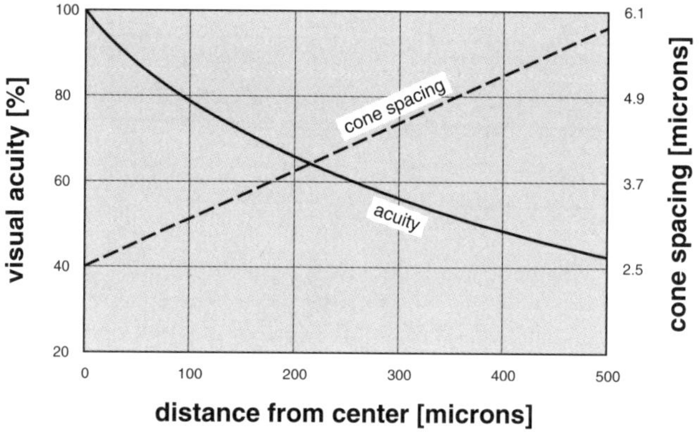

Figure 4.6: Visual acuity as a function of cone spacing. The center to center spacing increases almost linearly with increasing distance from the very center of the fovea. As a consequence, the angular resolution decreases and the acuity drops.

The image formation process can be described by a linear system, at least as a first approximation. Referring to Chapter 2, the object to be imaged is the input signal, the lens system functions as a linear system, and the image constitutes the output. The system response is entirely characterized by the impulse response. An impulse can be thought of as a point source in object space. Then, the object is a linear combination of scaled and shifted impulses. How does the lens transform an impulse? The image of a point through an aperture is the point spread function. For example, the image of a point obtained by a pinhole camera is the diffraction pattern. An image, then, is obtained by convolving the light intensity function of the object by the point spread function. The same operation can be performed in the frequency domain. The Fourier transform of the point spread function is the optical transfer function and its magnitude (modulus) is the modulation transfer function.

The modulation transfer function of the eye has been determined, for example by visual interferometry. Fig. 4.7 shows the theoretical curve, derived from the line spread function, as well as the actual curve obtained from measurements.

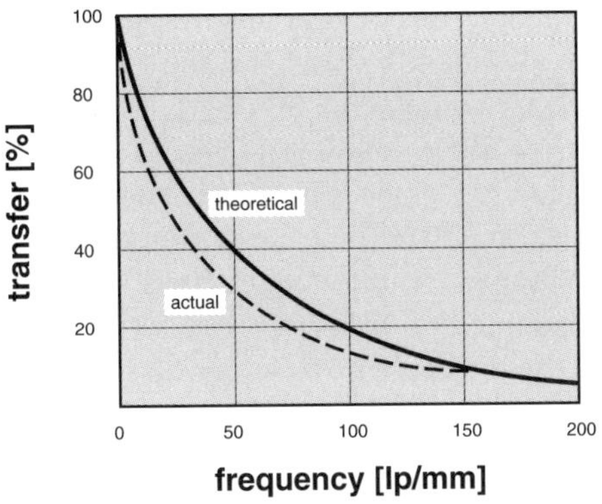

Figure 4.7: Modulation transfer function of the eye. The solid curve is the predicted MTF derived from the line spread function. The dashed curve is the actual MTF obtained from measurement using visual interferometry.

4.2.3 *Photometric Properties of the Photoreceptors*

The major differences between rods and cones are their different sensitivity, the spatial distribution, and the quantity. Rods are far more numerous than cones[2]. In Fig. 4.8 we also notice a difference in shape: rods are long and slender, cones are shorter and tend to be tapered. Only in the very center of the fovea are the cones thinner.

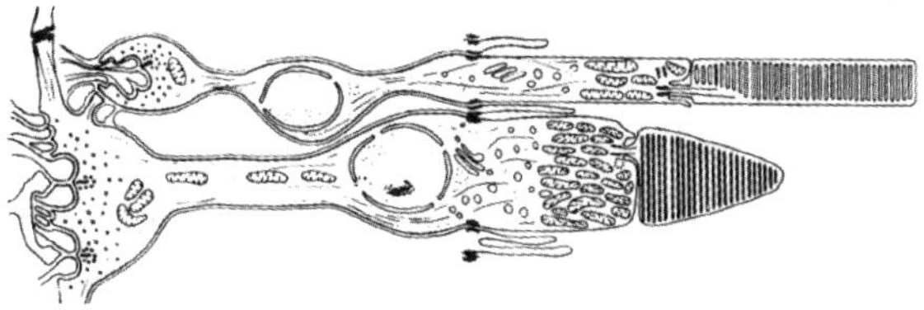

Figure 4.8: Rod and cone. Rods are slender and longer than cones. (From Levine, *Vision in Man and Machine*, 1985, McGraw-Hill, reproduced with permission of McGraw-Hill Companies).

Light passes through the inner to the outer segment of a photorecep-

[2]Estimates vary, some authors report 100 million rods and 5 million cones; others estimate the number of cones higher.

tor. The outer segment carries light-absorbing (visual) pigments. Such pigments are more abundant in rods and have been studied much more extensively than the visual pigments of cones.

In contrast to most other neural cells, the photoreceptors are depolarized in their resting state. That is, in the absence of light a current flows (dark current) across the membrane; sodium ions flow in and potassium ions leave the membrane. The photopigment of rods is rhodopsin.

If a photon is absorbed—a process called bleaching—a myriad of processes start. First, the molecule is decomposed into opsin and Vitamin A. This activates an enzyme, which in turn is responsible for the closing of many pores in the membrane. As a consequence, the dark current stops the flow and the potential across the membrane increases. In other words, the cell hyperpolarizes and the release of neurotransmitters at the synapse changes.

If a few neighboring rods absorb one photon each, the ganglion cell to which these rods are connected to, responds. This minuscule amount of energy constitutes the lower threshold of vision. *Scotopic vision* is initiated by rods and begins at an intensity of 10^{-6} mL (millilambert)[3].

Compare the sensitivity of rods with that of a high speed emulsion of photographic material. Here, a few photons are necessary to form a silver speckle on the silver halide crystal—just enough to trigger the decomposition of the entire crystal to silver during the development process. This is where the analogy stops. The emulsion can store one latent image while the retina receives several images per second, many hours a day and many days during the lifetime of a person.

Cones are less sensitive to light than rods because cone photopigments are present in much lower densities than rod pigments. More energy is necessary to cause a change in the cone's response to light. Vision initiated by cones, called *photopic vision,* spans approximately an intensity range from 10^{-3} to 10^5 mL. Thus, the dynamic range of vision is about 10^9 (see also Fig. 4.9).

The visual system cannot process simultaneously the entire dynamic range because the photoreceptors are only sensitive over one or two orders of magnitude. The photoreceptors adapt to a specific level, corresponding to the illumination of the environment (brightness adaptation). Fig. 4.9 depicts the ability of the eye to detect just noticeable brightness differences as a function of the absolute brightness. As one would expect, in scotopic vision only a few brightness differences are perceived while in photopic vision the eye becomes very sensitive to brightness differences.

The relatively flat part of the curve in photopic vision illustrates the

[3]For a summary of radiometric and photometric units refer to Chapter 6

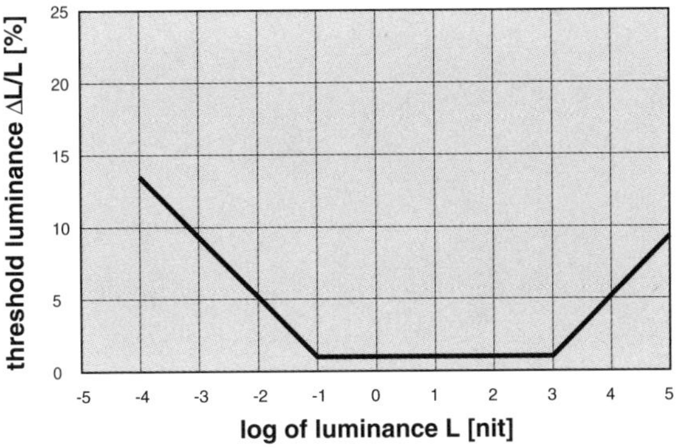

Figure 4.9: Brightness sensibility of human eye (Weber-Fechner law). Over a wide luminance range, the just noticeable brightness difference is approximately 2 % (see text for more information).

Weber-Fechner law which says that the ratio of perceived brightness differences to absolute brightness remains constant (approximately two percent). From this we may conclude that the eye can discriminate on the order of 50 to 100 different gray shades under good viewing conditions. This is in good agreement with the observation that for perceiving a smooth displayed image on a monitor up to 100 gray levels are required.

We conclude this section about photometric properties by pointing out the wavelength sensitivity of the photoreceptors. Fig. 4.10 is a plot of the wavelength sensitivity of rods and cones. Note that three types of cones are present[4]. The main difference between these three cone types is the sensitivity of their visual pigment, and their distribution in the retina.

Response of Ganglion Cells: Receptive Fields

Remember that there are many fewer ganglion cells than photoreceptors. The ratio is approximately 1 : 125. Only in the very center of the fovea is an individual cone passing its response to one ganglion cell. In the rest of the fovea the ratio is 1 : 3. So, what are the ganglion cells reporting back to the next processing centers?

First, we note that one ganglion cell receives impulses from several photoreceptors. Now, the receptive field of a ganglion cell refers to those receptors which are "connected" to it. The spatial extent of a receptive field can

[4]L-,M-,and S-cones. L- and M-cones are densely spaced and more numerous than S-cones.

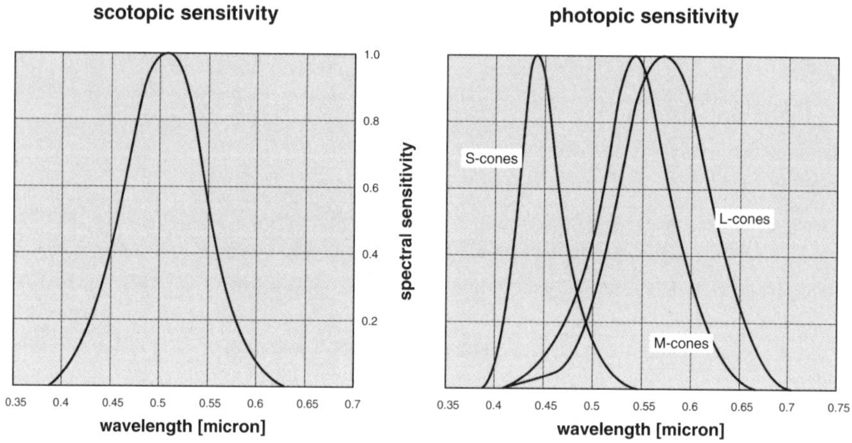

Figure 4.10: Spectral sensitivity of rods and cones. (Figure adapted from Wandell, 1995).

be found by positioning light stimuli (e.g. small flashes of light or moving bars) on the retina, and by determining the ganglion cell's response. The results of these intricate experiments and their thoughtful analysis are summarized as follows.

The receptive fields consist of a circular center and an annulus. An on-center ganglion cell increases its firing rate when the light spot falls inside the center of the receptive field. The frequency of impulses reaches a maximum when the spot size covers the entire center of the receptive field; the frequency drops when the spot stimulus becomes larger. By the time the spot size covers the entire receptive field, the cell fires at its resting state. When only the annulus is stimulated, the cell stops firing altogether—only to respond with a burst once the stimulus is turned off. Off-center ganglion cells exhibit exactly the opposite behavior. If the center is illuminated, the firing stops; if the annulus is stimulated, the firing increases.

This intricate interplay between exhibition and inhibition permits the ganglion cell to respond to contrast differences that may occur within its receptive field. Imagine illuminating the retina with a stimulus that is sharply divided into a bright and dark region. We can predict the outcome. Those ganglion cells, whose receptive fields are entirely in the dark or bright region, do not respond to the stimulus because the on and off responses cancel out. Only the cells whose receptive fields are partially overlapping the two regions will respond. Take this to the extreme and illuminate the entire retina with diffuse light: no response will occur since no noticeable brightness differences are present. If you ever have skied when clouds and fog move in, then you experienced the dramatic effect of diffuse light—the eerie feeling of loosing the ability to see the snow surface so important to plan and control your movements.

Fig. 4.11 schematically depicts the functional behavior of on- and off-center ganglion cells. The spatial organization of the receptive field and five different stimuli patterns A—E are shown in the top figure. In the first case of pattern A, the center is fully illuminated by a spot light, causing a maximum response of the on-center cell and no response in the off-center cell. If a strip is used, width equal to the diameter of the receptive field's center pattern (Fig. B), a slightly reduced response is to be expected, because a part of the center-surround area offsets the strong center response a bit. This trend continues with the third pattern where more than half of the center-surround almost neutralizes the center's response. Pattern D stimulates only the annulus, suppressing any response. Finally, pattern E depicts the situation where the entire receptive field is illuminated. Both, the on-center and off-center ganglion cell remain in their resting state.

We conclude that ganglion cells respond to brightness differences within their receptive fields, that is, to local intensity differences. Receptive fields differ in size. As one would expect, the size is smallest in the fovea and progressively increases further out in the visual field. The light intensity changes, transmitted by the optical nerve, are detected by biological filters of the retina. We will pursue this aspect further in Sec. 4.5.

4.3 Visual Cortex

4.3.1 Overview

The primary visual cortex is a complex substructure of the cerebral cortex, comprising approximately $1.5 \cdot 10^8$ neurons. The visual cortex is layered; six major layers have been identified. The layers differ regarding the density of neurons, axons and synapses, and interconnections to other brain areas.

The visual signals from the retina and the lateral geniculate body arrive at the visual cortex and other cortical areas. However, the visual cortex has been most intensely studied, presumably because of its location being more easily accessible than deeper brain structures.

The visual cortex is topographically organized: an area of about a two millimeter square has complete functionality. These areas—self-contained modules of the cortex—map out a portion of the visual field. Consequently,

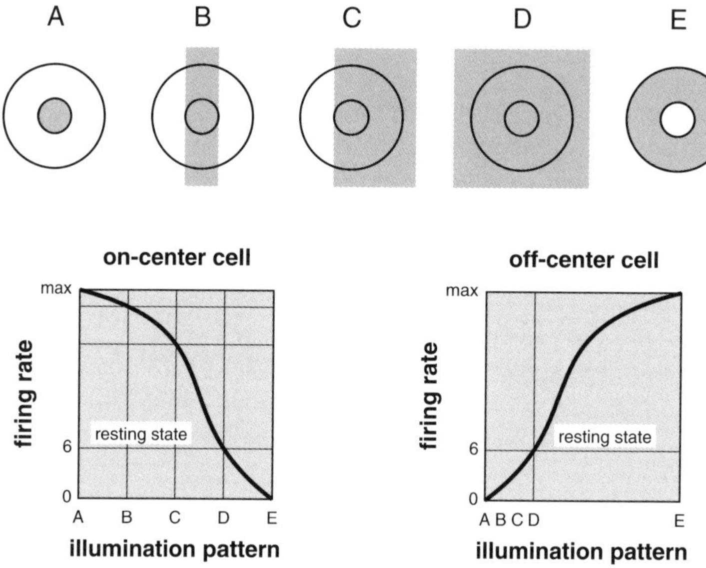

Figure 4.11: Schematic diagram of the response of on- and off-center ganglion cells. The top figure shows five different stimuli, A—E, including their position with respect to the receptive field. The firing rate of the on-center field is depicted in the lower left figure. The response of the off-center field is shown in the lower right figure.

if one such area is damaged, the corresponding part of the retinal image is not processed further and the result is local blindness. Neighboring modules do not compensate for the loss. However, the perceptual process of filling in completes the missing information by interpolating it from the surrounding area.

Following the signals from the two retinas, we notice that they merge at the optic chiasm (see also Fig. 4.1 on page 78). Here, information from both visual fields is mixed and transmitted back in such a fashion that the right hemisphere of the cortex receives impulses from both retinas and both visual fields. Likewise, the left hemisphere receives information from the right visual field through the left retina and from the left visual field through the right retina. The signals remain segregated, however. A cortical neuron receives input from one of the retinas, except for binocularly driven cells.

4.3.2 Cortical Receptive Fields

The specialization of cells in the cortex increases. So does the complexity of their receptive fields. Unlike cells of earlier levels, only a few cortical neurons have circular symmetrical receptive fields. Moreover, they respond quite differently to stimuli.

Cortical neurons can be broadly divided into simple cells and complex cells. A major distinction is their different response: simple cells function like an LSI system and their response to an input (stimulus) can be predicted based on their impulse responses. As the name suggests, complex cells exhibit a more complicated behavior.

Orientation Sensitive Cells

Some cortical cells respond best if a slit of light crosses their receptive fields at a specific angle. Changing the orientation and position only slightly evokes no response. The receptive fields consist of excitatory and inhibitory regions of rectangular shape with the longer side of the rectangle indicating the preferred orientation.

Other simple cells respond more strongly if one half of the receptive field is stimulated.

Direction Sensitive Cells

The most commonly found cells in the striate cortex are the complex cells. Like simple cells they respond to properly oriented stimuli. However, the cell's firing rate fades out rather quickly unless the stimulus is moved. So, complex cells are movement sensitive; they respond with a barrage of impulses if a properly oriented slit is swept across their receptive fields. Some complex cells are also direction sensitive. That is, it matters in which direction the slit is moved. That a large population of cells is highly sensitive to movements makes a lot of sense, at least from an evolutionary point of view. After all, to react properly and quickly to the environment, moving objects should be discovered promptly.

End-Stopped Cells

End-stopped cells are further specialized in that they are sensitive to the length of the stimulus. They respond much more strongly if the slit of light ends or changes direction within their receptive field. Thus, they respond best to corners and curvature.

Binocular Sensitive Cells

So far, information from the two eyes has been treated separately, even though one cortical hemisphere receives information from both eyes. As photogrammetrist we are professionally interested in stereopsis. Within the superficial layers of the visual cortex many neurons respond well if both retinas are stimulated.

Binocular cells respond to depth. Some of these cells only fire if the stimulus is roughly as far away as the distance on which the two eyes are focused (fixation point). Other cells evoke a brisk barrage of impulses if the stimulus is nearer or farther away from the fixation point. To re-phrase this in the photogrammetrist's vocabulary: some binocular sensitive cells respond best to stimuli with zero-parallax, others are more sensitive to stimuli with positive or negative x-parallaxes.

Another characteristic feature of these disparity-tuned cells is that they are also orientation and movement sensitive. As one would expect, they do not respond at all if only one eye is stimulated. Though disparity-tuned cells undoubtedly contribute to stereovision they are just a partial explanation of how we perceive depth. One should bear in mind that stereopsis is only one of several depth cues.

Let us pause for a moment in our journey through the visual system and recapitulate. First and foremost, we recognize that what reaches the brain is not an image, but information about changes in the scene, for example, light intensity differences, their orientation, and movement. The specialization of cells and the complexity of their receptive fields increase. How far will this specialization go? After cells were discovered in the visual area of a monkey that responded to the shape of paws, the notion of a *grandmother cell* arose. Is there a cell that would respond to grandmother's face? And, if indeed such a cell were to be found, would it help us in understanding how we achieve the rich, constant, 3-D representation of the world from ambiguous, distorted, incomplete responses of specialized cells?

4.4 Visual Perception

The universe consists of particles, atoms, molecules, electromagnetic fields, and a lot of empty spaces. We experience a very different world, however; a 3-D world of objects and their relationships. Objects have dimensions, they consist of colored surfaces; they may smell. The perceptions, such as colors, tones, and smells do not exist in the 'real' world—they are the result of internal constructions based on information recorded by our senses. You and I have different mental constructions of the physical world, but from describing these constructions and from reacting to them in a very similar way we infer that our perceptions are indeed very similar and consistent.

From studying the senses of animals we know that their senses are attuned to record other information of the real world. Some respond to frequencies of the electromagnetic spectrum not available to us, others sense odors that have no perceptual meaning for us.

The cardinal question of visual perception is: *how* do we all arrive at a similarly constructed world from data that is ambiguous, distorted and incomplete?

Over the centuries philosophers have attempted to explain perception. Philosophers of the associationism argued that the visual input is insufficient for correct perception. To achieve correct perceptions, we must learn how to interpret visual signals, and we learn it by way of association. The famous scientist Hermann von Helmholtz argued in a similar way. To him, perception is based on inference. The nature of an object is unconsciously inferred from visual stimuli.

The Gestalt psychologists rejected the idea that perception could be explained by associating simple sensations. Their basic tenet was that "the whole is more than the sum of its parts". For example, two dots exhibit a sense of orientation, which is not present in a single dot. They argued that the form and structure of sensations and their interrelationships should be taken into account. The Gestaltists thought that this synergism is accomplished by magnetic force fields between brain events. The Gestalt psychology has fallen into disrepute, mainly, because no evidence was found for the force fields in the brain.

Yet another way to explain perception was put forward by researchers in psychophysics. Contrary to the belief of Empiricists and Gestaltists, they argue that all of the information needed to explain perception is available from the real world. Perceptions, such as, shape, size, depth, motion or color are embedded in the signals we receive; perception, then, is just a matter of properly recording and encoding visual stimuli.

A fourth approach toward explaining perception comes from the field of cognitive psychology. Here, a more information theoretical approach is adopted where computer models of perceptual processes are legitimate goals for establishing psychological theories. This, together with a more quantitative approach in research paves the way for "computational perception" and results that can be converted to algorithms.

4.4.1 *Perceptual Organization*

The neurophysiological approach to vision left us with the image decomposed into simple local features, such as edges, corners and some depth information. Such low-level descriptions must be organized into larger perceptual structures. Perceptual organization is the first process of perception. It detects groupings and structures from an image without prior knowledge of its content. Such groupings and structures are believed to be the input for object recognition and image understanding.

The human visual system is remarkably adept at detecting many different patterns and significant groupings of elements in an otherwise random set of elements. Grouping processes include clustering, detecting single and parallel lines and curves, and segmenting the image into regions which share similar image characteristics.

Perceptual organization is not only a late vision or cognitive process. Preattentive vision experiments (response of human visual system to stimuli without focus of attention) indicate that perceptual organization occurs very early on in the visual pathway. Quite often, grouping processes are recursive. The output of one process becomes input to another grouping process, perhaps together with other image information. An example is edge fragments that are clustered to edge elements which, in turn, are grouped to lines.

The following are examples of a set of criteria for grouping the image and finding associations. Most of these principles have been advocated by the Gestalt psychologists and are known as the *Gestalt laws of organization.*

proximity groups local features together which are close together. In Fig. 4.12(a) we perceive columns because the vertical spacing of dots is smaller. Depth is a very strong factor for proximity. Things with similar disparity values are grouped together and perceived as belonging to the same surface.

similarity groups similar features together. In Fig. 4.12(b) the horizontal spacing of the solid dots is smaller than the vertical spacing, yet we perceive columns. This is an example where similarity overrides proximity.

common fate groups things together which appear to move together. It can be demonstrated graphically by generating random dots, which are then rotated and superimposed on the original.

good continuation is demonstrated in Fig. 4.12(c). We perceive two smooth crossing lines rather than two V-shaped forms. Smooth continuity is preferred over abrupt changes.

closure emphasizes a preference for closed figures. Fig. 4.12(d) is seen as a rectangle and not as braces.

symmetry groups features together that are similar. In Fig. 4.12(e) curves 1 and 2 are grouped together because of symmetry. However, the perception of curves 4 and 5 is stronger because of reflection around the vertical axis. Clearly, curve 3 is not grouped with any other curve.

figure ground separation The smaller of two areas is usually perceived as a figure against a background. In Fig. 4.12(f) a black propeller is seen against a white background.

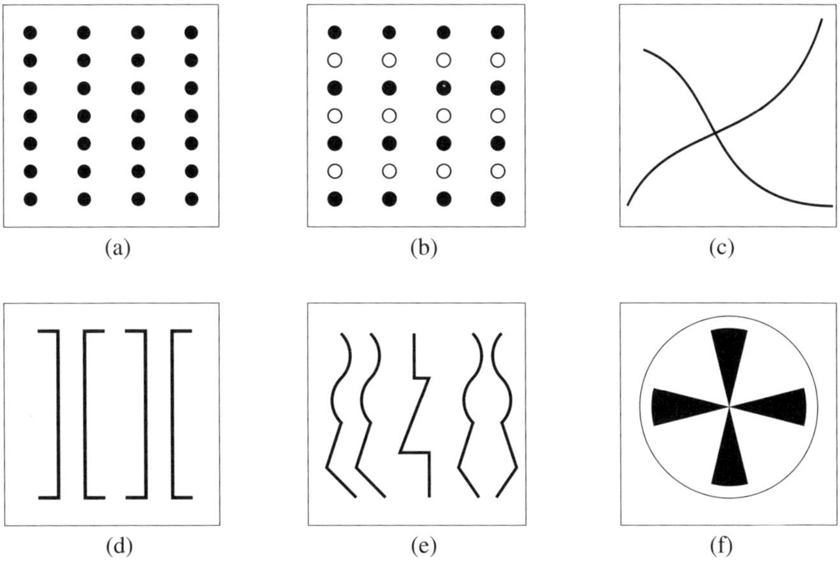

Figure 4.12: Examples of the Gestalt laws. An illustration of *proximity* is shown in (a) while (b) is depicting *similarity*. An example of *good continuation* is shown in (c); (d) illustrates *closure*, (e) *symmetry* and (f) exemplifies *foreground/background separation*.

4.4.2 Other Perceptual Processes

Here, we mention some other powerful perceptual processes, which could be used in computational vision.

filling in or completion is responsible for us to not perceive the world as a patchwork of edges and blobs (as might be concluded from the neurophysiological discussion about vision). A very illustrative example is the blind spot (see Sec. 4.2.1). Close one eye and fix a point with the open eye. Move a pencil with one hand so that it crosses the visual field. When the pencil is imaged at the blind spot, it disappears, as expected. However, you are not left with a black spot; rather the hole in the retinal image is covered (filled in) by the surrounding background. Filling in appears to belong to a more general perceptual process called surface interpolation.

virtual lines are imaginary lines, linking nearby tokens. Fig. 4.13 is an example. A similar phenomenon is illusionary contours, investigated by Kanizsa. In Fig. 4.13(b) we perceive the structure of a square. The four corners are lying on circles. Another (unlikely) interpretation of this figure is four partial circles.

texture is a very important but not well understood perceptual process. Texture is strongly related to surfaces. Slowly changing texture patterns give a strong perception for surface normals. Julesz studied texture segmentation intensively. He concludes that textured regions cannot be segregated if their first and second order statistics are identical. Later, Julesz introduced the notion of textons and argued that they play a complementary role in human texture segregation.

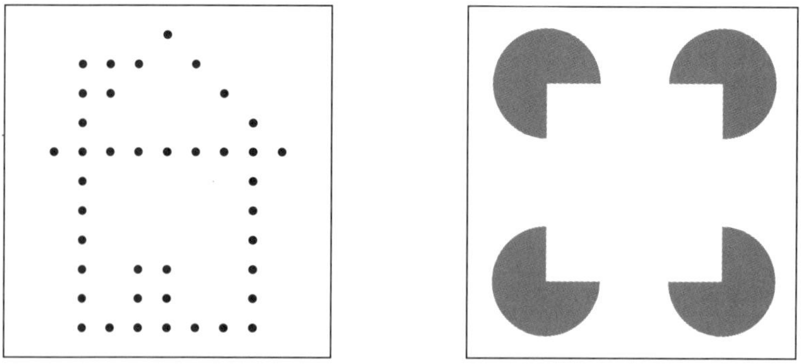

(a) (b)

Figure 4.13: Example for virtual lines (a). Fig (b) demonstrates the phenomenon of illusionary contours. The figure is perceived as a square and not as four partial circles.

4.5 *Marr's Theory about Vision*

The physiological approach to vision answered the question: what happens where? How something happens cannot be fully explained unless the cell's behavior can be described by a complete wiring diagram. For answering the question why single cells respond the way they do, a broader view must be adopted. As Marr put it

> ... trying to understand perception by studying only neurons is like trying to understand bird flight by studying only feathers: It just cannot be done. In order to understand bird flight, we have to understand aerodynamics; only then do the structure of feathers and the different shapes of bird's wings make sense.

Marr's theory about vision has a strong information processing underpinning. He argues for understanding an information process - vision - at three different levels.

computational theory specifies what the visual system must do. It answers the question about the purpose of the computation and the strategy for solutions.

representation and algorithm investigates the representation of input and output and the algorithm that transform one into the other.

hardware implementation answers the question how the representation and the algorithm can be physically implemented by neurons.

The tenet of Marr's theory is that the shapes and positions of things can be made explicit from images without knowing what these things are and what role they play. However, this cannot be accomplished in one step, rather in a sequence of representations designed to facilitate the subsequent construction of physical properties of objects. The three main steps are briefly discussed.

4.5.1 *Primal Sketch*

The purpose of the primal sketch is to make intensity changes in the image explicit. Intensity changes, or edges for short, are an important physical property of objects. In the real world, edges occur over a wide range of spatial extents. A sharp edge, for example, is manifest within a small area, comprising a few pixels only. On the other hand, a fuzzy edge can only be detected by looking at a much larger area. Marr and Hildreth (1980) propose a sequence of LoG operators to detect edges at various scales. The

LoG operator (Laplacian of a Gaussian) is obtained by taking the second derivative of a Gaussian filter. The Laplacian (∇^2) is particularly suited because it is direction independent. By varying the standard deviation σ of the Gaussian, the desired sequence, also called multi channel implementation, is obtained. Obviously, the parameter σ determines the spatial extent within which an edge is detected. Edges are identical with the zero-crossing contours that result from intersecting the convolution surface with a plane, whose convolution value is zero. Thus, a sharp edge is obtained by convolving the image with a small σ (fine channel), and fuzzy edges result from coarser channels.

There is much evidence that the human visual system performs the same operations. Cells exist in the cortex that respond to different spatial frequencies. Spatial information is processed in each part of the visual field by five independent channels. Actually, the LoG operator is approximated by the difference of two Gaussians of slightly different σ. The two coarser channels have transient properties, responding to fluctuating patterns, while the finer channels respond to stationary objects. The finest channel is related to acute vision.

The primal sketch is more than just an agglomeration of zero-crossings. Perceptual processes operate on the image as well as on the edges, resulting in a curvilinear organization, virtual lines and groupings. Zero-crossings from different channels are combined, governed by the rule that edges in different channels are localized in space.

4.5.2 2.5-D Sketch

Its purpose is to make explicit the orientation and depth of visible surfaces as well as discontinuities. The name of this sketch derives from the assumption that it captures a great deal about the relative depths and surface orientations, and local changes and discontinuities, but some aspects are more accurately represented than others.

> Very locally we can easily say from motion or stereopsis information whether one point is in front of another. But if we try to compare the distances to two surfaces that lie in different parts of the visual field, we do very poorly and can do this much less accurately than we can compare their surface orientations.

The 2.5-D sketch is built up from the primal sketch, augmented with information from stereopsis, texture, analysis of motion, and shading. The surface orientation is much more accurate than depth. Only local changes in depth have a comparable accuracy. Discontinuities in depth may arise from stereopsis and occlusion. Occlusion may be specified by the presence of occluded edges in the primal sketch, or by analyzing motion patterns.

The 2.5-D sketch is represented as a set of primitives, depicted as "needles". The length of each needle describes the degree of tilt of that part of the surface, while the orientation of a needle reflects the direction of slant. The distance from the viewer is represented by a scalar quantity.

Interpolation procedures are invoked in areas of insufficient information. In areas of low contrast, no edges are present and therefore no depth information. The missing depth information is interpolated from surrounding areas where contrast is present. Another example for an interpolation process is illusionary contours (see Fig. 4.13(b)).

The 2.5-D sketch is the end product of early vision processes, solely derived from images, without support from late vision or knowledge of the scene. The early vision processes are modular, they work in parallel and independent of one another. The segmentation problem is implicitly solved by making explicit the discontinuities between different surfaces.

4.5.3 3-D Model Representation

The purpose of this last step is to describe shapes and their spatial organization in an object centered coordinate system. Marr and Nishihara suggest a modular organization of shape descriptions in a coordinate frame which is determined by the shape itself (canonical coordinate frame). The modular organization allows a description that is independent of the degree of details with which an object is described.

The theory is restricted to a set of generalized cones. A generalized cone is obtained by moving a cross section of constant shape but variable size along an axis. A vase is a good example of a generalized cone. An object may consist of several generalized cones, each with its own axis. All axes of one object form the component axes of that object.

A library of 3-D model descriptions at different levels of specificity is generated for objects that may possibly appear in a scene. The same model description must be derived from the image. Object recognition then entails comparing these descriptions with the library.

Occluding contours of an image provide strong clues for finding the axes of generalized cones. Occluding contours are the silhouettes of objects. Even though most silhouettes are ambiguous, humans interpret them in a particular way. Marr hypothesizes that additional information is used to constrain the perception of 3-D shapes to silhouettes as we see them. These constraints are general and do not require a priori knowledge of the scene.

4.6 *To Probe Further*

Most of the neurophysiological section in this chapter is based on the two books *Eye, Brain, and Vision,* by D.H. Hubel (Scientific American Library, 1988), and *Foundations of Vision* by Brian Wandell (1995). Hubel's book is written for non-specialists. He writes "In writing this book I have the astronomer in mind as my prototypical reader—someone with scientific training but not an expert in biology, let alone neurobiology." Wandell's *Foundations of Vision* (Sinauer Associates, Inc., 1995) offers a wealth of data and explains many vision processes as linear systems. In *Vision in Man and Machine* by M. D. Levine (McGraw-Hill Book Company, 1985), the author even goes a step further by explaining vision from a signal processing view. Marr's book *Vision* (W.H. Freeman, San Francisco, 1982) is based on a strong information processing view of vision. His theory about vision, especially early vision, is still widely accepted by researchers from different fields. An interesting approach to visual perception is presented in *Perception* by I. Rock (Scientific American Library, 1984).

Chapter 5

Computer Vision

Like the human visual system, computer vision is concerned with reconstructing the 3-D world from 2-D projections, and interpreting and describing it in a useful manner. Digital photogrammetry has a similar goal and is confronted with the same fundamental problems. Thus, it stands to reason to analyze the concepts, methods developed, and solutions found in computer vision. By this I do not mean that we should blindly copy whatever has been done in computer vision. To judge what concepts may apply for solving photogrammetric problems, we must develop a reasonable understanding of what problems are solved in computer vision, what approaches have been proven as successful, and on what theoretical underpinnings the solutions are based.

This chapter provides a short overview of computer vision, its paradigm, basic concepts and major issues. We build on the previous chapters, particularly on Chapters 3 and 4. There is no neat separation between computer vision, image processing, and pattern recognition. In fact, many early vision processes are plain image processing tasks. That is where image processing stops, however, and that is where the real challenge of computer vision begins: make structures in images explicit, perceptually organize and match them with models of scene objects.

Rather than touching loosely on all sorts of aspects, I have opted to focus on some of the key concepts, followed by edge detection and tracking features through scale-space as examples of early vision processes. The section about object recognition serves as a late vision example.

5.1 *Background*

Computer vision is a relatively new and rapidly changing field. Many of the essential concepts have only evolved during the last two decades. The goal is to reconstruct the 3-D world from 2-D images, and to interpret and describe it in a useful manner. The term "useful" is vague: what might be a useful representation for one application may be totally inadequate for another one.

Several fundamental problems make the development of vision systems difficult and limit their applications. Reconstruction, as such, is an ill-posed problem. The digital image may contain not only a lot of useless, but even misleading data. At the outset, information is only implicitly available. There is a huge representational gap between the visible input (iconic) and the world models (highly abstract, symbolic). This gap cannot be bridged in one giant step; intermediate representations are necessary. Then, the basic tasks of computer vision are making information explicit; designing reasonable representations; and describing objects in a form that is independent of different illumination and viewpoint conditions.

By and large, computer vision pursues the same goal as human vision: generate descriptions about the scene from images. The descriptions must be explicit and meaningful to allow other system components to carry out a task. In that aspect, computer vision is part of an entire system that interacts with the environment, say a robot. But tasks such as decision making, planning, and executing decisions are not part of computer vision. Since humans perform vision tasks with such great ease, without conscious effort, it is tempting to emulate human vision. As discussed in the previous chapter, the human visual system is not well understood, particularly its cognitive part. What we can conclude from studying biological vision is that very effective solutions of the vision problem exist. Computer vision is faced with the challenge to reinvent the solution.

Fig. 5.1 schematically depicts a computer vision paradigm. Don't take it too literally: there is no such thing as a uniformly accepted paradigm. Most everyone agrees that vision must be solved in a modular fashion, for the difference between input and output is so huge that it requires a modular approach. Exactly what these modules (canonical decomposition of vision) and their interrelationships are, is still a matter of considerable debate.

Vision begins with image formation. Since computer vision may be viewed as the inverse process of image formation, it only makes sense to obtain a thorough understanding of image formation, a subject treated in more detail in Chapter 6.

Low level image processing tasks, such as gray level modifications, provide the transformation from the raw to the preprocessed image. At the

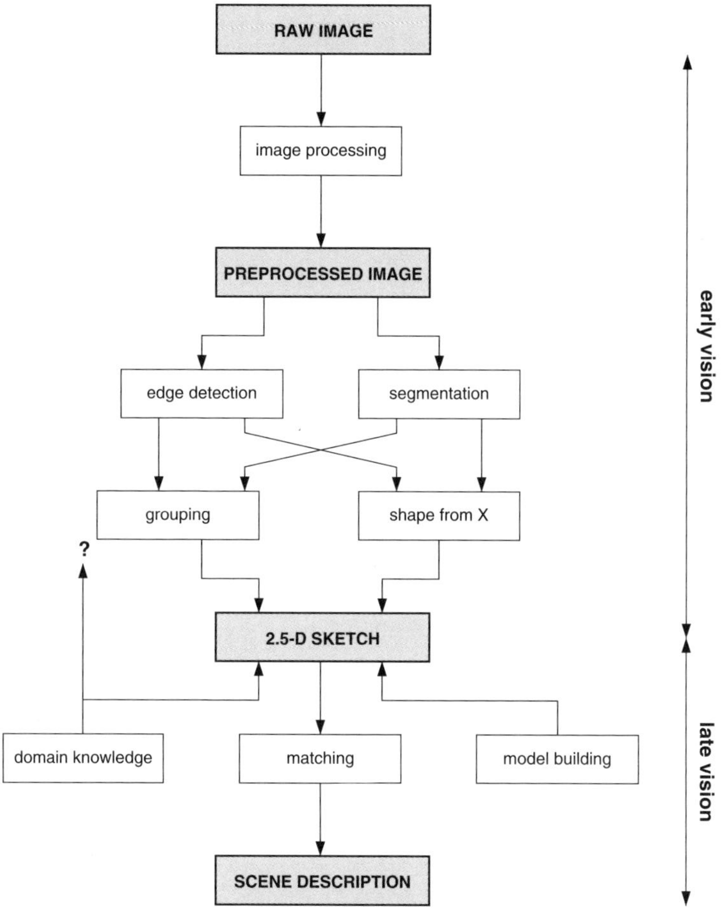

Figure 5.1: Paradigm overview of computer vision.

same time, defects of the raw image, caused by the image acquisition system, may be removed.

Now the process of extracting useful information begins. Examples include the detection of corner points and edges. Both are discontinuities of the image function. Edges are likely to have been caused by structures in the scene, such as object boundaries, markings, and surface discontinuities. The edge pixels are linked to edges and eventually grouped into higher-level entities, such as straight lines, arcs, and parallel lines.

Another way to find structures in the image that—hopefully—correspond to scene events, is segmentation. Here, image regions are sought, based on similar characteristics of spatially coherent pixels. Texture is an example of defining image regions (see Sec. 3.5). Aggregates of pixels with similar properties may be grouped into higher-order regions, perhaps by taking into account domain knowledge.

It has long been recognized that shape plays a fundamental role in recognizing objects (think of the enormous feat of the human visual system to recognize faces based on a few artist's strokes). The shape of surfaces is a vital prerequisite for object recognition. Different cues provide shape information. Probably, stereopsis comes to mind first to the photogrammetrist, but there are equally important cues, such as shading, motion, texture, color—collectively called shape from X.

The 2.5-D sketch[1] is a representational collection of the results of early vision processes. In Marr's vision paradigm, the 2.5-D sketch should be obtained without any knowledge of what the scene may look like and what the potential application of the vision system is (e.g., recognizing buildings from an aerial scene, inspecting parts on a conveyor belt, navigating a robot). This request is quite often violated in real vision systems, for example, segmentation and grouping are performed with a particular application in mind. As you realize, this makes a system specific and application dependent which is acceptable as long as the system is not (mis)used for another application. It may not come as a surprise that the most successful vision systems are application dependent—the quest of building general purpose vision systems remains a dream.

The 2.5-D sketch contains less data than the raw image, but more important, it is more explicit. For example, an edge could be an object boundary or a shadow; a single pixel can be everything. Depth and 3-D shape information is particularly important. The 2.5-D sketch is the transition from image space to object space. Subsequent processes, termed late vision, are scene oriented rather than image oriented.

The process of grouping, that is, the perceptual organization of extracted features, may also occur after the 2.5-D sketch. In fact, some researchers call grouping 'middle level vision' to emphasize the importance of perceptual organization and parameterization of features.

If the application of the vision system is object recognition then a data base with models of objects is generated (model building). The grouped and parameterized features are now matched with the object library. Usually, there is no perfect agreement between model and extracted features and—in a more complex setting—an inference process starts with the goal to minimize and to explain remaining differences.

[1] The name refers to Marr's vision paradigm. For a brief review see Sec. 4.5.

5.2 Key Concepts, Key Issues

5.2.1 Vision is Modular and Hierarchical

At the heart of the computer vision paradigm lies the recognition that vision (biological and artificial) is modular. The modules can be viewed as a canonical decomposition of vision (Haralick and Shapiro, 1990). The solution of visual tasks (e.g., object recognition, navigation) usually requires processing on all levels. Useful information that may lead to scene descriptions is not explicitly available in raw images. Physical phenomena, such as discontinuities of surfaces, depth, reflectance and illumination, are confounded in the intensity values of the pixels by the imaging process. The raw image cannot be used directly to interpret the image function. The problem is to define intermediate levels (modules) and corresponding descriptions. Local extrema and their derivatives are appropriate primitives because they are quite often related to object space events.

One of the problems of obtaining intensity discontinuities is the scale because physical phenomena differ vastly in size and extent. Moreover, finding all discontinuities may not be desirable; noise and unnecessary details should be suppressed. This can be achieved by smoothing the raw image, for example by convolving it with a Gaussian filter. The size σ of the Gaussian filter determines the scale in which the intensity discontinuities are represented. We obtain an infinite number of representations if σ is varied continuously. This representation is known as the scale space image. An attractive property of the Gaussian over other convolution kernels is that no new object space events occur when σ is increased. Descriptions found at a coarse scale also appear on finer scales, though not exactly at the same location.

The concept of representing the descriptive primitives in a continuum of scale levels is appealing because it circumvents the problem of specifying at what scale a particular object space event occurs. However, most practical implementations are discrete. That is, the scale space is discretized. The result of smoothing a raw image with a finite set of smoothing operators, e.g. Gaussian with discrete σ values, is called an image pyramid. Each level of the pyramid corresponds to a different resolution of the raw image, hence the term multi-resolution. A fundamental problem of the discrete scale space is tracking the descriptive primitives from one level to the next. The smoothing operation displaces the primitives. Tracking features through the discrete scale space is difficult.

5.2.2 Vision is Ill-Posed

As is evident from the paradigm, early vision processes are aimed toward reconstructing the scene from images. It is the inverse problem of image formation. Like many inverse problems, early vision is ill-posed. Take edge detection, for example. It requires numerical differentiation and that is clearly ill-posed. Hadamard defined a mathematical problem to be well-posed when its solution exists, is unique, and is robust against noise.

Regularization theories have been proposed to solve ill-posed problems. The basic notion is to restrict the space of acceptable solutions by choosing a function that minimizes an appropriate functional. One method for finding x from $y = Ax$ is to minimize $\|Ax - y\|^2 + \lambda\|Px\|^2$ where Px is a stabilizing functional and λ is the regularization parameter. In essence, an ill-posed vision process is regularized by imposing physically plausible constraints, for example, smoothness of surfaces.

Despite the mathematical elegance of regularization theory, it remains a partial solution only. There are problems of applying the theory if the quantity to be computed is a discontinuous function. The world we want to reconstruct from images is full of discontinuities. Another problem is the degree of smoothness for the unknown function to be recovered, for example, when interpolating surfaces. This and other problems must be solved before regularization theory can serve as a general method for solving ill-posed vision processes.

Of course, vision processes were solved before the framework of regularization theory became available. This was accomplished by introducing assumptions and constraints, sometimes in a rather ad hoc fashion. The secret is to strike a fine balance between constraining the problem just enough to obtain a solution and at the same time to remain as general as possible. Quite frequently, assumptions are too restrictive and algorithms fail when applied to slightly different scenes.

Yet another approach to make vision well-posed is to integrate visual processes. The individual shape-from-X processes are computationally underconstrained and in themselves not robust. By combining different processes, the parameters will be more constrained and the solution becomes more robust and unique. It is important to realize that it is the mere combination of processes which makes the solution more stable and *not* the introduction of additional constraints imposed on individual processes.

The combination (or integration) can be realized on two levels. The results of individual shape-from-X processes may be considered as contributions toward the 3-D shape description of surfaces. The integration can be performed within the framework of testing a hypothesis.

Another approach is to integrate information from different cues in one process. Examples include the combination of stereo and shading, or the simultaneous computation of the illuminant direction and shape from shading. The concept of adding the radiometric model to the geometric model in least-squares matching falls into the category of combining stereo with shading.

5.2.3 Transition from Early Vision to Late Vision

The 2.5-D sketch in the computer vision paradigm is the interface between early and late vision. Early vision processes, such as edge detection, grouping edge segments, stereopsis, texture, motion and color, are data-driven processes with the goal to derive physical properties of the object space (scene) from implicit information in the images. The most important information in the 2.5-D sketch is shape, depth and boundaries of surfaces, where boundaries are manifest in discontinuities of depth or surface normals.

According to Marr's vision theory, this information can be captured from images only, without a priori knowledge about objects in the scene. Surface reconstruction is the result of early vision. Random dot stereograms prove that the human visual system reconstructs surfaces from parallax information only. If Fig. 5.2 is viewed under a stereoscope, the center is clearly perceived as a surface floating above the surrounding background.

Figure 5.2: Random dot stereogram. When viewed stereoscopically the center square is floating above the background.

While early vision is associated with reconstructing visible surfaces, late vision is primarily concerned with interpreting this information by way of symbol processing. The notion is to segment (group) the 2.5-D sketch into higher-level tokens and associate a meaning to it. The vague term "meaning" is chosen purposely for it strongly indicates that late vision is application dependent. The further we move from the 2.5-D sketch the more application dependent late vision becomes. You realize that the perceptual organization may continue after the 2.5-D sketch, perhaps now application dependent.

Late vision processes depend on knowledge about the scene (domain knowledge) and general knowledge (world knowledge). An example for general, high-level knowledge is the fact that a shadow cast on a smooth surface does not cause people to perceive two different surfaces, despite the fact that the shadow boundary is a very strong edge. Knowledge of this kind is readily accessible to people, and this is part of the reason why human vision is so flexible and capable of dealing with a variety of different scenes. Computer vision systems, on the other hand, lack this flexibility and versatility.

A crucial issue in any vision system is the proper interaction between early and late vision. Where exactly should data-driven processes end and goal-driven modules begin? How far should scene independence be preserved? How can knowledge be brought in without limiting the scope of applications?

5.3 Edge Detection

Physical boundaries of objects play an important role in the human visual system's ability to recognize objects. Think of the few strokes it takes an artist to portray a face. Often, object boundaries show up as rapid changes in the gray levels within a small area of an image. Such discontinuities of the gray level function are commonly called edges. Not all object boundaries create an edge, nor do all edges correspond to boundaries. To infer boundaries from local intensity changes is a rather complex process, involving the following steps.

detecting edge pixels is the process of determining local discontinuities in the gray levels. What is a discontinuity, or more specifically, how large must the difference in gray levels between neighboring pixels be (threshold), and how large should the neighborhood be (spatial extent)?

linking edge pixels to edges is the identification of edge pixels that belong to the same edge. Edge linking may be viewed as a labeling process that makes an edge explicit.

grouping edges is the process of identifying straight line segments, polylines, conics, and parallel lines, for example. It may be application dependent. Grouped edges are essential elements of a meaningful scene description.

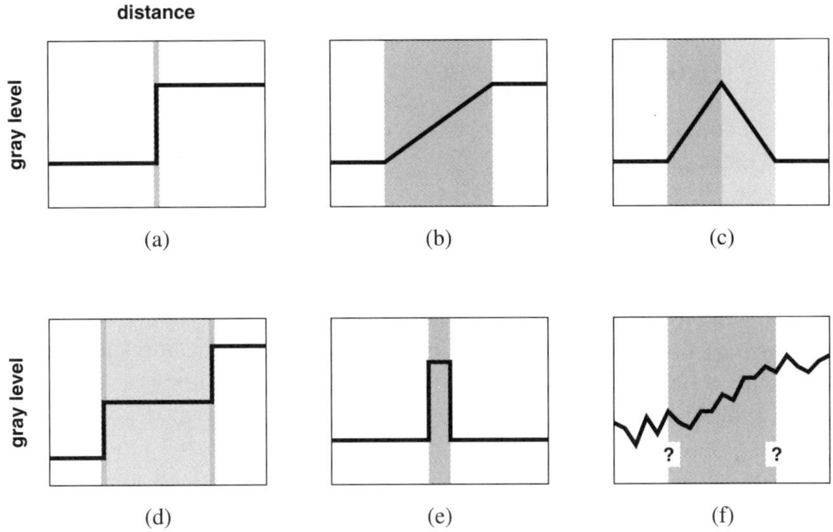

Figure 5.3: Examples of different edges. (a) step; (b) ramp; (c) roof; (d) stair; (e) spike; (f) noisy. The spatial extent over which the gray levels change is shaded. See text for more detailed explanations.

Fig. 5.3 schematically depicts different types of edges, including their spatial extent. Consider each graph as an intensity profile across an edge. The spatial extent, over which the intensity change takes place, is shaded in gray. In (a) we have a step edge, characterized by a sudden change in the gray levels. The bigger the change the more distinct we perceive the edge. The spatial extent is one pixel. The ramp edge in (b) has a spatial extent of several pixels. Obviously, the larger the extent and the smaller the change in the gray levels, the fuzzier the edge appears. The roof edge in (c) shows an interesting problem: do we have three edges or are we considering the roof as one phenomenon? A similar problem arises with stair edges depicted in (d): is the stair one phenomenon or do we have two individual step edges? The spike in (e) symbolizes what some researchers call blobs or lines. It is an edge that is several pixels wide.

So far we discussed ideal edges. In real images, the gray levels are contaminated with all sorts of noise, and the ramp edge now looks more like (f) where it is clearly more difficult to identify the spatial extent and the location of the edge. Note that we can associate a 'sign' to an edge, determined by the transition from a darker to a brighter region or vice versa.

5.3.1 Detecting Edge Pixels

An edge operator detects a local edge, here called an edge pixel. From a good edge operator you expect that it locates edges accurately (good localization), that it finds all (or most) edges, and that it does not signify wrong edges. Detecting intensity changes amounts to analyzing the differences of gray levels within a small spatial extent. Taking differences amplifies noise. Consequently, an edge detection scheme should reduce the impact of noise. One possibility is to smooth the image before edge detection. Some edge operators combine the two steps of smoothing and sharpening.

The requirement that an edge operator detect all edges is a non-trivial one; comparing Fig. 5.3 reveals that there are many different kinds of edges. A challenging issue is to detect edges with a wide spatial extent. Over how many pixels do we consider intensity changes still as an edge and how significant should the gray level difference be? You realize that the notion of an edge defies a rigorous mathematical definition; ultimately, it is application dependent. Not surprisingly then, many edge operators have been developed. There is no single best or optimal operator and quite often several edge detectors may be used in concert to solve a particular application.

Several criteria may be used to classify edge operators, for example, first or second (or higher order) derivative operators and directional or non-directional operators. As indicated in Fig. 5.4, the first derivative becomes a maximum at the location of an edge; second derivative operators indicate an edge by a zero-crossing. The following classification groups edge operators into three broad categories.

gradient edge operators determine the gradient of the image function. In Sec. 3.3.3 we have discussed various ways in which the gradient is approximated in the (discrete) digital image. Examples of gradient operators include Roberts, Prewitt, Sobel, Kirsch—all devised in the sixties and seventies.

template matching operators fit a model edge to the image. When a good fit is found, the pixel is labeled with the model edge. Examples include the Kirsch operator.

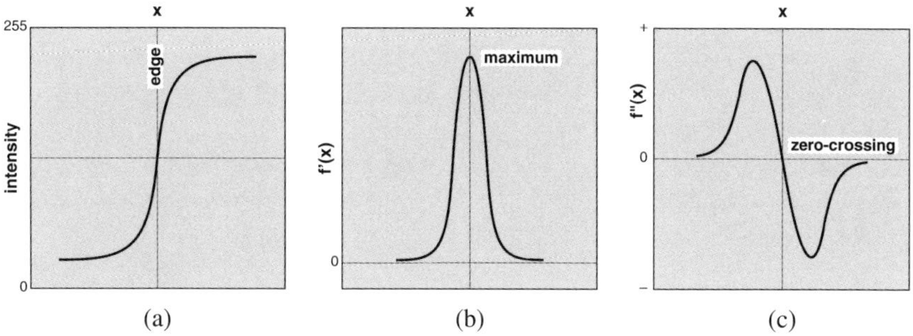

Figure 5.4: First and second derivative of a function. The edge in (a) is identified as a maximum in the first derivative (b) and by a zero-crossing in the second derivative (c).

second derivative operators indicate an edge pixel by a zero-crossing (see Fig. 5.4). If the derivative is taken symmetrically in all directions then the operator becomes non-directional.

Directional edge operators may use differentiation in any orientation. Preferred directions are either perpendicular to the edge orientation or in the direction of the gradient. Such operators generally locate edges more precisely than non-directional detectors. However, the derivative direction must first be determined, usually in a separate step. Moreover, directional operators cannot combine the operations of smoothing and differentiation in one step since smoothing is a scalar quantity but the gradient is a vector.

Rotation invariant edge operators employ non-directional differentiation. Now let us pop the cardinal edge detection question: should you use a directional or non-directional operator? Apart from personal preference, the decision depends on further processing tasks which is sort of driven by the application.

For educational purposes, the next section will present the non-directional LoG operator. The operator is rather simple and well studied. Its analysis elucidates some key concepts of edge detectors.

Laplacian of Gaussian (LoG) Edge Operator

The Laplacian operator, defined as

$$\nabla^2 = \frac{\vartheta^2}{\vartheta x^2} + \frac{\vartheta^2}{\vartheta y^2} \tag{5.1}$$

is independent of the orientation of the coordinate system. Since it is a scalar operator (the dot product of two gradients), it can be combined with the smoothing operation. The latter is necessary to reduce the effect of noise on the differentiation.

The LoG operator uses a Gaussian as a smoothing operator. We introduced the Gaussian in Sec. 3.3.2 and encountered its remarkable property of being localized in the spatial and frequency domain.

$$g(x,y) = \frac{1}{\sqrt{2\pi}\sigma} e^{-(x^2+y^2)/2\sigma^2} \tag{5.2}$$

where σ is the scale, or space constant, of the Gaussian.

Let us first perform smoothing and differentiation sequentially, say on image I. We have symbolically

$$S = G * I \tag{5.3}$$

$$C = \nabla^2(S) = \nabla^2(G * I) \tag{5.4}$$

with S the smoothed image and C the convolved image. The convolution property allows rewriting Eq. 5.4 as follows

$$C = (\nabla^2 G) * I = \text{LoG} * I \tag{5.5}$$

which is remarkable in that it combines smoothing and differentiation in one step. The second derivative of the Gaussian leads to the definition of the LoG operator

$$\boxed{\text{LoG} = \nabla^2 G = \frac{1}{\sigma^4}\left(\frac{x^2+y^2}{\sigma^2} - 2\right) e^{-(x^2+y^2)/2\sigma^2}} \tag{5.6}$$

Fig. 5.5 shows a cross-section of the LoG operator. The width of the central lobe, w, is related to σ by

$$w = 2\sqrt{2}\sigma \tag{5.7}$$

Like the Gaussian, the LoG also asymptotically reaches zero at infinity. For practical implementations, the size is limited, however. The operator size is usually chosen as $3 \cdot w$; coefficients outside this size are truncated.

The convolution of an image with the LoG operator yields scalar quantities. They are easier to examine than the vector quantities of directional

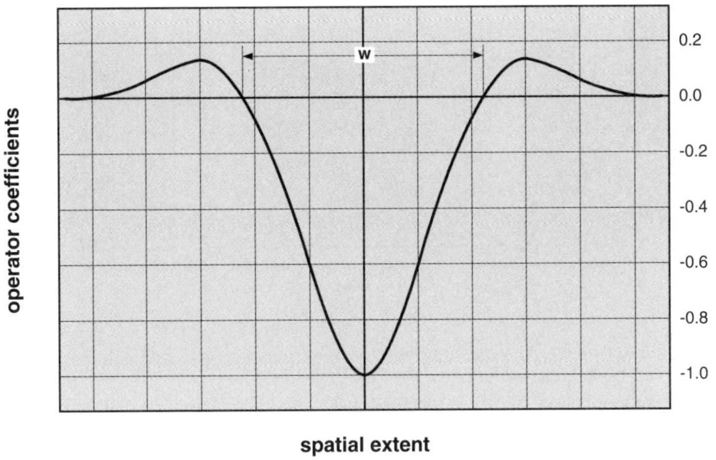

Figure 5.5: Cross-section through LoG operator. The negative central lobe has a diameter of $w = 2\sqrt{2}\sigma$. The infinite size operator is usually truncated at a radius of $1.5 \cdot w$.

operators. Edge pixels are found where the convolution values are zero. Since we deal with real numbers we will hardly find a convolution value that is precisely zero. It is more feasible to look for neighboring pixels of opposite sign and infer that somewhere in between is a zero-crossing.

Fig. 5.6 on the following page depicts some artifacts of edges obtained with the LoG operator. In Fig. 5.6(a) two edges intersect at a Y-junction. Zero-crossings cannot properly represent this situation because they are closed contours, causing a phantom edge. However, this arbitrary edge is very weak and disappears if the edges are thresholded. But then a gap will appear. Fig. 5.6(b) demonstrates the dislocalization problem associated with corners. Zero-crossings only precisely mark an edge if it is straight within the operator size. Fig. 5.6(c) shows a step edge. The LoG operator delineates the two edges properly but also generates two phantom edges (dashed lines). You can predict this rather strange behavior by looking at the shape of the operator in Fig. 5.5. Imagine the operator is placed somewhere on the leftmost edge in Fig. 5.6(c). Here we have a transition from bright to dark; the convolution values will be positive on one side and negative on the other. Now repeat the same process on the rightmost edge. Again, the transition is from bright to dark, but the convolution values are reversed. To cope with this oddity, two phantom edges are introduced. With proper thresholding of the edge strength they disappear, however.

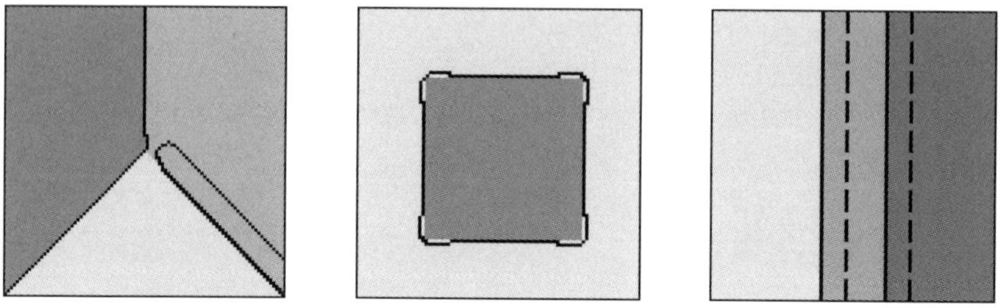

(a) (b) (c)

Figure 5.6: The synthetic images demonstrate three common problems with edges obtained with the LoG operator. Due the fact that zero-crossings are closed contours, it cannot properly represent the situation depicted in (a), where two edges intersect at a Y-junction. (b) demonstrates the dislocalization problem associated with corners. Whenever an edge is not straight within the operator size, the zero-crossing at that position won't mark the edge properly. The more acute a corner, the larger the dislocalization. The step edge in (c) causes two phantom edges (dashed lines), resulting from the need to reverse the convolution sign in the region of the step. Both (real) edges are transitions from brighter to darker regions (see text for further explanations).

5.3.2 Linking Edge Pixels

The process of edge detection left us with an indication of where edge pixels are or where we should look for them. To obtain a meaningful description, the individual edge pixels must be linked together. In subsequent tasks, the interest shifts from edge pixels to edges. Edges are entities with a starting and end point and a sorted list with pixels in between. Linking edge pixels may be viewed as a process of making an edge explicit. The procedure depends on the edge operator used. We continue with the LoG operator.

The convolution of an image with the LoG operator results in an image that contains real numbers. Finding edge pixels amounts to finding zero-crossings, that is, transitions from positive to negative convolution values or vice versa. In the scan line approach, every row is checked for a zero-crossing within that row, and the pixel closer to the zero-crossing is marked as an edge pixel and connected to a neighboring edge pixel from the previous row. If it cannot be connected then it is considered as the beginning of a new edge. As you correctly guessed, this simple approach is not without problems as there may be ambiguous situations on how to continue an edge.

Another approach is to follow an edge. Suppose we have identified an edge pixel. The continuation must be found within a three by three area. An interesting problem arises when more than one candidate is found. Guided by the Gestalt law of good continuation (see Sec. 4.4.1), we continue with that pixel which causes the least change in orientation.

Conceptually, the most pleasing approach to determine zero-crossing contours is briefly explained next. Consider the convolved image as a DEM, that is, as the discrete representation of a topographic surface. This DEM has positive and negative elevations. If we were to intersect the DEM by a horizontal plane at elevation zero then we would trace the zero-crossing contours. A possible implementation of this concept is to determine all regions with a positive convolution value. Let us first seek the maximum peak and then grow a region with this seed pixel until we meet a negative convolution value which then becomes a boundary pixel. After having found all boundary pixels that belong to the region with the maximum peak we proceed in the same fashion with the next highest peak and so forth. Depending on subsequent tasks, it may be beneficial to add attributes to the boundary pixels, such as the magnitude and orientation of the slope.

Fig. 5.7 shows an example of zero-crossings obtained with different size LoG operators.

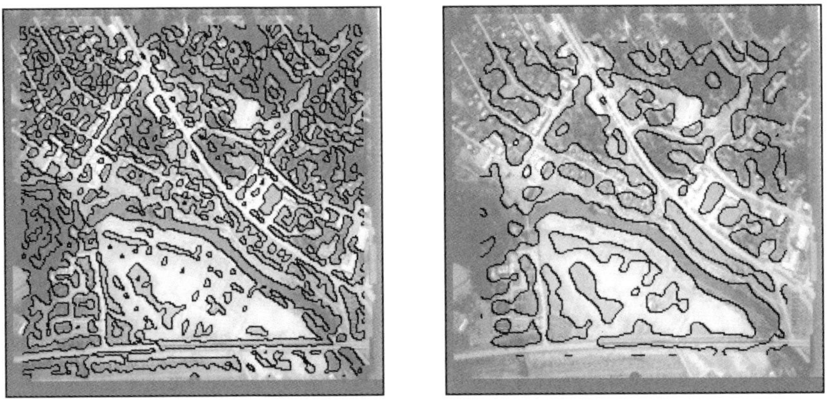

Figure 5.7: Example of zero-crossings obtained with the LoG operator. In the left image, w is two pixels and in the right figure, $w = 4$ pixels. Note the wider separation of the zero-crossings as the size of the LoG operator increases.

5.4 *Scale-Space Theory and Image Pyramids*

Real world objects can be described at various levels of detail, ranging from the molecular structure to the composition of parts, and on to the universe in which objects exist. This leads to the notion of scale. However, only a fairly narrow range of scales is meaningful for our applications. For recognizing buildings from aerial scenes, for example, it is meaningless to describe the building at the micron level or at kilometer level. The image of objects, too, exists only over a limited range of scales. Obviously, the resolution of the image defines the lower boundary while the size of the image, or a sub-image that contains the object, constitutes the upper scale limit.

Maps are familiar examples of multi-scale representations. Other examples include magnification by optical means and its electronic equivalent, zooming in and out on graphics displays. As we have discovered in the previous chapter (Sec. 4.2.3), receptive fields are a biological implementation of multiple scales. The response of receptive fields on the retina and in the visual cortex is very similar to that of Gaussian derivatives.

Only recently has the concept of multiple scales been formalized into a mathematical theory, known as scale-space theory. The theoretical underpinning is the representation of a signal at all scales by a family of derived signals with the property that details are suppressed at coarser scales. Image pyramids are another realization of the multi-scale concept. Here, the objective is to reduce the number of pixels on every level, while in the scale-space representation the sampling remains constant. Fig. 5.8 illustrates these concepts with a one dimensional signal. The scale-space representation in (a) is obtained by smoothing the original signal with an operator of ever increasing size. The pyramidal representation in (b) requires a resampling of the smoothed signal at every level at an ever increasing sampling size. Since the process of smoothing reduces the amount of information contained in the original signal, the scale-space representation is highly redundant. An important feature characteristic is their range of scales in the scale space. Feature A in Fig. 5.8(a) occurs throughout the scale space—hence it is a very dominant feature. On the other hand, feature B exists only on three levels. Fine details disappear early.

5.4.1 *Scale-Space Generation*

The scale-space representation of a signal, say $f(x, y)$, is a set of signals $\{f_s^n(x, y, n)\}$ that represents the original signal at all scale levels as a function of the scale parameter n. These derived signals constitute the scale space family of $f(x, y)$. The generation of the scale space is not an arbi-

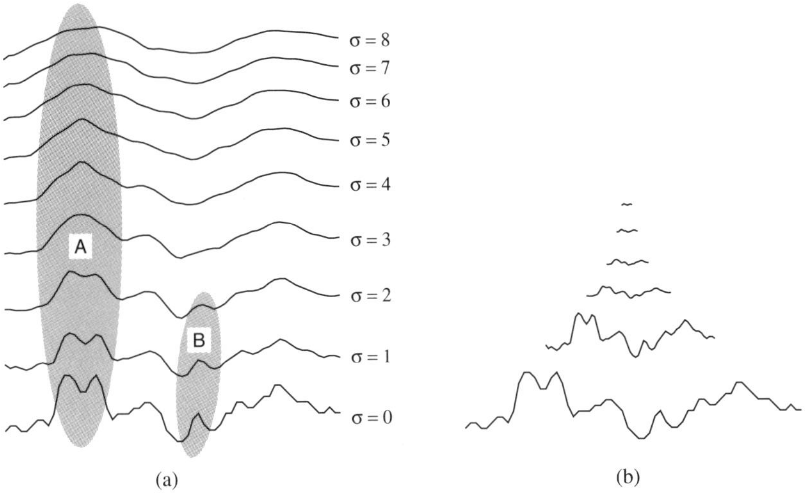

Figure 5.8: Multi-scale representation of a 1-D signal. The original signal at the bottom in both figures is convolved with a Gaussian smoothing filter of increasing σ. The scale-space representation in (a) shows the smoothed signal with the original tesselation, while the pyramidal representation in (b) represents the smoothed signal with decreasing resolution. Note how fine details disappear within a relatively small range of scales, as example B in (a) demonstrates. Coarser features persist through several levels, see example A.

trary process. To meet the objective of making information explicit that is only implicitly available in the original signal, two criteria must be satisfied.

1. For generating the scale-space family, the original signal must be convolved by a single scale-generating function $s(x, y; n)$.

2. The scale-generating function must have the property that details corresponding to higher frequency phenomena are suppressed on coarser levels. Moreover, no new information may occur on any level that is not already present on previous levels.

The first condition can be formally stated as

$$f_s^n(x, y; n) = f(x, y) \star s(x, y; n) \tag{5.8}$$

where $f_s^0(x, y; 0)$ obviously is the original signal $f(x, y)$.

A scale-generation operator that satisfies the second condition is the Gaussian

$$G(x, y; \sigma) = \frac{1}{2\pi\sigma} e^{-\frac{x^2+y^2}{2\sigma^2}} \tag{5.9}$$

Now, the scale-space family f_s^n of signal $f(x, y)$ can be created by

$$f_g^\sigma(x, y; \sigma) = G(x, y; \sigma) \star f(x, y) \tag{5.10}$$

The scale-space theory provides a framework for detecting significant structures as a function of their scale. The information is only implicitly available, however. Information that is embedded in the scale-space can be made explicit by tracking features. Edges are examples of features that can be extracted in and tracked through the scale-space. Extracted edges at slightly different scales are similar and can be traced from coarser levels to finer levels with increasingly improved localization. That is, coarser scales render significant edges while finer scales provide their exact position. Tracking features offers the additional advantage of making relationships explicit—a property that can be exploited in relational image matching.

5.4.2 Image Pyramids

There are two steps in the generation of pyramids: smoothing followed by resampling to a coarser grid. Consequently, the image size rapidly decreases. A common approach is to double the grid size from one level to the next. Suppose an original image size of $2^N \times 2^N$ pixels. Then, the pyramidal representation is a sequence of images with sizes of $2^k \times 2^k, k = N, N-1, \ldots, 0$. Does resampling to a coarser grid cause loss of information? If the smoothed version of the image is obtained by a low-pass filter that suppresses all frequencies higher than half the maximum frequency, then resampling to a grid twice as large is theoretically loss-less.

The simplest, but not the best way, to generate a pyramid is to assign to the pixels on level $k-1$ the average gray level obtained from the corresponding 2×2 neighborhood on the previous level k. Here, smoothing is achieved by an averaging filter in the spatial domain. We recall from the discussion in Section 2.4 the aliasing and ringing effects of the averaging filter.

A better choice of a smoothing kernel is a low-pass filter. It satisfies the criterion of equal contribution of every pixel to the generation of the next coarser level. Unfortunately, an ideal low-pass filter requires infinite support in the spatial domain. Different solutions have been proposed to compromise between ideal low-pass property and finite support. For a 3×3 size, often the binomial filter with the constants

$$\frac{1}{16}\begin{vmatrix} 1 & 2 & 1 \\ 2 & 4 & 2 \\ 1 & 2 & 1 \end{vmatrix}$$

is used. It satisfies the criteria of symmetry and equal contribution. When applied recursively, the binomial filter increasingly approximates the Gaussian filter.

Yet another approach to generating pyramids is to represent the difference between smoothed and unsmoothed images. Such pyramids are also known as Laplacian pyramids. A Laplacian image L^k on level k is obtained from $L^k = I^k - I^k \star S$, where I^k is the image at level k and S is the smoothing kernel. The computation of the difference image is followed by resampling L^k to the next coarser level $k-1$. Obviously, the Laplacian pyramid emphasizes differences in gray levels, such as edges.

Fig. 5.9 depicts a normal image pyramid and the Laplacian pyramid of the same original image. Note that the two pyramids capture different aspects of the original image.

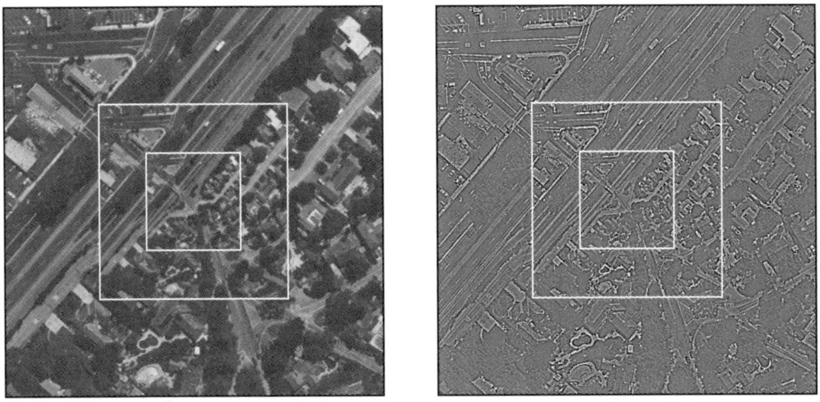

(a) (b)

Figure 5.9: Example of an image pyramid (a) and Laplacian pyramid (b). Only the first three levels are shown.

The advantage of image pyramids is their rapidly decreasing size that greatly facilitates computations (cost) as well as their representations (memory). Pyramids are used in many applications, for example image matching (hierarchical approach), surface representation, and image storage for fast displays at different resolution levels. On the negative side, we note that pyramids are not invariant to translations. Such operations require a new pyramid generation.

5.5 Perceptual Organization

For an overview of visual perception refer to Sec. 4.4.

5.5.1 Background

Perceptual organization is the ability to derive relevant groupings and structures from an image without prior knowledge of its content. The human visual system is remarkably adept at detecting many different patterns and significant groupings of elements in an otherwise random set of elements. Such grouping processes include clustering, detecting single and parallel lines and curves, and segmenting the image into regions which share similar image characteristics.

Perceptual organization is a prerequisite for object recognition. It must be largely independent of lighting conditions, viewer point and scale. Thus a perceived grouping is an intrinsic property of the image. It must be derived by examining regions since a single gray level of an individual pixel contains almost no perceptual information.

Perceptual organization is not only a late vision or cognitive process. Preattentive vision experiments (response of human visual system to stimuli without focus of attention) indicate that perceptual organization occurs very early on in the visual pathway. Quite often, grouping processes are recursive. The output of one process becomes input to another grouping process, perhaps together with other image information. An example is edge fragments that are clustered to edge elements which in turn are grouped to lines.

Since vision is ill-posed, one image may have many different interpretations. In order to reduce the solution space, assumptions about the world we see have to be made. Perceptual organization and, in particular, the Gestalt principles of organization provide the human visual system with sensible assumptions about physical and biological objects. Generally, the shapes of natural objects vary smoothly and quite often they are symmetrical. Matter is cohesive, so adjacent regions are likely to belong to the same structure.

In computer vision, perceptual organization is known as grouping, or more generally, as segmentation.

5.5.2 *Analysis of Blockworld Scenes*

More than twenty years ago many researchers in artificial intelligence tried to solve the segmentation problem by dividing a visual scene into a number of distinct objects. The complexity of natural images was avoided, and the experiments were restricted to what is known as "blockworld" in AI: white, prismatic solids which are evenly illuminated.

The best known program is probably from Waltz. He set out to recognize objects from a collection of lines. Crucial elements in the analysis are junctions (where two or more lines meet). There is a direct relationship between the junction type and the possible arrangement of surfaces. Another important relationship is how the different types of edges (convex, concave, occluding) are combined.

Despite the success of some of the scene analysis programs, their practical importance is limited. The principles embodied in the segmentation of blockworld scenes fail to reveal significant structures in natural images.

5.5.3 *Curvilinear Segmentation*

Boundaries of objects are one of the most important structures derived from images. Naturally, many computer vision programs group edge elements into boundaries. They usually begin with linking edge points (edge following, edge linking) followed by approximating the contour by straight lines and curves. A number of methods have been proposed for accomplishing this task. Techniques for solving the edge following problem include graph search and dynamic programming.

Hough Transform

One of the most popular methods for line and curve detection is the Hough transform, initially proposed by Hough and later modified by several researchers. The method is an interesting example of seeking a representation that is much better suited for solving the segmentation problem. This representation is called the parameter space. Hough used the slope-intercept equation $y = mx + q$ for creating the parameter space m, q. That is, a line in the x, y coordinate system is represented as a point in the m, q space. Likewise, all possible lines through a point in x, y are represented in the parameter space by a straight line. The strategy is to compute the line in the m, q space corresponding to every edge element in the x, y space. Then, straight lines in x, y are found at intersecting positions in m, q. The parameter space is discretized and implemented as an accumulator array. Consequently, a straight line in the m, q space must be rasterized and all cells that are turned on are incremented.

A better parameterization of the line is by the polar equation $x \cos\phi + y \sin\phi = r$, because m may be infinite in the slope-intercept representation. With the polar form, the parameter space is r, ϕ. This form is suitable for other curves, and the Hough method can be generalized. For example, $(x-a)^2+(y-b)^2 = r^2$ would allow finding circles in the three-dimensional parameter space a, b, r. We will employ this technique in Chapter 13. A generalized form of the Hough transform will be presented in the advanced image matching chapter.

As elegant and straightforward as the Hough transform is, it has a number of disadvantages. Apart from the requirement for huge accumulator array sizes[2], the distance between points is not considered. It will group points from far regions if they happen to be on a straight line, and will fail to consider points in close proximity if they are only slightly off the line. This is in strong contrast to the way humans perform collinearity grouping.

Virtual Lines

Fig. 5.10(a) shows a random-dot flow pattern, created by superimposing a pattern of random dots on an identical pattern which was rotated slightly. The human visual system groups the dots in such a way that a rotational flow is perceived. A similar effect is obtained when a copy of the random dots is expanded in a radial direction and then superimposed on the original random dots. The grouping process results in a radial flow pattern as vividly demonstrated in Fig. 5.10(b).

The virtual line algorithm links points based on proximity and orientation. Take a point and determine for all points closer than d_{min} the orientations. Repeat this procedure for every point and determine the distribution of all orientations. In a typical flow pattern, certain directions are dominant. The preferred directions emerge from the orientation distribution as peaks. A simple thresholding operation determines what orientations are allowed. Now the procedure is repeated by assigning a virtual line to the closest point which satisfies the orientation criterion. A modification of this general schema is necessary if all directions are equally likely. Such is the case in the rotational flow patterns of Fig. 5.10. However, preferred directions are clearly obtained on a local basis, for example, in one quadrant.

---

[2]For n-parameter curves each quantized to m levels, m^n accumulator cells are required.

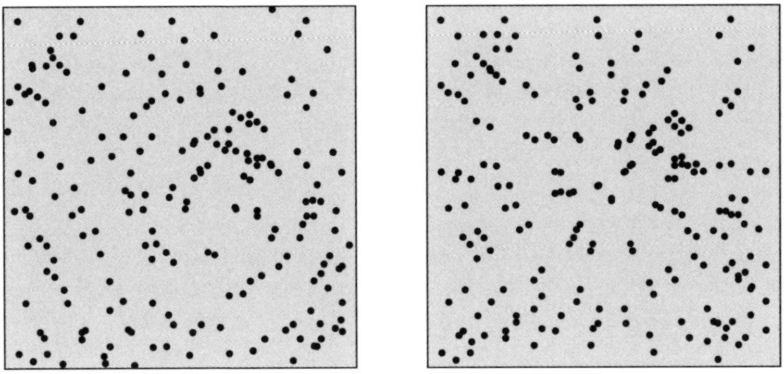

(a) (b)

Figure 5.10: Virtual lines. Randomly generated dots are rotated in (a) to give a sense of rotational motion. The same dots are scaled and superimposed on the original set in (b). The virtual lines give a perception of radial motion.

5.5.4 Region Segmentation

Region segmentation divides the image into units that are homogeneous with respect to one or more characteristics. From a perceptual point of view, those regions should be meaningful as far as the interpretation is concerned. Thus, region segmentation is to a certain degree application dependent and one issue is how to influence the process by a priori knowledge about the scene.

One approach to region segmentation is region growing. Properties of local groups of pixels are aggregated. In the simplest case, pixels are grouped to regions based on their properties and neighborhood. Global techniques consider the properties of a much larger population of pixels, for example, by thresholding the histogram. This amounts to a foreground/background separation.

Another approach is called split and merge. Here, the image is regarded as a discrete state and every pixel is initially a region. The state changes when a boundary between regions is removed (merging) or introduced (splitting).

No doubt, texture plays an important role in perceptual organization. Texture segregation divides an image into regions of similar texture primitives, sometimes referred to as texels. A texel has certain invariant properties which occur repeatedly in different positions and orientations. Julesz performed extensive studies of human texture perception. He claimed that two regions cannot be discriminated if their first- and second-order statistics are identical. First-order statistics refer to overall brightness while second-order statistics express differences in granularity (spatial distribution of texels) and slope (orientation).

While Julesz tried to arrive at a universal mathematical formula to explain the perception of texture boundaries, Marr developed an information processing theory which explained how a set of descriptive principles can be used to recover texture. We recall that the primitives in the primal sketch comprise edges, bars and blobs with associated orientations, contrast, dimensions and positions. Structures are recovered by clustering place tokens into small structures which, in turn, are grouped to larger units on the basis of proximity and changes in spatial density, a process Marr called theta aggregation.

5.5.5 More General Approaches

Although many computer vision programs have incorporated some aspects of perceptual organization, the approaches remain rather rudimentary. Most methods are ad hoc and treat grouping aspects in isolation. Obviously, perceptual organization as an information processing system is not understood on the theoretical level[3]. Another likely reason for the lack of an integrated approach is the fact that the outputs of some perceptual processes do not immediately lend themselves to a physical interpretation of the scene.

McCafferty describes a computational approach to structuring and grouping. He sets out to formulate a mathematical model for including the Gestalt laws of perceptual organization in a vision system. In his work, grouping is posed as an energy minimization problem.

As described earlier, the solution space of the ill-posed visual reconstruction problem is reduced by introducing stabilizing functionals (regularization theory). Since the Gestalt laws of organization constrain the human visual system to physically plausible solutions, they provide the basis for the stabilizing functionals. To enforce the grouping due to proximity, similarity, closure, continuation, symmetry, figure-ground separation, and common-fate, the following minimization problem is formulated.

$$\begin{aligned} E(g) &= \lambda_1 \|E_1(g)\|^2 + \lambda_2 \|E_2(g)\|^2 + \cdots \\ & \quad \lambda_7 \|E_7(g)\|^2 \end{aligned} \tag{5.11}$$

[3]Refers to Marr's approach of understanding visual processes on the three levels of theory, algorithms, implementation.

where $\lambda_1, \ldots, \lambda_7$ are regularization parameters and $E_1(g), \ldots, E_7(g)$ are stabilizing functions, and $E(g)$ is the total energy to be minimized.

The grouping problem, cast as an energy minimization problem, involves finding the global minimum energy state. The energy distribution across the state space is multimodal; thus the global minimum cannot be found by a simple threshold. It is difficult if not impossible to know if the lowest energy state found is in fact identical with the global minimum.

For computing the energy of a grouping, ways must be found to compute the individual energy terms. For example, proximity is a measure of the spatial separation between place tokens. It can be computed as:

$$\frac{1}{i} \sum_{p=1}^{i} \frac{1}{n_i} \sum_{q=1} n_i d_{pq} \tag{5.12}$$

where i is the number of tokens, n_i is the number of neighbors of tokens, and d_{pq} is the distance between token p and q. That is, proximity is calculated as the average distance between each token and its neighbor.

Finally, proper values for the regularization factors λ must be found. They essentially determine the weight of an individual grouping in the combination of the Gestalt laws of organization. The weighting factors are either constants, in which case the relative importance of groupings never changes, or they could also be functions of some other information, such as raw image data, or a priori knowledge.

5.6 Object Recognition

For a person to respond properly to the environment, he must analyze, interpret and understand visual stimuli. Ideally, the same feat must be accomplished by a robot or an autonomous vehicle. To pick a part from a bin or to navigate through a cluttered environment, the robot must understand its environment from sensor data and stored knowledge. The net result of image understanding is a fully interpreted scene. Image understanding, or image interpretation, is application dependent.

Object recognition is an important subtopic of image understanding. While image understanding is essential for many computer vision applications, its role in digital photogrammetry is less important. Here, the emphasis is on recognizing and locating objects. Subsequent tasks, such as analyzing objects and their interrelationships, are typically performed by GIS.

Object recognition is a broad and very active research area. Many different methods and aspects have been reported. It is, of course, impossible to describe in one section the different approaches or to make proper reference to the huge body of literature. The purpose here is to provide an overview and to touch on some of the most critical issues, such as reducing the search space, modeling the objects, and matching data features with object models.

5.6.1 Global Model-Based Recognition

The traditional approach to object recognition is to search for a correspondence (match) between object models and sensory data (e.g., image). Matching establishes an interpretation of image data consistent with objects and phenomena of the world. The single most critical factor in this endeavor is to find suitable representations, both for the objects and input data. Most recognition systems describe objects and data as geometric or relational structures.

Matching Geometric Structures

A common solution is to parameterize object models and data. Matching then, is an attempt to find the best fit between two parametric representations. The fit is measured with a merit function $M(p, q)$ where p is the parameter vector describing the object, and q is the parameter vector derived from the data. There are several techniques available to optimize the parameters including direct analytical solution and hill-climbing (gradient) techniques.

The parameters may refer to shape, such as curvilinear boundary description, or to other global properties of the object, such as area, elongateness, Euler number, etc. Useful properties are those that remain invariant under transformation or projection.

Matching $\psi - s$ Curves

The $\psi - s$ curve represents a line as a function of $\psi(s)$ where s is the parameter of the tangent ψ. Fig. 5.11(a) shows a line in the x, y domain and Fig. 5.11(b) depicts the same line in the parameter space s, ψ. Horizontal straight lines transform to horizontal straight lines on the s - axis ($\psi = 0$) and straight lines with a slope α are represented as horizontal lines, $\psi = \alpha$. Thus, vertices cause discontinuities in the $\psi - s$ domain and the amount of the discontinuity is proportional to the angle of the vertex. Since the tangent of a circle is a constant, a circle will be represented by a straight line with the slope proportional to the radius.

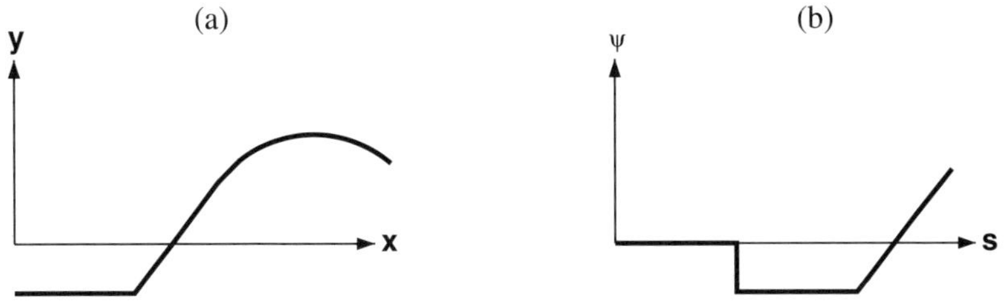

Figure 5.11: A line consisting of straight line segments and a partial circle is represented in the original x, y space (a) and in the parameter space ψ, s (b).

The $\psi - s$ representation for describing the shape of object models and data models is advantageous because it is rotation and translation independent. Chapter 11 will elaborate further on this method.

Matching Moment Invariants

Moment invariants are useful for recognizing patterns. An image is represented by the spatial moments of its intensity function. The ij^{th} generalized moment is

$$m_{ij} = \int f_{ij} I(x, y) dx dy \tag{5.13}$$

where $I(x, y)$ is the image intensity function. The function $f_{ij}(x, y)$ may represent the density in which case m_{ij} corresponds to the standard definition of moments; or it may involve sine and cosine functions where m_{ij} would represent the coefficients of a Fourier series expansion of $I(x, y)$.

Linear, quadratic and cubic polynomials of the moments are used as shape descriptors. Linear regression is used to distinguish these nonlinear functions by treating the powers of the moments as new linear variables.

A simple pattern recognition model with two moment invariants has been proposed to recognize airplanes. For each aircraft the two moment invariants

$$\begin{aligned} X &= \mu_{20} + \mu_{02} \\ Y &= \sqrt{(\mu_{20} - \mu_{02})^2 + 4\mu_{11}^2} \end{aligned} \tag{5.14}$$

are computed. The pattern is represented as a point in the X, Y parameter space. A system built on this approach outperformed human observers in a test of identifying airplanes in a library of 132 different types.

5.6.2 Feature Matching With Geometric Constraints

In contrast to global matching methods, feature-based matching approaches the object recognition problem in several steps. The basic idea is to extract local data features, which are matched with spatially localized object features. Such features include distinct points, edges, curves and surface patches. The following tasks must be solved.

1. Build a library of objects (model base). The geometric description of every object depicts shape characteristics in a local object coordinate system.

2. Extract features from the images. Group and segment them such that data features correspond to one object *(segmentation problem).*

3. From the model base select those objects which are likely to correspond to a set of given data features *(indexing problem).*

4. Find instances of objects in the data by establishing a correspondence between data features and object features *(correspondence problem).*

5. Find instances of objects in the data by transforming the object to the image for checking global consistency *(hypothesize-and-test problem).*

Extracting features, organizing and representing them in a suitable form is a typical perceptual organization problem. In the simplest form, it would involve edge detection, edge formation and edge segmentation. There is a plethora of edge detecting operators available. Edge forming depends on the edge detector used. Curvilinear edge segmentation results in straight line segments and curved parts of an edge. Yet another processing stage may analyze consecutive straight line segments for rectangular polygons.

A straightforward but naive solution to the indexing problem is simply to take every object from the model base and perform steps 3 and 4. A more intelligent approach would be to consider additional attributes such as color. Also, the number of data features may be used to select only those objects that have more features, assuming that the data features are only a subset of all features describing an object. Finally, domain-specific knowledge may be used. For example, it may well be that certain objects do not occur in certain parts of the image or in the neighborhood of other objects.

Table 5.1: Angles in grads between object features.

	F_1	F_2	F_3	F_4	F_5
F_1	0	150	250	150	250
F_2	250	0	100	0	100
F_3	150	300	0	300	0
F_4	250	0	100	0	100
F_5	150	300	0	300	0

Table 5.2: Angles in grads between data features.

	f_1	f_2	f_3
f_1	0	300	150
f_2	300	0	250
f_3	250	150	0

Correspondence Problem

We explain the correspondence problem and the significance of geometric constraints by way of a simple 2-D example. Suppose the feature extraction process (e.g. edge detection, grouping and segmenting) generated the three data features f_1, f_2, f_3 (see Fig. 5.12(a)) which we assume correspond to the boundaries of an object. Shown in Fig. 5.12(b) is an object selected from the object library. The problem is to establish a correspondence between data features and object features without performing the costly transformation object to image.

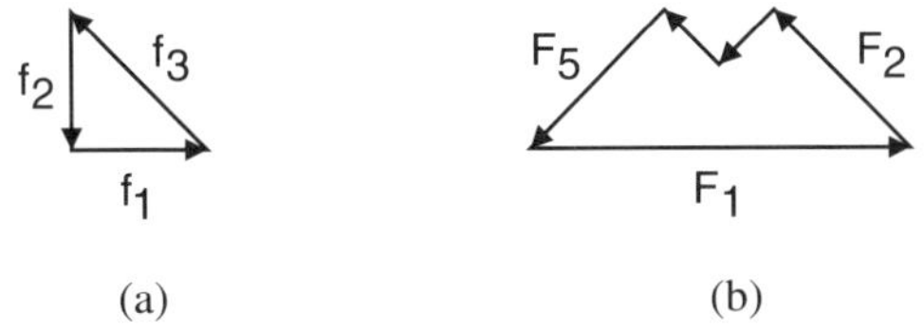

Figure 5.12: Data features extracted from image (left) and object selected from the object library (right).

In a brute force approach we could pair each data feature with all object features and check all possible combinations for consistency. This would amount to an exhaustive search of the correspondence space (see Fig. 5.14 on page 131). The n−dimensional correspondence space (n = number of data features) is tesselated by the number of object features. Referring to our example we obtain $5^3 = 125$ assignments most of which do not make any sense. Thus, the goal is to determine only likely pairings.

Instead of exploring all pairings, geometric constraints are used to eliminate those which make no sense. However, the method should require much less computing time than the rigorous solution of transforming the object to the image and checking if it matches. In our example angles are used as geometric constraints. In Table 5.1 all angles between object features are listed. Table 5.2 contains the same for the data features.

As indicated in Fig. 5.13, data feature f_1 is assigned to every object feature F_i. Next, data feature f_2 is assigned to those object features that satisfy the angle constraint. From Table 2 we take the angle between f_1 and f_2 (300^g) and check in the appropriate rows of Table 1 for the same angle. For the assignment f_1 to F_3 two pairings satisfy the constraint. The same holds for f_1 to F_5. Now, only these pairings need be pursued further. Again, the assignment for f_3 is performed by considering the angular constraint $\angle(f_2, f_3) = 250^g$. We end up with the four sets of possible pairings for f_1, f_2, f_3, namely, $(F_3, F_2, F_1), (F_3, F_4, F_1), (F_5, F_2, F_1), (F_5, F_4, F_1)$. With additional constraints, for example, distances, the potential pairings could be further pruned.

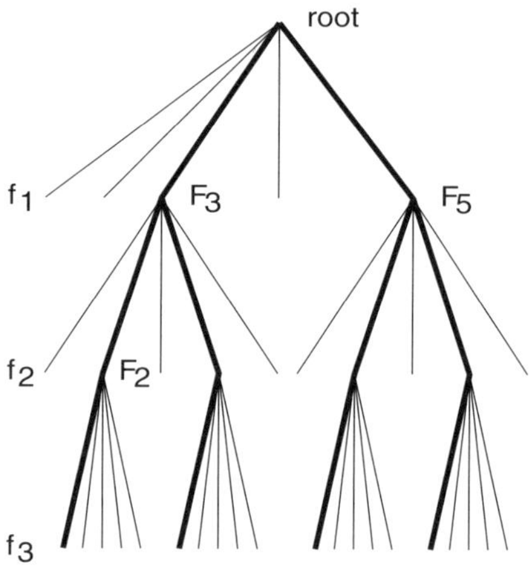

Figure 5.13: Correspondence problem cast as a tree search problem.

We note that the four solutions found are not globally consistent. The geometric constraints only assure local consistency. The angle constraint is a binary constraint. If enforced, two consecutive nodes in the tree are consistent. A unary constraint would ensure a single node to be consistent. With unary and binary constraints, the consistency between three or more nodes is not guaranteed. Therefore, the transformation of the object to image is necessary.

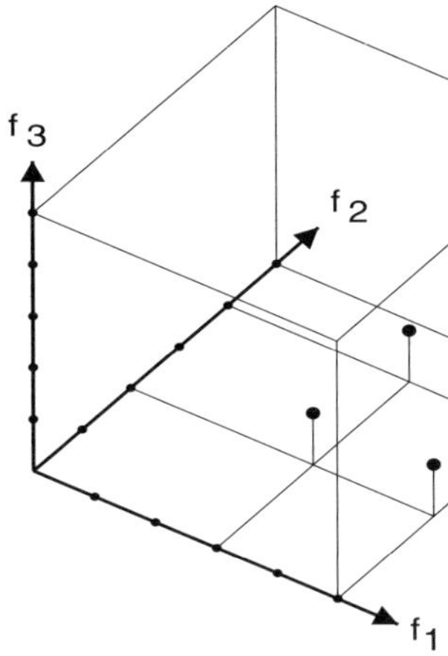

Figure 5.14: Correspondence space. Solid dots indicate possible correspondences found by applying geometric constraints.

Every point in the discrete correspondence space constitutes a hypothesis for an assignment of data features to object features. Obviously, most assignments can be ruled out immediately. For example, two distinctly different data features cannot be assigned to one and the same object feature (assuming rigid objects). Unary and binary constraints further reduce the assignments to a feasible set.

As indicated in Fig. 5.13 the search for feasible assignments is cast as a tree search problem. On the first level of the tree, data feature f_1 is paired with all object features $F_i, i = 1, ...n$, but only if unary constraints are satisfied. An example of a unary constraint is the length. The same procedure is repeated on the next levels of the tree. Only those nodes are further expanded which satisfy the constraints. Every valid leaf node defines a path in the tree, indicating feature assignments which must now be examined for global consistency.

Verifying Hypotheses

Every leaf of the tree is a hypothesis about the correspondence of data features to object features. These hypotheses must be tested by transforming the object into the image. Global consistency is reached if all data features match the corresponding object features. Usually, there are more object features than data features. That is, there are some unpaired object features. They are now used to examine the gray levels for any evidence of data features which remained undetected in the feature extraction process.

In the general case of a 3-D transformation, seven parameters must be determined. This transformation should be formulated in a general fashion; for example, it should not be restricted to points only, rather it should allow including straight line segments, curves or surface patches.

A hypothesis can be tested before the leaf of a tree is reached. In fact, as soon as enough data features are assigned to perform a 3-D transformation, it may well be advisable to verify the hypothesis immediately. If successful, the search in the correspondence space can be terminated. This variation of the general approach may be justified if the tree is very deep (e.g., many object features) or if a particularly salient matching is found.

5.6.3 Nonmodel-Based Object Recognition

The approaches to object recognition described so far are based on the assumption that objects can be described geometrically, for example, by defining their boundaries or surfaces. This may be impossible, however, if it comes to recognizing objects with complex shapes. Such is the case with aerial scenes. Even rather simple structures, such as buildings, come in such a variety of different sizes and shapes that it is a fruitless attempt to precisely describe and store all of them in a model library. For outdoor scenes other solutions must be found.

5.7 To Probe Further

There are numerous books on the subject of computer vision. Although fairly old compared to the fast development of the field, *Computer Vision* by Ballard & Brown (Prentice-Hall, Inc., 1982) is still relevant. It covers a broad range. More mathematically inclined readers may enjoy the book *Computer and Robot Vision* by Haralick & Shapiro (Addison-Wesley Publishing Company, 1992). It comes in two volumes and includes chapters about pattern recognition and mathematical morphology. Also notable is *Robot Vision* by Horn (MIT Press, 1986). Photogrammetrists may enjoy the chapter about Photogrammetry & Stereo. It also covers the subject of shape from shading. An artificial intelligence view on computer vision can be found in the easy-to-read Chapter 3 of *Artificial Intelligence* by Charniak & McDermott (Addison-Wesley Publishing Company, 1985).

As the field matures, more specialized texts appear. Sarkar and Boyer's *Computing Perceptual Organization in Computer Vision* (World Scientific, 1994) makes a valuable contribution to middle level vision. In *Scale-Space*

Theory in Computer Vision (Kluwer Academic Publishers, 1994), Lindeberg provides a theoretical framework about the scale-space. In addition to the theoretical treatment, the author also offers interesting applications of the scale-space theory.

Chapter 6

Radiometry and Photometry

This chapter provides some background information about radiometry and photometry. Understanding camera and scanner specifications requires familiarity with radiometric and photometric concepts and quantities. Moreover, we should gain some insight into the working principles of detectors. This, in turn, requires knowledge about flux transfer which concerns the way electromagnetic radiation interacts with matter.

The first section summarizes the most important properties of electromagnetic radiation. The section about radiometry begins with the definition of radiometric quantities and their relationship, and it ends with thermal radiation. After briefly describing photometric quantities, the chapter concludes with a discussion of radiative energy transfer. The goal is to gain a basic understanding of the energy transfer through a lens.

Optics is not treated in this book. I assume that the reader is familiar with the concepts of pupils, image formation, aberrations and diffraction, and vignetting.

6.1 *Electromagnetic Radiation*

Electromagnetic radiation (EMR) is a dynamic form of energy made manifest only by its interaction with matter. It propagates with a velocity of $3 \cdot 10^8$ m/sec through a vacuum. You may regard it as a convenient high-speed link between source and sensor, free of charge. The properties of EMR change when it interacts with matter (e.g. reflectance, absorptance). Detected changes may be used to infer with what kind of material EMR interacted.

6.1.1 Wave Characteristic of EMR

James Clerk Maxwell derived equations to describe the propagation of electromagnetic waves. An electromagnetic wave consists of an electric field **E** and magnetic field **H**. The fields are perpendicular to each other and also perpendicular to the direction of propagation. The speed of propagation is constant everywhere in the same medium. In a vacuum it is the speed of light. Fig. 6.1 depicts these properties.

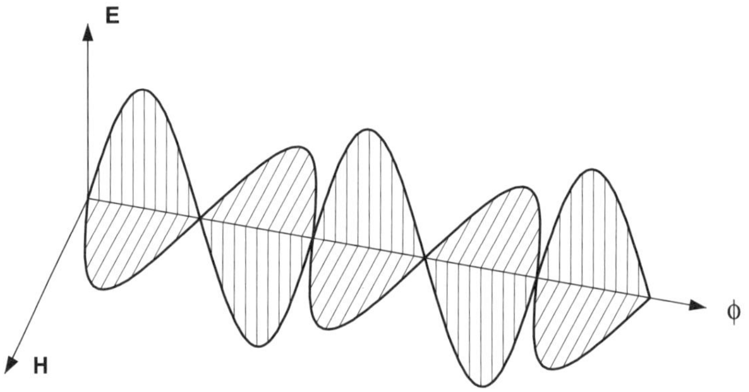

Figure 6.1: The electric (E) and magnetic (H) components of an electromagnetic wave are perpendicular to each other and to the direction of propagation (Φ).

The rate and direction with which EMR radiates across a unit area is expressed by the Poynting vector Π as the result of the cross product of the electric and magnetic field.

$$\Pi = \mathbf{E} \times \mathbf{H} \tag{6.1}$$

This equation can be derived from Maxwell's equations. The Poynting vector represents the flow across an isolated area. It is normal to the plane of **E** and **H**, confirming the fact that the energy travels perpendicular to the two fields.

6.1.2 Quantum Nature of Radiation

From observations, we know that when light interacts with matter it behaves as though it consists of particles. For example, the photoelectric effect can be explained as the emission of an electron due to the absorption of a quantum. On the other hand, electrons behave like waves, as is evident from interference and polarization. The dichotomy cannot be resolved—the dual nature cannot be unified by one model.

Planck derived the quantum theory of electromagnetic radiation. The generation of electromagnetic waves occurs in short wave trains (bursts). That is, radiation is thought of as discrete packets (quanta) of energy, propagating with the same speed and direction as the continuous waves. Each quantum of radiation has an energy Q_ν associated with it. This energy is called a *photon*. Note that the subindex indicates frequency dependency.

$$Q_\nu = h \cdot \nu \tag{6.2}$$

where h is Planck's constant ($6.625 \times 10^{-34} J \cdot s$). The wavelength, λ, the frequency, ν, and the speed of light, c are related by

$$c = \lambda \cdot \nu \tag{6.3}$$

Substituting ν in Eq. 6.2 leads to

$$\boxed{\begin{aligned} \lambda &= \frac{h \cdot c}{Q_\nu} \\ Q_\lambda &= \frac{h \cdot c}{\lambda} \end{aligned}} \tag{6.4}$$

Eq. 6.4 can be thought of as providing the correspondence between the wave nature of light and its particle nature (wave-particle duality of EMR).

Example 6.1:

With Eq. 6.4 we can determine the wavelength of radiation of a certain energy. The wavelength of a photon carrying the energy of one electron volt ($1eV = 1.59 \times 10^{-19}$ joules) is $1.25\ \mu m$. Likewise, the wavelength of a photon with 1 MeV is $1.25\ pm$. The more energetic the radiation the shorter the wavelength.

6.1.3 Electromagnetic Spectrum

Electromagnetic energy spans an impressive range of wavelengths from 10^{-10} to 10^{10} μm, or, equivalently, a frequency range from 10^{24} to 10^4 Hz.

As indicated in Fig. 6.2, the spectrum is divided into different portions, beginning with cosmic rays at the high frequency end to broadcast waves at the long wavelength end. The most interesting portion for remote sensing and photogrammetric applications is the optical spectrum, which occupies wavelengths from 0.3 to 15 μm. In this range, electromagnetic energy can be manipulated by optics. The microwave range between 1 GHz and 10 GHz (30 mm to 300 mm) is of great interest because microwave instruments operate over a wide range of weather conditions.

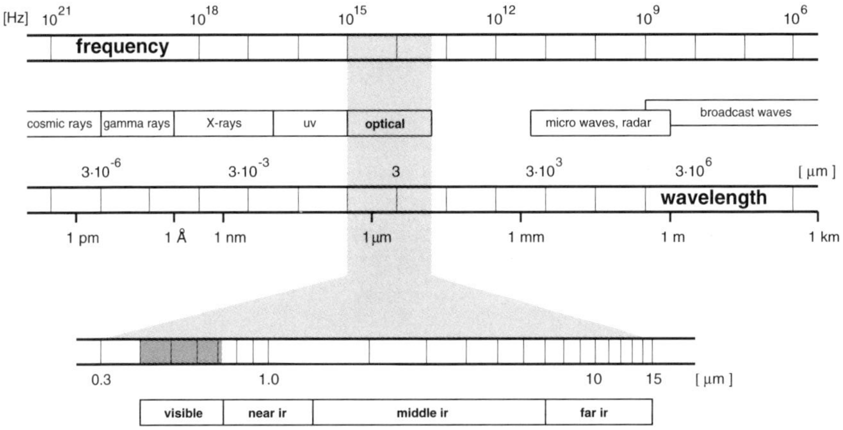

Figure 6.2: Electromagnetic spectrum.

The optical portion of the spectrum is further subdivided into the visible and infrared spectrum. The visible portion extends approximately from 0.38 to 0.76 μm, followed by near infrared (0.76 to 3.0 μm), middle infrared (3.0 to 7.0 μm), and far infrared (7 to 15 μm).

Energy from these portions of the spectrum interacts differently with different materials. For example, in the visible and near infrared range, reflected electromagnetic radiation depends on the status of vegetation (moisture content, cellular structure, pigmentation), and status of soils (moisture and mineral content). Thermal properties of the surface are most distinct in the far infrared. Electrical properties and roughness of surfaces may be inferred from analyzing the microwave portion of the electromagnetic spectrum.

Consequently, different sensing systems have been developed to record one or several portions of the spectrum for inferring properties of surfaces and material from reflected radiation. Since most of the wavelengths shorter than 0.3 μm do not penetrate the atmosphere, this portion of the spectrum is not exploited in remote sensing and photogrammetry.

6.2 Radiometry

Radiometry is the science of the measurement of electromagnetic radiation. Radiation refers to the propagation of energy through space. The previous section summarized some of the most important properties of EMR, such as the quantum property and the distribution of radiation over the electromagnetic spectrum.

In this section we introduce the definition of the quantities central to radiometry and photometry. This is followed by describing some useful relationships between these quantities. We then summarize some of the laws of thermal radiation necessary to understand the photon flux.

Units and naming conventions are in accordance with the International Commission on Illumination. Radiometric quantities are perfectly adequate to deal with all quantitative aspects of electromagnetic radiation.

6.2.1 Solid Angle

Central to the definition of some of the radiometric and photometric quantities is the solid angle. The solid angle, Ω, is defined as the area of a spherical surface, A, divided by the square of the radius r of the sphere. The unit is the *steradian*, $[sr]$, a dimensionless quantity. There are 4π steradians of solid angle in a sphere ($4\pi r^2/r^2$). Fig.6.3 depicts the concept of a solid angle.

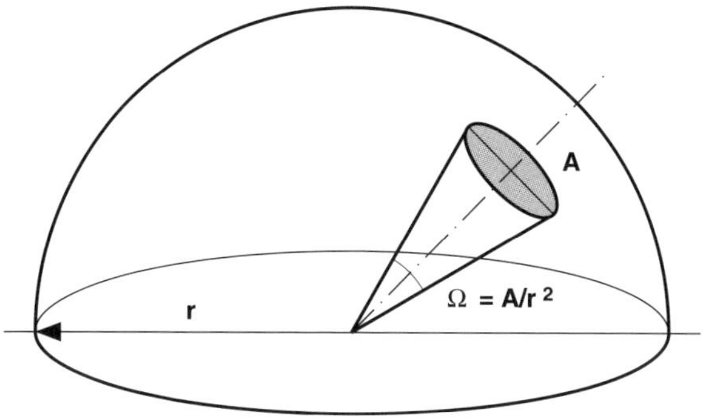

Figure 6.3: Concept of a solid angle. The area A on the spherical surface is divided by the square of the sphere's radius. The unit is the steradian.

Example 6.2:

> *Suppose the sun is a point source. What is the solid angle between the sun and earth? The area A is the earth's disc with a diameter of diameter of* $\approx 12,730$ *km. Approximating the distance earth/sun by 150 million km, we obtain a solid angle of about* $5.7 \cdot 10^{-9}$ *sr.*

6.2.2 Radiometric Quantities

Radiant Energy, Q

The energy carried by electromagnetic radiation is called *radiant energy*. It is capable of performing physical work when interacting with matter. It is measured in joules. All wavelengths contained in the radiation are included, except in the case of *spectral radiant energy*, Q_λ which is defined as the radiant energy per unit wavelength interval at a specific wavelength λ (joule per nm).

$$\boxed{Q_\lambda = \frac{dQ}{d\lambda}} \tag{6.5}$$

Radiant Flux, Φ

Radiant flux, sometimes also called power, is the time rate with which radiant energy passes a spatial position. Conceptually, it corresponds to the flow rate of water passing along a pipe. The defining equation is

$$\Phi = \frac{dQ}{dt} \tag{6.6}$$

The flux is measured in joules per second $[J \cdot s^{-1}]$ or in watts $[W]$. When the radiant flux is constant then the total radiant energy passing a spatial position during the time interval Δt is simply $Q = \Phi\Delta t$. When the flux changes in time, Q is found by integrating the flux within the time interval, that is

$$Q = \int_{t_1}^{t_2} \Phi(t)dt \tag{6.7}$$

Spectral radiant flux, Φ_λ, is the radiant flux per unit wavelength interval at a specific wavelength λ (Watt per nm).

$$\boxed{\Phi_\lambda = \frac{dQ_\lambda}{dt} = \frac{d\Phi}{d\lambda}} \tag{6.8}$$

Radiant Flux Density, E and M

The radiant flux intercepted by a plane surface, divided by the area of the surface, is defined as flux density

$$E = \frac{d\Phi}{dA} \tag{6.9}$$

Flux density is measured in watts per square meter, $[W \cdot \mathrm{m}^{-2}]$. When radiant flux is emitted from a surface (e.g. reflected radiant flux), it is called *radiant exitance M.* When radiant flux is incident on a surface, it is termed *irradiance E.*

When the radiant flux incident on a surface changes from point to point, the total flux per area is found by integrating the irradiance over the area, that is

$$\Phi = \int E(A)dA \tag{6.10}$$

Spectral irradiance, E_λ, is the radiant flux per unit wavelength interval at a specific wavelength, λ, at a point in a specified surface ($W \cdot \mathrm{m}^{-2} \cdot \mathrm{nm}^{-1}$, Watts per square meter and unit wavelength interval).

$$\boxed{E_\lambda = \frac{dE}{d\lambda} = \frac{d^2\Phi}{dAd\lambda}} \tag{6.11}$$

Radiant Intensity, I

Radiant intensity refers to a point source. It is the radiant flux per unit solid angle leaving the point source in a given direction. The units are flux per steradian $[Wsr^{-1}]$ and the definition of radiant intensity is

$$I = \frac{d\Phi}{d\Omega} \tag{6.12}$$

Whether or not the source of radiation can be considered a point source depends not only on the physical extension of the source, but also on the particular configuration, such as the distance between source and receiving surface and the size of the receiving surface area.

Spectral radiant intensity, I_λ, is the radiant flux per unit solid angle per unit wavelength interval at a specific wavelength λ incident on or emerging from a point in space in a given direction.

$$\boxed{I_\lambda = \frac{dI}{d\lambda} = \frac{d\Phi}{d\Omega d\lambda}} \tag{6.13}$$

Radiance, L

Radiance refers to an extended source of radiant flux. It is defined as the radiant flux per unit solid angle leaving an extended source in a given direction, or

$$L = \frac{d^2\Phi}{d\Omega dA\cos\theta} \tag{6.14}$$

As indicated in Fig. 6.4, the term $dA\cos\theta$ refers to the projected (foreshortened) area of the extended source in the direction of the solid angle. Radiance is measured in watts per square meter per steradian, $[W\,m^{-2}\,sr^{-1}]$. The two angles θ, ϕ define the orientation of the solid angle. Note that radiance also depends on the position. That is, L depends on two angles and two coordinates (position and orientation of the solid angle). This is usually indicated by $L(x, y, \theta, \phi)$.

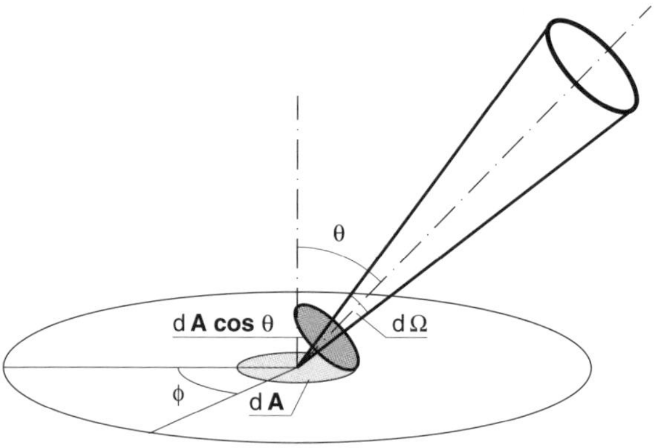

Figure 6.4: The concept of radiance. Radiant flux leaves an extended source in a given direction (ϕ, θ) within a solid angle Ω.

The concept of radiance is very frequently used. A case in point is the radiant flux leaving an extended area in the direction toward a sensor.

Spectral radiance, L_λ, is the spectral concentration of radiance

$$\boxed{L_\lambda = \frac{dL}{d\lambda} = \frac{d^3\Phi}{\cos\theta d\Omega dA d\lambda}} \tag{6.15}$$

6.2.3 Lambertian Radiators

A *Lambertian radiator* (also called *Lambertian surface*) has a radiance which is independent of direction. This can be expressed by the following equation:

$$L(\theta, \phi) = \text{const} \tag{6.16}$$

where θ and ϕ define the direction of the solid angle (see Fig. 6.4). You can visualize a Lambertian radiator best by an imaginary rotation of the cone (solid angle) in Fig. 6.4 about its surface point S. The radiance propagating into the solid angle is the same, regardless of its orientation.

What is the irradiance onto a plane that is parallel to the plane containing the radiation source? Let E_o be the irradiance corresponding to a solid angle whose axis is perpendicular to the radiation source ($\theta = 0$). Now let the solid angle point in a direction different from $\theta = 0$. The same amount of radiance is now incident on an area that is larger by a factor of $1/\cos\theta$. Consequently, the irradiance decreases and we have

$$\boxed{E = E_o \cos\theta} \tag{6.17}$$

which is known as Lambert's cosine law.

A good approximation for a Lambertian surface is a white matte paper, illuminated by diffuse sunlight. The paper appears at the same brightness, regardless from what direction we look at it.

6.2.4 Relationship of Radiometric Quantities

Relationship Between Radiance and Irradiance

We are seeking a functional relationship of the form $E = f(L)$. Substituting $d\Phi/dA$ in Eq. 6.14 on the facing page by the defining equation for the radiant flux density (Eq. 6.9) renders

$$L = \frac{E d\Phi}{d\Omega \cos\theta} \quad \text{or}$$

$$E = \int_\Omega L(\theta, \phi) \cos\theta d\Omega \tag{6.18}$$

Radiance describes the angular distribution of radiation and irradiance integrates this angular distribution over a specified solid angle Ω. Eq. 6.18 expresses this fundamental relationship for a point in the surface.

The elemental solid angle $d\Omega$ is conveniently expressed by spherical coordinates. Substituting $d\Omega$ in Eq. 6.18 by $\sin(\theta)d\theta d\phi$ (see also Fig. 6.5) leads to

$$E = \int_0^{2\pi} \int_0^{\pi/2} L(\theta, \phi) \cos(\theta) \sin(\theta) d\theta d\phi \tag{6.19}$$

In case of a Lambertian radiator ($L(\theta, \phi)$ = const), L can be removed from the integral and Eq. 6.19 evaluates to π, resulting in the simple relationship

$$\boxed{E = \pi L} \tag{6.20}$$

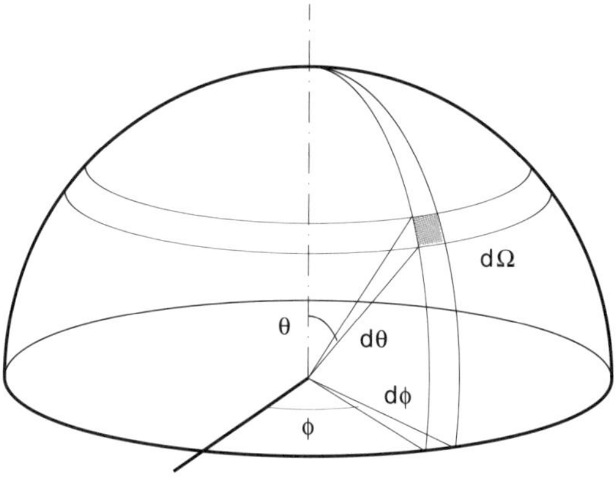

Figure 6.5: Element of solid angle expressed by spherical coordinates.

Relationship Between Radiance and Flux

Quite often one is interested in determining the flux from radiance. From Eq. 6.14 on page 142 we obtain

$$\begin{aligned} d^2\Phi &= L \cos\theta dA d\Omega \\ \Phi &= \int_A \int_\Omega L(x, y, \theta, \phi) \cos\theta d\Omega dA \end{aligned} \tag{6.21}$$

The flux is found by integrating the radiance over both the area A of the surface and the solid angle. Again, an elemental solid angle is denoted by $d\Omega$. To consider the positional dependency of the radiance, the flux must be obtained by integrating the above equation over the entire surface.

6.2.5 Thermal Radiation

The random motion of particles in matter results in particle collisions, which in turn cause a change in electron orbits or in vibrations of molecules and atoms. The kinetic energy of particle motion is heat energy. If the electron orbits change, say from a high energy band to a low energy band, heat energy is transformed to radiant energy. The difference in kinetic energy of an electron is carried by the photon. Likewise, a photon incident on matter may cause an electron to change from a low energy orbit to a higher energy level, thereby absorbing the photon's radiant energy.

Planck described this process based on thermodynamic laws. An ideal thermal emitter is called a blackbody. It transforms heat energy to radiant energy according to the maximum possible rate permitted by thermodynamic laws. Any real body (gray body) transmits or absorbs at a lower rate.

Planck's blackbody radiation law is expressed as

$$\boxed{M_\lambda = \frac{2\pi hc^2}{\lambda^5\left(e^{\frac{hc}{\lambda kT}} - 1\right)}} \tag{6.22}$$

where:

M_λ	spectral exitance in W m^{-2} μm^{-1}
T	absolute temperature in Kelvin
λ	wavelength in meters
h	Planck's constant $= 6.6256 \times 10^{-34}$ W s^2
k	Boltzmann's constant $= 1.38054 \times 10^{-23}$ W s K^{-1}
c	speed of light $= 2.997925 \times 10^8$ m$\cdot s^{-1}$

The numerical value of the numerator in Eq. 6.22 is usually called the first radiation constant $c_1 = 3.74151 \times 10^{-16}$, units $[W\cdot\text{m}^2]$. The second radiation constant is $c_2 = c \cdot h/k = 1.43879 \times 10^{-2}$, units $[\text{m} \cdot K]$.

Fig. 6.6 shows the spectral exitance distribution for blackbodies at different temperatures. The spectral exitance, M_λ, is a function of the wavelength. It reaches a maximum, which is of shorter wavelengths for blackbodies radiating at higher temperatures. The change of λ_{max} as a function of temperature is expressed by Wien's displacement law

$$\lambda_{max} = C/T \tag{6.23}$$

with $C = 2.898 \times 10^{-3}$ m K. You can derive this equation by setting the first derivative of Eq. 6.22 to zero and solving for λ.

Example 6.3:

The sun radiates close to a blackbody of 5900 *K. Using this temperature in Eq. 6.23 renders* $\lambda_{max} = 0.49\,\mu m$ *which corresponds, not surprisingly, to the blue/green portion of the visible spectrum.*

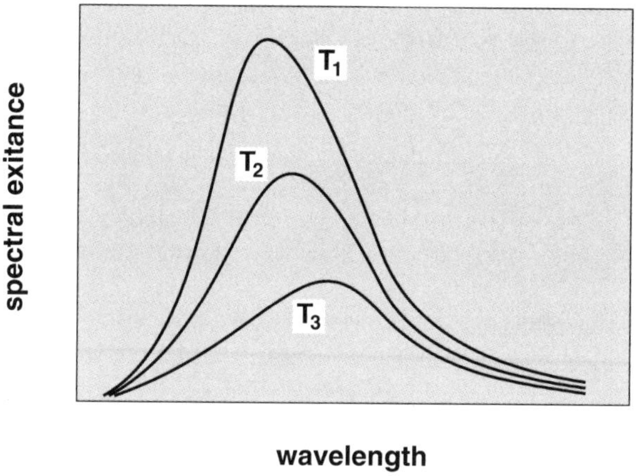

Figure 6.6: Spectral exitance distribution of blackbodies at different temperatures, $T_1 > T_2 > T_3$.

The energy of a blackbody (exitance) within a spectral band from λ_l to λ_u is found by integrating Planck's law within the boundaries l, u. The evaluation of the integral for the entire spectrum, that is the total radiant energy at a temperature T, is found by the Stefan-Boltzmann law

$$M_T = \sigma T^4 \tag{6.24}$$

with $\sigma = 5.669 \times 10^{-8}$ W m^{-2} K^{-4}.

Any real body radiates according to its temperature and material characteristics. The radiation is less than that of a blackbody, which is an ideal thermal emitter. The ratio of exitance of a graybody, M_λ^g, to that of a blackbody, M_λ^b, is the spectral emissivity, ϵ_λ.

$$\epsilon_\lambda = \frac{M_\lambda^g}{M_\lambda^b} \tag{6.25}$$

According to Kirchhoff's law, the spectral emissivity of a material is equal to its spectral absorptance, provided the material is in thermal equilibrium. The spectral emissivity is commonly determined by measuring the spectral absorptance of the material.

Spectral emissivity depends on the material, the wavelength, and the temperature. The influence of temperature is almost negligible unless the material undergoes a change of state (e.g. from solid to liquid). Also, in certain wavelength ranges, the emissivity is almost constant for any material. Such is the case for the 8 to 14 μm infrared band, where $0.85 < \epsilon < 0.95$.

6.2.6 Photon Flux

The radiant quantities defined in previous sections refer to the wave nature of radiation. For example, the radiant flux is associated with the square of the amplitude of an oscillating electric and magnetic field. The Poynting vector, introduced in Eq. 6.1 on page 136, expresses the rate at which electromagnetic energy passes through a unit area.

For some applications in radiometry it is advantageous to relate flux to the photons. An example is a detector that counts the number of photons impinging on its sensor. Therefore, the radiant quantities are replaced by *photon flux quantities.*

The number of photons emitted by a radiation source at all wavelengths is called the *photon number*, N_p. It can be readily determined from the radiant energy divided by the energy of one photon:

$$\boxed{N_p = \int \frac{Q_\nu}{h \cdot \nu} d\nu = \int \frac{\lambda \cdot Q_\lambda}{h \cdot c} d\lambda} \tag{6.26}$$

The photon flux, Φ_p is defined as the time rate of flow of photons, or

$$\Phi_p = \frac{dN_p}{dt} \tag{6.27}$$

with the unit $[s^{-1}]$.

All other photon flux quantities are the same as the radiant quantities, except that radiant flux must be replaced by the photon flux.

6.3 Photometry

Photometry deals only with the visible portion of the spectrum. Photometric quantities refer to the response of the human eye to electromagnetic radiation. Only radiation in the visible spectrum causes a photochemical change in the retina. Radiant fluxes with wavelengths shorter than 0.4 μm are absorbed by the cornea. Wavelengths in the near infrared penetrate the eye lens and are focused on the retina, but they do not cause a visual sensation. However, excessive irradiance in the spectrum 0.7 to 0.9 μm may damage the retina.

6.3.1 Photometric Quantities

Photometric quantities, also termed luminous or visual quantities, such as brightness, quantity of light, and luminance, are only defined for the visible portion of the spectrum. The same symbols are used for equivalent radiometric and photometric quantities. For clarity it may be important to differentiate them. In that case subscript e is used for the radiometric quantity and subscript v for the photometric quantity. Although the definition of radiant and photometric quantities may be the same, the units are quite different. Note that the attributes *radiant* and *luminous* are always used to express what quantities are meant.

Luminous Energy, Q

It is the quantity of light and measured in *lumen-seconds*, also known as *talbot* $[lm \cdot s]$.

Luminous Flux, Φ

Luminous flux is the time rate with which light passes a spatial position. It is measured in lumens, unit symbol $[lm]$. A lumen is the flux of a unit solid angle of a point source with a luminous intensity of one candela.

Luminous Flux Densities, E and M

Same definition as radiant flux densities. Different kinds of units are used, such as *lux* (lumen per square meter), $[lx]$; *footcandle* (lumen per square foot), $[lm \cdot \mathrm{ft}^{-2}]$; and *phot* (lumen per square cm), $[lm \cdot \mathrm{cm}^{-2}]$.

Luminous Intensity, I

Luminous intensity is the luminous flux from a point source proceeding into a unit solid angle in a given direction. It is measured in *candelas*, $[cd]$ where one candela is the luminous intensity of a point source that operates like a blackbody at the temperature of solidification of platinum ($T = 2042\ K$). For all practical purposes *candela* and *candle* are the same, even though they have a slightly different definition.

Luminance, L

Luminance is also called photometric brightness. The definition is the same as that for radiance. Again, units vary. The preferred unit is *nit* (candela per square meter), $[nt]$. Other units include *stilb* (candela per square centimeter), $[sb]$ and *lambert L* ($[cd \cdot \pi^{-1} \cdot \mathrm{cm}^{-2}]$).

6.3.2 Relationship Between Radiant and Photometric Quantities

The conversion of radiant flux of the visible spectrum to a visual stimulus is dependent on the wavelength. Fig. 6.7 depicts the *spectral luminous efficiency*, $V(\lambda)$. In a broad sense, $V(\lambda)$ can be considered as the averaged wavelength sensitivity of the L-, M-, and S-cones of the human eye (compare Fig. 4.10 on page 88). The maximum is reached at about 0.55 micron.

The visual response Φ_v to radiant flux Φ_e can be computed by multiplying the flux with $V(\lambda)$.

$$\boxed{\Phi_v = c \int_0^{\infty} \Phi_e(\lambda) V(\lambda) d\lambda} \tag{6.28}$$

where $c = 680\ [lm\, W^{-1}]$.

6.4 Radiative Energy Transfer

A frequently occurring problem in radiometry and photometry is the determination of the quantity of flux transferred from one surface to another. The principles derived in this section apply to both radiometric and photometric (luminous) flux.

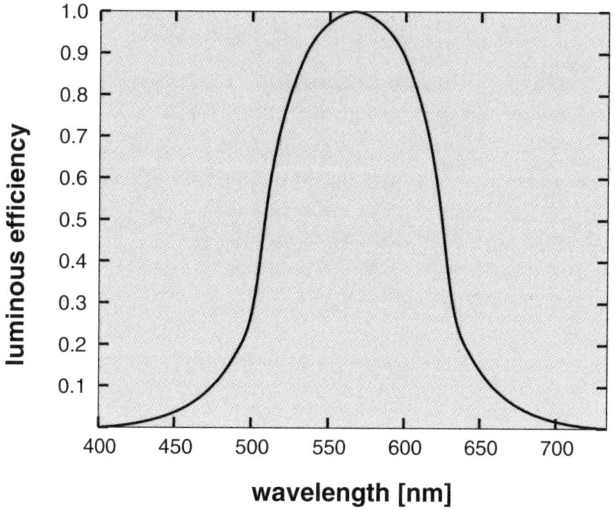

Figure 6.7: Spectral luminous efficiency, $V(\lambda)$.

6.4.1 General Case

Fig. 6.8 illustrates the general case of flux transfer between two surfaces. The surface on the left side contains the radiation source (source surface). Suppose that at a distance R from the source is a point in the receiving surface with the element dB. The angle between the surface normal in A and R is denoted by θ_A. Similarly, the angle between the surface normal in B and R is denoted by θ_B.

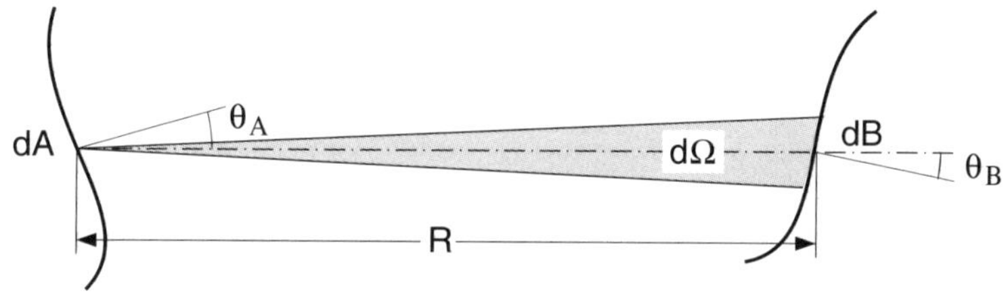

Figure 6.8: Radiation transfer between source and receiving surface.

Let us now determine the flux incident on dB from the elemental radiance source dA. The starting point for this derivation is the definition of radiance (Eq. 6.14 on page 142). The solid angle element $d\Omega$ is determined by dB and the distance R, or

$$d\Omega = \frac{dB \cos\theta_B}{R^2} \tag{6.29}$$

Substituting the above expression for a solid angle element in Eq. 6.14, we find

$$d^2\Phi = \frac{L}{R^2}\cos\theta_A \cos\theta_B dAdB \tag{6.30}$$

$$\Phi = \int_{A_o}\int_{B_o} \frac{L}{R^2}\cos\theta_A \cos\theta_B dAdB \tag{6.31}$$

This fundamental equation describes the flux transfer from a source to a receiving surface. Note that in the most general case L, R, θ_A, θ_B depend on the positions in both surfaces, as the double integration over both surfaces indicates.

If one is only interested in the flux per unit area at a particular point in the receiving surface, then the irradiance E at that point is obtained and Eq. 6.31 reduces to

$$E = \int_{A_o} \frac{L}{R^2}\cos\theta_A \cos\theta_B dB \tag{6.32}$$

6.4.2 Radiative Energy Transfer Through Lens System

A special application of Eq. 6.31 is the radiative energy transfer through a lens system. Fig. 6.9 illustrates how radiant energy from an elemental source element dA (object point) is imaged at dC (image point). Note that in contrast to Fig. 6.8 more energy is received at dC because the lens considerably increases the solid angle into which energy propagates[1]. For the following discussions we assume a planar, circular entrance pupil (aperture).

What is the irradiance E_B at the image point dC? This question is answered in two steps. First, the total flux across the aperture is determined. The irradiance E_B is then easily obtained by following its definition.

Flux Across Aperture

Let $d^2B = r \cdot d\phi \cdot dr$ be the surface element of a circular aperture (Fig. 6.10). The flux incident on d^2B is obtained by Eq. 6.31. Integrating over the entire aperture results in

[1]The solid angle is determined by the entrance pupil.

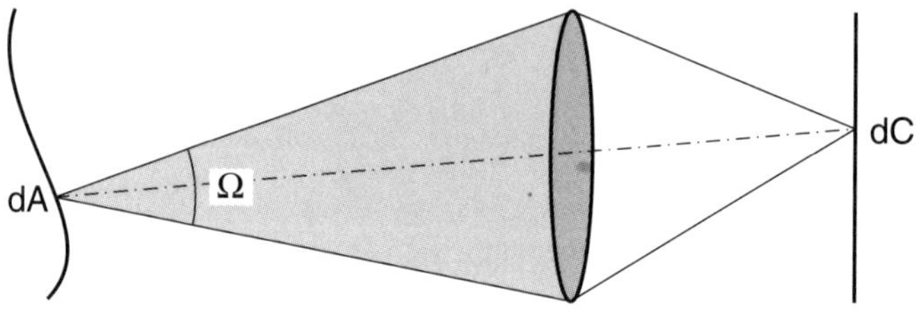

Figure 6.9: Radiative energy transfer through lens system.

$$\begin{aligned} dB &= \int_0^{2\pi} r \cdot dr \cdot d\phi \;=\; 2\pi \cdot r \cdot dr \\ B &= 2\pi\, r \int_0^{r_0} dr \;=\; \pi\, r_0^2 \end{aligned} \tag{6.33}$$

where r_0 the radius of the aperture.

With Eqs. 6.33 and 6.31 we obtain the flux incident on the aperture as follows

$$d\Phi_B = \pi r_o^2 \int_A \frac{L}{d^2} \cos\theta_A \cos\theta_B \, dA \tag{6.34}$$

Substituting d by $d_0/cos\theta_B$ in the above equation leads to

$$d\Phi_B = \pi r_0^2 \int_A \frac{L}{d_0^2} \cos^3\theta_B \cos\theta_A dA \tag{6.35}$$

Irradiance at Image Plane

The irradiance E_C at the elemental image area dC is readily obtained from the defining equation 6.11 on page 141

$$E_C = \frac{d\Phi_B}{dC} \tag{6.36}$$

Eq. 6.35 must be integrated over a small area ΔA to account for the fact that "point sources" do not exist. At the same time we approximate the small area ΔA by a planar Lambertian radiation source and obtain for the irradiance E_C

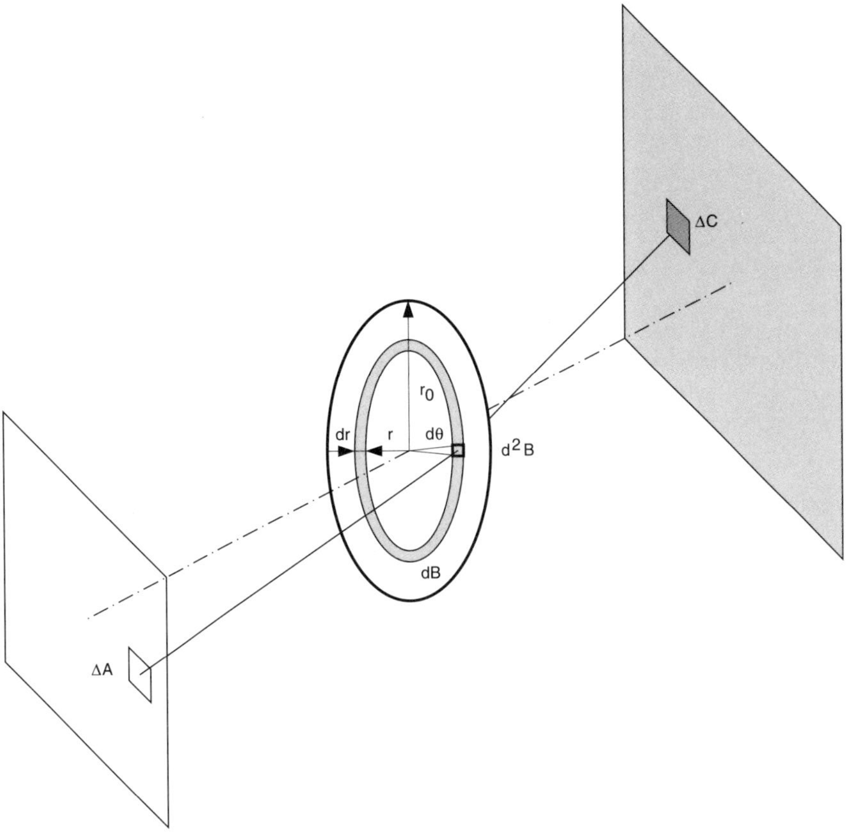

Figure 6.10: Irradiation on aperture B and image plane C.

$$E_C = \frac{L_A \, B \, \Delta A \cos\theta_A \cos^3\theta_B}{d_0^2 \, \Delta C} \tag{6.37}$$

where

θ_A	angle between radiation source and optical axis
θ_B	angle between image and optical axis
ΔA	area of radiation source
ΔC	area of image
$B = r_0^2 \pi$	area of aperture
d_0	distance between radiation source and aperture

Now ΔA and ΔC are related by the square of the lateral magnification, or the square ratio of the object distance d_0 to the focal length f. Taking this into account and assuming that the small radiation source is perpendicular to the optical axis ($\theta_A = \theta_B = \theta$) we find the rather simple equation

$$\boxed{E_C = \frac{L_A B}{f^2} \cos^4 \theta = 2\pi r_0^2 \frac{L_A}{f^2} \cos^4 \theta} \qquad (6.38)$$

Eq. 6.38 expresses the fact that the image irradiance is proportional to the scene radiance and that it drops off by a factor of $\cos^4$ from the optical axis (cosine to the fourth law).

6.5 To Probe Further

There are several books on the subject of radiometry and photometry. Two relatively easy-to-read books are *Introduction to Radiometry and Photometry* (Artech House, 1994) by W. R. McCluney, and *Radiometry and the Detection of Optical Radiation* (Wiley & Sons, 1983) by R. W. Boyd.

Many books on detectors have introductory chapters dealing with radiometry. Examples are *Optical Radiation Detectors* (John Wiley & Sons, 1984) by E. L. Dereniak & D. G. Crowe; *Solid-State Imaging with Charge-Coupled Devices* (Kluwer Academic Publishers, 1995) by A. J. P. Theuwissen; and *Physical Detectors of Optical Radiation* (Academic Press, 1983) by W. Budde.

Part II

Fundamentals of Digital Photogrammetry

Digital photogrammetry uses digital imagery rather than photographs. This is the chief distinction to traditional photogrammetry and almost all other differences are a consequence of this. Of paramount interest—I presume—are the questions on how to obtain digital images, how to work with them effectively, and how to use the computer to automate photogrammetric procedures.

Digital images used in photogrammetry are either acquired directly by digital cameras or indirectly by scanning aerial photography. Digital cameras have not yet equaled aerial cameras in terms of performance, resolution, and robust operability. Although manufacturers of aerial cameras have announced work on digital cameras, it will take some time before data acquisition will be all digital. Following the spirit of this book, systems and procedures in digital photogrammetry are compared with their counterparts in traditional photogrammetry. Hence, after discussing principles of digital cameras, a comparison is made with aerial cameras and film-based material, adding another facet to the problem of resolution and to the quest to determine an appropriate pixel size.

The only way to effectively work with digital imagery in an operational environment is by using digital photogrammetric workstations—or softcopy workstations, if you prefer. Let's face it, automatic procedures have not yet reached the level of generality and robustness that would be required to execute tasks in a black box environment, without the (occasional) interaction of a human operator. Considering the rapid change in the functionality and performance of digital photogrammetric workstations, system specific details are omitted for the benefit of discussing more general principles.

One of the most fundamental processes in photogrammetry is to identify and to measure conjugate points in two or more overlapping photographs. In fact, stereo photogrammetry relies entirely on conjugate points. In digital photogrammetry, rigorous attempts are made at automating stereopsis. We could easily devote an entire book to the subject of *image matching*, but have opted to present the material in two chapters. Again, emphasis is placed on general principles and to a much lesser extent on a review of existing matching methods. The first matching chapter examines the fundamental problems and elucidates solutions. This rather general discussion is followed by describing more advanced matching methods, including feature-based, symbolic (structural), and template matching, as well as matching with Generalized Hough Transforms. Subsequent chapters in Parts III and IV (second volume), will address additional aspects, for example, multiple image matching, matching in object space, and matching by trial and error.

Chapter 7

Electronic Imaging Systems

The schematic diagram in Fig. 7.1(a) shows the traditional way of obtaining imagery with film-based frame cameras. Here, *scanners* are used to convert analog photography, such as diapositives or negatives, to digital imagery (see Chapter 8). This chapter is concerned with the direct way of obtaining digital images, as depicted in Fig. 7.1(b).

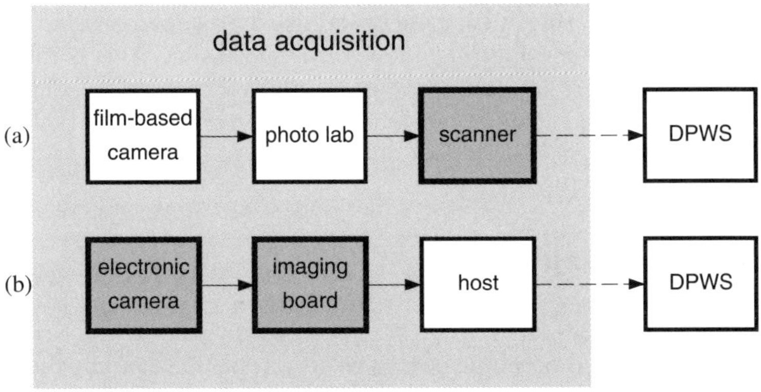

Figure 7.1: Data acquisition. In (a) the traditional way of obtaining images is sketched. Film-based cameras are replaced in (b) by electronic cameras. Scanners will be discussed in Chapter 8.

The interdisciplinary science of electronic imaging is concerned with designing and developing imaging systems. Early systems had quite limited applications. Vidicon-tube cameras were not very accurate, proprietary HW and SW increased the cost and offered little flexibility. The appearance of solid-state cameras in the early eighties eliminated these shortcomings. Electronic cameras based on charge-coupled devices (CCD camera) greatly changed electronic imaging systems and spawned an explosion of applications, ranging from consumer camcorders and electronic still cameras, to document scanners, machine vision, and scientific cameras.

The success of electronic imaging systems is not only related to CCD cameras, however. Recent advances in computer technology, such as powerful CPUs, 32-bit operating systems, and the PCI bus bring electronic imaging into the realm of personal computers. The result is increased performance and more flexibility at lower cost, sparking a new interest in machine vision applications, e.g. industrial inspection and metrology, surveillance, and medical imaging.

Electronic imaging distinctly influences digital photogrammetry. It makes *real-time photogrammetry* possible; we find electronic imaging systems at the heart of *mobile mapping systems*; in general, close-range photogrammetry undergoes a definite change.

Considering the rapid change in electronic imaging technology, I will emphasize principles and skim over descriptions and specifications of systems as they may soon become obsolete. The goal of this chapter is to reach a basic understanding of the working principles and main characteristics of digital cameras. Since CCD cameras are used predominantly, we begin with some background information about CCD sensors. This is followed by a discussion of solid-state cameras, including a comparison with film-based cameras. The chapter concludes with a brief description of frame grabbers.

7.1 Introduction

Fig. 7.2 depicts how differently the basic functionality of an electronic imaging system can be implemented in real systems. The functionality in Fig. 7.2(a) is divided into several modules. Image capture refers to the process of obtaining an image during the time of exposure. In the case of a CCD camera, the accumulated charge in every semiconductor capacitor constitutes the captured image[1]. Since the captured image is inherently analog, an analog to digital (A/D) conversion must take place at some stage. The result is a raw digital image that is processed by low level signal processing tasks before it is transferred to the main processing unit where further processing and analyzing takes place, depending on the application

[1]One is tempted to draw an analogy between the charge accumulated at the capacitors and the latent image that is obtained with film-based cameras during the time of exposure.

at hand. By and large, the conceptual block diagram sketched in Fig. 7.2(a) is only slightly affected by technological developments.

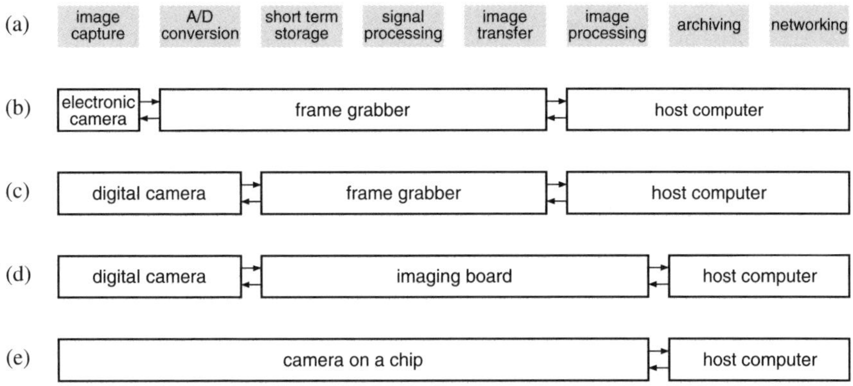

Figure 7.2: General functionality of electronic imaging systems (a) and various ways to implement it in real systems (b)-(e), beginning with traditional imaging systems consisting of electronic camera (b) or digital camera (c), frame grabber and host computer.

Fig. 7.2(b)-(e) are examples of four different implementations of the basic functionality. The first example depicts the traditional electronic imaging system comprising an electronic camera (e.g. solid-state camera), a frame grabber that provides for the A/D conversion, processing and transfer, and a host computer. There is a subtle but important difference between the first and second implementation. As shown in Fig. 7.2(c), the A/D conversion is now performed by the camera and the output is a digital signal. Consequently, such cameras are called *digital cameras.*

The implementation shown in Fig. 7.2(d) is again only slightly different from the previous one. Its main purpose is to demonstrate another technology trend. The processing power of frame grabbers increases due to new signal processing chips, fast memory at affordable prices, and miniaturization of microprocessors. Some manufacturers try to reflect the increased functionality of the frame grabber by a new name: *imaging board.* The distinction between the two products is blurred, however.

The last example shows an extreme case of miniaturization: the camera on a chip. It basically consists of a single chip and a lens supplemented with signal processing chips for image compression, color interpolation, and interface functions. Its major advantage is size and easy packaging.

7.2 CCD Sensors: Working Principle and Properties

There is a rich literature about solid-state and CCD sensors, for example *Budde* (1983), *Dereniak and Crowe* (1984), *Theuwissen* (1995), and *Yang* (1988).

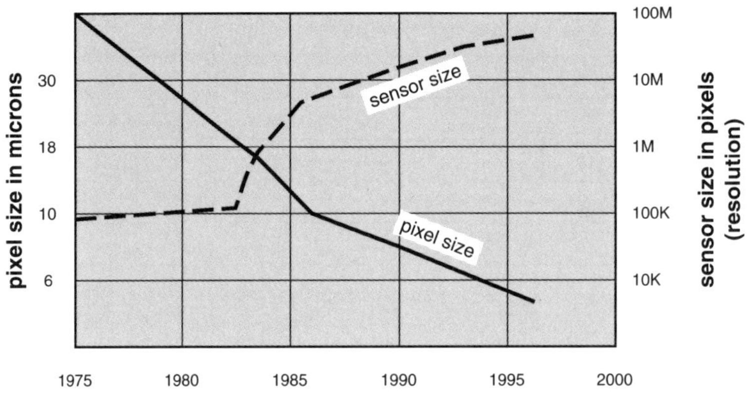

Figure 7.3: Development of CCD arrays over a period of 25 years.

The *charge-coupled device (CCD)* was invented in 1970. The first CCD line sensor contained 96 pixels; today, chips with over 50 million pixels are commercially available. Fig. 7.3 illustrates the astounding development of CCD sensors over a period of 25 years. The sensor size in pixels is usually loosely termed *resolution*, giving rise to confusion since this term has a different meaning in photogrammetry[2].

7.2.1 Working Principle

Fig. 7.4(a) is a schematic diagram of a semiconductor capacitor—the basic building block of a CCD. The semiconductor material is usually silicon and the insulator is an oxide (MOS capacitor). The metal electrodes are separated from the semiconductor by the insulator. Applying a positive voltage at the electrode forces the mobile holes to move toward the electric ground. In this fashion, a region (depletion region) with no positive charge forms below the electrode on the opposite side of the insulator.

[2]Resolution refers to the minimum distance between two adjacent features, or the minimum size of a feature, which can be detected by photogrammetric data acquisition systems. For photography, this distance is usually expressed in line pairs per millimeter (lp/mm).

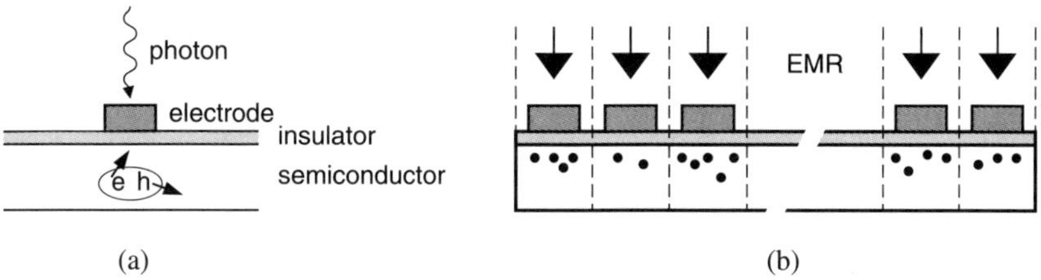

Figure 7.4: Schematic diagram of CCD detector. In (a) a photon with an energy greater than the bandgap of the semiconductor generates an electron-hole pair. The electron e is attracted by the positive voltage of the electrode while the mobile hole moves toward the ground. The collected electrons together with the electrode form a capacitor. In (b) this basic arrangement is repeated many times to form a linear array.

Suppose EMR is incident on the device. Photons with an energy greater than the band gap energy of the semiconductor may be absorbed in the depletion region, creating an electron-hole pair. The electron—referred to as *photon electron*—is attracted by the positive charge of the metal electrode and remains in the depletion region while the mobile hole moves toward the electrical ground. As a result, a charge accumulates at opposite sides of the insulator. The maximum charge depends on the voltage applied to the electrode. Note that the actual charge is proportional to the number of absorbed photons under the electrode.

The band gap energy of silicon corresponds to the energy of a photon with a wavelength of 1.1 μm (see Eq. 6.4 on page 137 for the relationship between wavelength and energy). Lower energy photons (but still exceeding the band gap) may penetrate the depletion region and be absorbed outside. In that case, the generated electron-hole pair may recombine before the electron reaches the depletion region. We realize that not every photon generates an electron that is accumulated at the capacitor site. Consequently, the quantum efficiency is less than unity (see also Fig. 7.8 on page 166).

An ever increasing number of capacitors are arranged into what is called a *CCD array*. Fig. 7.4(b) illustrates the concept of a one-dimensional array (called a linear array) that may consist of thousands of capacitors, each of which holds a charge proportional to the irradiance at each site. It is customary to refer to these capacitor sites as *detector pixels*, or pixels for short. Two-dimensional pixel arrangements in rows and columns are called full-frame or staring arrays.

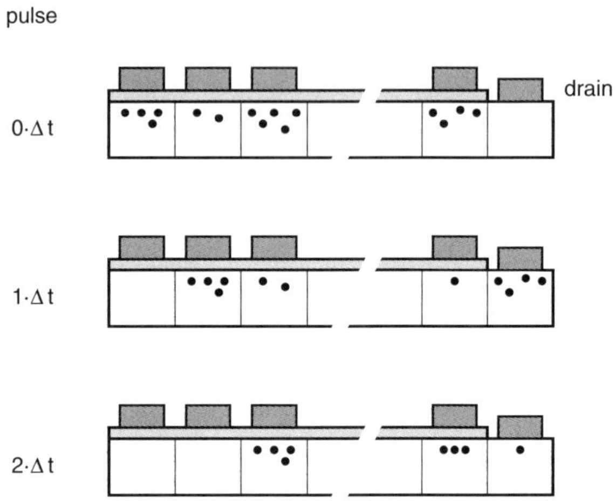

Figure 7.5: Principle of charge transfer. The top row shows a linear array of accumulated charge packets. Applying a voltage greater than V_1 of electrode 1 momentarily pulls charge over to the second electrode (middle row). Repeating this operation in a sequential fashion eventually moves all packets to the final electrode (drain) where the charge is measured.

The next step is concerned with transferring and measuring the accumulated charge. The principle is shown in Fig. 7.5. Suppose that the voltage of electrode i+1 is momentarily made larger than that of electrode i. In that case, the negative charge under electrode i is pulled over to site i+1, below electrode i+1, provided that adjacent depletion regions overlap. Now, a sequence of voltage pulses will cause a sequential movement of the charges across all pixels to the drain (last electrode) where each packet of charge can be measured. The original location of the pixel whose charge is being measured in the drain is directly related to the time when a voltage pulse was applied.

Several ingenious solutions for transferring the charge accurately and quickly have been developed. It is beyond the scope of this book to describe the transfer technology in any detail. The following is a brief summary of some of the methods.

Linear Array With Bilinear Readout

As sketched in Fig. 7.6, a linear array (CCD shift register) is placed on both sides of the single line of detectors. Since these two CCD arrays are also light sensitive, they must be shielded. After integration, the charge accumulated in the active detectors is transferred to the two shift registers during one clock period. The shift registers are read out in a serial fashion as described above. If the readout time is equal to the integration time, then this sensor may operate continuously without a shutter. This principle, known as *push broom*, is put to advantage in line cameras mounted on moving platforms to provide continuous coverage of the object space (see 7.3.3).

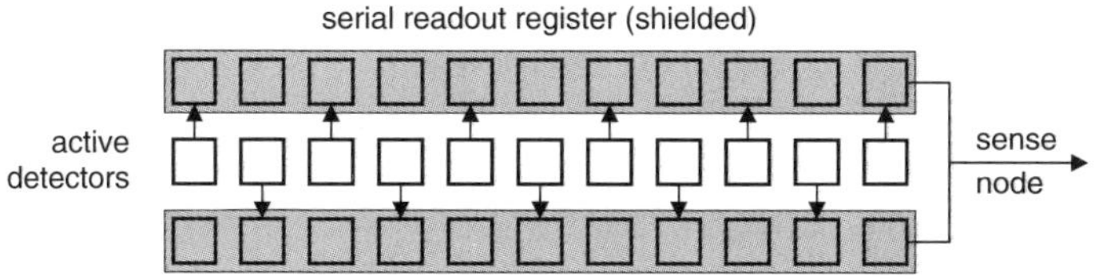

Figure 7.6: Principle of linear array with bilinear readout. The accumulated charge is transferred during one pixel clock from the active detectors to the adjacent shift registers, from where it is read out sequentially.

Frame Transfer

You can visualize a frame transfer imager as consisting of two identical arrays. The active array accumulates charges during integration time. This charge is then transferred to the storage array, which must be shielded since it is also light sensitive. During the transfer, charge is still accumulating in the active array, causing a slightly smeared image.

The storage array is read out serially, line by line. The time necessary to read out the storage array far exceeds the integration. Therefore, this architecture requires a mechanical shutter. The shutter offers the advantage that the smearing effect is suppressed.

Interline Transfer

Fig. 7.7 on the next page illustrates the concept of interline transfer arrays. Here, the columns of active detectors (pixels) are separated by vertical transfer registers. The accumulated charge in the pixels is transferred at once and then read out serially. This again allows an open shutter operation, assuming that the read out time does not exceed the integration time.

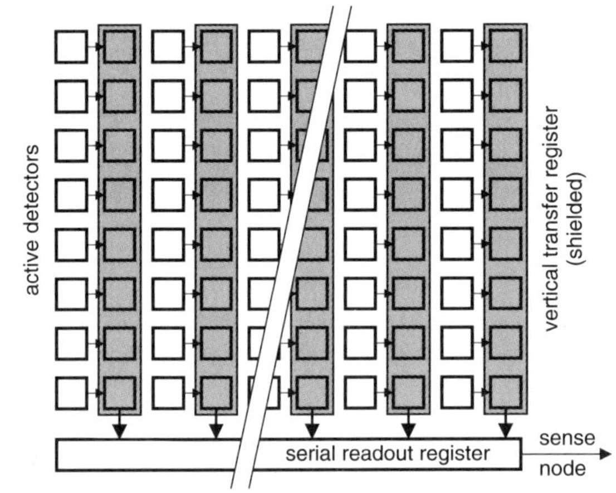

Figure 7.7: Principle of linear array with bilinear readout. The accumulated charge is transferred during one pixel clock from the active detectors to the adjacent shift registers from where it is read out sequentially.

Since the CCD detectors of the transfer register are also sensitive to irradiance, they must be shielded. This, in turn, reduces the effective irradiance over the chip area. The effective irradiance is often called *fill factor*. The interline transfer imager as described here has a fill factor of 50%. Consequently, longer integration times are required to capture an image. To increase the fill factor, microlenses may be used. In front of every pixel is a lens that directs the light incident on an area defined by adjacent active pixels to the (smaller) pixel.

Time-Delay and Integration (TDI) Transfer

Time-delay and integration is an interesting CCD array architecture that resembles closely what is known in photogrammetry as forward motion compensation for aerial cameras. Suppose either an object or the camera moves at a constant velocity, causing an image blur along the trajectory. The idea of TDI transfer is to subdivide the integration time into n intervals and to transfer the charge accumulated during one such interval to the next column of pixels where charge continues to accumulate during the next interval. Conceptually, this approach corresponds to moving the film by an amount proportional to the distance the object traveled during the exposure time.

7.2.2 *Major Properties*

Noise

Unfortunately, not all of the electrons being measured in the drain are generated from incident radiant flux. The detector, including the electronic components, also produces electrons, collectively called noise. The systematic component of noise can be determined, for example, by measuring the unexposed array, and subsequently be eliminated from the signal (like a systematic error). However, random noise remains and corrupts the signal.

Random noise originates from different sources. Due to the discrete nature of electromagnetic radiation, incident flux on the CCD sensor has a random component (radiation noise). The generation of electrons by photons (photoelectrons) in the detector is also a random process. Thermal motion of charged particles and thermal current fluctuations in the detector and other electronic components of the CCD sensor cause what is called *dark current.* You can imagine that temperature greatly affects dark current. Indeed, dark current doubles with every 8^0 C temperature increase. Dark current fluctuates with a standard deviation equal to its square root. Noise related to dark current is expressed in number of electrons.

The total amount of noise electrons depends on several factors. For example, dark current strongly depends on temperature and integration time. On the other hand, the clock rate (read-out frequency) affects the noise level of the output electronics. For example, the minuscule charge transferred to the sense node must be greatly amplified so that it can be measured.

Spectral Response

Silicon is the most frequently used semiconductor material. In an ideal silicon detector, every photon exceeding the band gap ($\lambda < 1.1\ \mu$m) causes a photon electron that is collected and eventually measured. The quantum efficiency is unity and the spectral response is represented by a step function. As indicated in Fig. 7.8, the quantum efficiency of a real CCD sensor is less than unity for various reasons. For one, not all the incident flux interacts with the detector (e.g. reflected by the electrode in front illuminated sensors). Additionally, some electron-hole pairs recombine. Photons with longer wavelengths penetrate the depletion region and cause electron-hole pairs deep inside the silicon. Here, the probability of recombination is greater and many fewer electrons are attracted by the capacitor. The drop in spectral response toward blue and UV is also related to the electrode material that may become opaque for $\lambda < 0.4\ \mu m$.

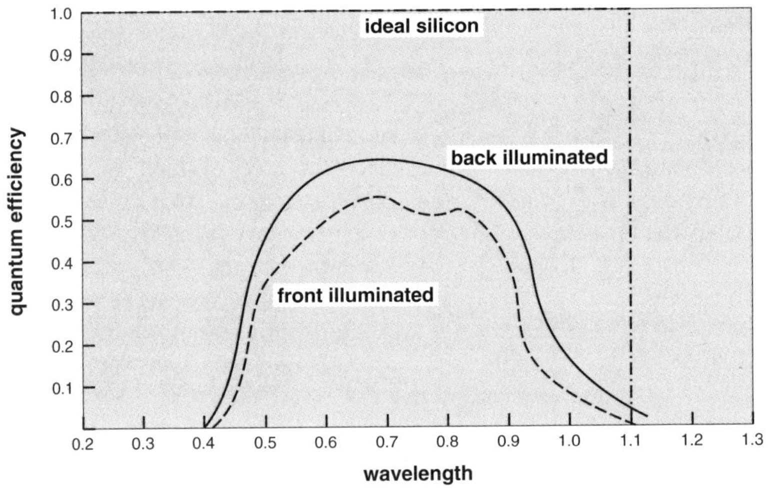

Figure 7.8: Spectral response of CCD sensors. In an ideal silicon detector all photons exceeding the band gap energy generate electrons. Front illuminated sensors have a lower quantum efficiency than back illuminated sensors because part of the incident flux may be absorbed or redirected by the electrodes (see text for details).

Sensors illuminated from the back avoid diffraction and reflection problems caused by the electrode. Therefore, they have a higher quantum efficiency than front illuminated sensors. However, the detector must be thinner, because high energy photons are absorbed near the surface—opposite the depletion region—and the chances of electron/hole recombination are lower with shorter diffusion length.

In order to make the detector sensitive to other spectral bands (mainly IR), detector material with the corresponding bandgap energy must be selected. This leads to hybrid CCD arrays where the semiconductor and the CCD mechanism are two separate components.

Linearity and Dynamic Range

Fig. 7.9 illustrates the functional relationship between irradiance and accumulated charge at a pixel site. The noise floor refers to randomly generated electrons, for example, by the dark current. Photon electrons are generated by incident flux, constituting the signal. What is actually measured is the signal superimposed with noise. As the irradiance is increased, more photon electrons are generated (as well as noise electrons) until the total amount of electrons reaches the sensor's saturation. Irradiance beyond E_{max} causes the generated photons to spill over to neighboring capacitors, a process known as *blooming*.

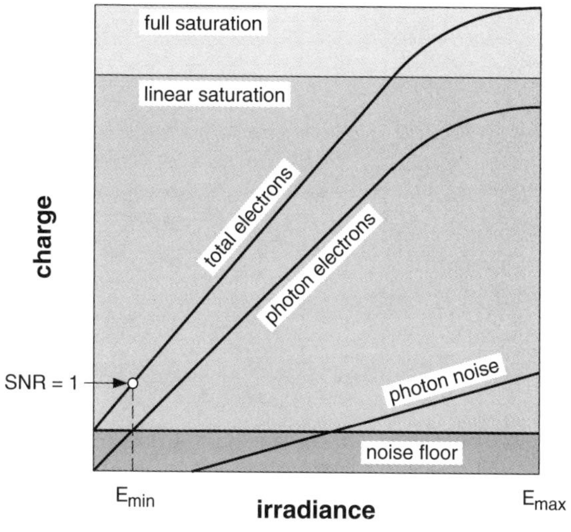

Figure 7.9: Relationship between irradiance and accumulated charge at a pixel site.

The response of the sensor is linear over a wide range of irradiance. Sensor manufacturers specify linear and full saturation in number of electrons. E_{min} in Fig. 7.9 refers to the minimum irradiance necessary to generate a signal that is just as large as the noise (Signal to Noise Ratio, SNR = 1). Full saturation depends on the effective pixel size and on the sensor architecture. Saturation values range between 10,000 and 500,000 electrons, with 50,000 a representative value for imagers used in photogrammetric applications.

The dynamic range, DR, is defined as the ratio of the maximum signal to the standard deviation of noise, or

$$DR = \frac{n_{sat}}{n_{noise}} = \frac{E_{max}}{E_{min}} \quad (7.1)$$

$$DR = 20 \log \frac{n_{sat}}{n_{noise}} \quad (7.2)$$

with n_{sat} the number of electrons at saturation, n_{noise} the number of noise electrons, and E_{max} and E_{min} the maximum and minimum irradiance, respectively (see also Fig. 7.9). Eq. 7.2 expresses the dynamic range in decibels [dB].

Example 7.1:

With a well capacity of 200,000 electrons and a noise floor of 60 electrons, the dynamic range of this array is 3,333 or 70 dB. Suppose the output signal is quantized by an A/D converter. How many bits are required to keep the quantization noise comparable to the signal noise? The dynamic range of the ADC is $2^N\sqrt{12}$ with N the number of bits. It follows that a 12 bit converter with a dynamic range of 14,189 or 83 dB does not significantly increase the overall signal noise. However, an 8 bit converter with a dynamic range of 887 (59 dB) would not be adequate.

7.3 Solid-State Cameras

Electronic cameras have been used for special photogrammetric applications since the early seventies. However, vidicon-tube cameras available at that time were not very accurate because of the inherent instability of imaging tubes. The appearance of solid-state cameras in the early eighties eliminated this severe shortcoming. Charge-coupled-device sensors are well suited for photogrammetric applications because they provide excellent stability and uniformity. Consequently, solid-state cameras based on CCD sensors, termed *CCD cameras*, are used predominantly in digital photogrammetry. *Holst* (1996) provides a recent account of CCD-cameras, including sensor architecture, camera performance, and image quality. If you are more interested in the calibration aspects of CCD-cameras you will find interesting details in *Beyer* (1992), *Chen and Schenk* (1992), *El-Habrouk et al.* (1996), *Seedahmed et al.* (1998), *Lenz and Fritsch* (1988).

Traditionally, the output of a solid-state camera is an analog video signal. More recently, cameras with a digital output have become available. The popular term "digital camera", sometimes informally used for electronic cameras, should be used more restrictively, referring to solid-state cameras with digital output.

The classical film-based cameras used in photogrammetry are often divided into aerial and terrestrial (close-range) cameras. The same principle can be applied to solid-state cameras. A solid-state camera with a resolution comparable to a classical frame camera must have on the order of $20,000 \times 20,000$ sensing elements. Such image sensors do not exist and will not likely become available commercially in the near future[3]. Alternative solutions to frame cameras are the so-called *line cameras*, which are briefly discussed in Sec. 7.3.3.

The chief advantage of solid-state cameras over the classical film-based cameras is the instant availability of images for further processing and analysis. This is essential in real-time photogrammetry where the time delay between image capture and analysis may range from 20 msec to several seconds or even minutes.

Another advantage of solid-state cameras is the increased spectral flexibility. The major drawback is the limited resolution or limited field of view.

7.3.1 Camera Overview

Fig. 7.10 depicts a functional block diagram of the major components of a solid-state camera.

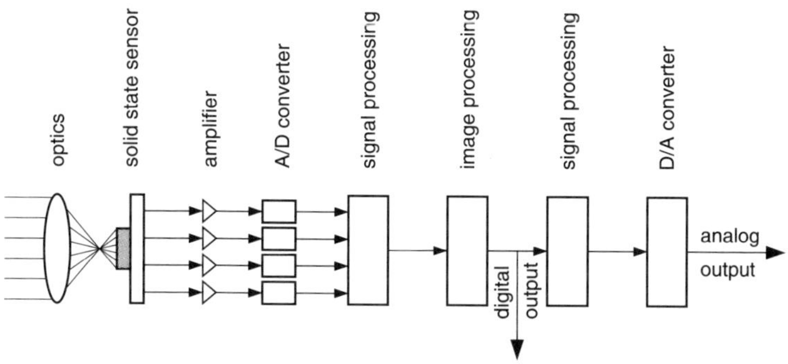

Figure 7.10: Functional block diagram of a solid-state camera. A real camera may not have all components. The diagram is simplified, e.g. external signals received by the camera are not shown.

The optics component includes lens assembly and filters, such as an infrared blocking filter to limit the spectral response to the visible spectrum

[3]A possible solution is to butt smaller sensor chips together.

(see also Fig. 7.8 on page 166). Many cameras use a C-mount for the lens. Here, the distance between mount and image plane is 17.526 mm. As an option, the optics subsystem may comprise a shutter[4].

The most distinct characteristic of an electronic camera is the image sensing device. Section 7.2 provides an overview of charge-coupled devices.

The solid-state sensor, positioned in the image plane, is glued on a ceramic substrate. The sensing elements (pixels) are either arranged in a linear array or a frame array. Linear arrays are used for aerial cameras while close range applications, including mobile mapping systems, employ frame array cameras.

The accuracy of a solid-state camera depends a great deal on the accuracy and stability of the sensing elements, for example on the uniformity of the sensor element spacing and the flatness of the array. From the manufacturing process we can expect an accuracy of $1/10^{th}$ of a micron. Considering a sensor element, size 10 μm, the regularity amounts to 1/100. Camera calibration and measurements of the position and the spacing of sensor elements confirm that the regularity is between $1/50^{th}$ and $1/100^{th}$ of the spacing.

The voltage generated by the sensor's read out mechanism must be amplified for further processing, which begins with converting the analog signal to a digital signal. This is not only necessary for producing a digital output, but also for signal and image processing. The functionality of these two components may range from rudimentary to very sophisticated in a real camera.

You may consider the first two components (optics and solid-state sensor) as image capture, the amplifiers and ADC as image digitization, and signal and image processing as image restoration. A few examples illustrate the importance of image restoration. The dark current can be measured and subtracted so that only its noise signal component remains; defective pixels can be determined and an interpolated signal can be output; the contrast can be changed (Gamma correction); and image compression may be applied. The following example demonstrates the need for data compression.

Example 7.2:

> *Consider a CCD camera with an array size of 5K $\times$ 5K and a dynamic range of 12 bits. This requires 300 Mbits of storage for each image. Now suppose the camera operates at 30 frames/sec. This will generate a pixel data rate of about 768 Mpixels/sec.*

---

[4]Not to be confused with the *electronic shutter*, which is functionally represented by the read-out mechanism of the sensor.

The digital output shown in Fig. 7.10 may, in fact, consist of multiple parallel ports. Large CCD arrays can be divided into sub-arrays, each with its own read-out mechanism. That is, all sub-arrays work in parallel and are also output in parallel to reduce the bandwidth.

The camera electronics may have additional components which serve the purpose of increasing the camera's functionality. An example is the acceptance of external sync that would allow the synchronization of the camera with other devices. For example, multiple camera setups with uniform sync may be required for certain close-range applications. More and more of the traditional frame grabber functionality is built into the camera.

Another example of an auxiliary component is the shutter control. Cameras with mechanical (or LCD) shutters need appropriate electronics to read an external signal to trigger the shutter.

7.3.2 Analog Output

The analog output follows the video standards. There is no uniform, worldwide accepted video standard, however. The first standard dates back to 1941 when the National Television Systems Committee (NTSC) defined RS 170 for black-and-white television. Today, the monochrome standard is often called EIA 170 and the color format is known as NTSC. These standards are used in North America, parts of South America, in Japan and the Philippines.

European countries developed other standards, e.g. PAL (phase alternate line) and SECAM (sequential color and memory). Yet another standard for black-and-white television was defined by CCIR (Commité Consultatif International des Radiocommunications). It differs only slightly from the NTSC standard, however.

The primary consideration for developing standards was displaying images with a visually pleasing appearance under the constraint of a rather limited video transmission bandwidth. The result of this optimization process is interlace and scan times, which reduce flicker to an acceptable level.

Fig. 7.11 on the following page shows the concept of interlace. In the NTSC and EIA 170 (former RS 170) standards, a frame of 525 scan lines consists of two fields, called *odd* and *even field.* However, to provide time for the vertical retrace, only 485 lines are displayed. That is, every field has 242.5 active lines. The first line of the odd field begins in the left upper corner (solid line). After the horizontal retrace, the procedure repeats until 242 lines are displayed. Now, the remaining half line is displayed at the bottom, followed by a vertical retrace to the center of the very top scan line where the display of the even field begins.

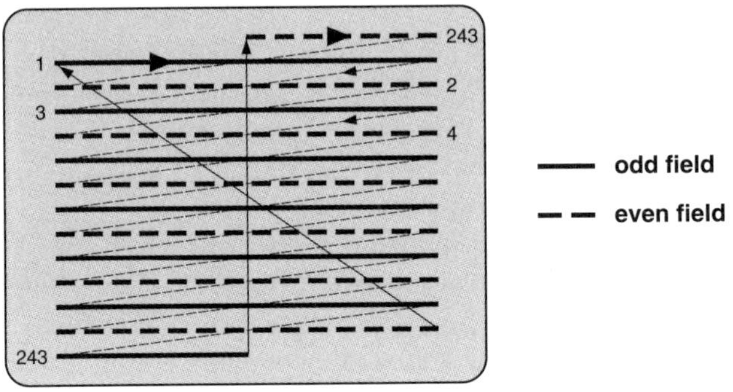

Figure 7.11: Diagram of video display according to the EIA 170 standard.

The video standards specify the maximum bandwidth, the line frequency (number of frames/sec × lines per frame), and the blanking period. All other parameters are derived from these specifications. The relevant numbers for NTSC are: nominal bandwidth of 4.2 MHz, total bandwidth[5] of 6 MHz, line frequency of 15.7343 kHz, and 11.1 μs blanking period. The total line time is simply the reciprocal of the line frequency, or 63.555 μs for NTSC. Subtract the blanking period and you obtain a minimum active line time of 52.455 μs.

Example 7.3:

> *A more interesting exercise is to determine how many pixels a video signal can transmit. Theoretically, the smallest pixel size is given by the Nyquist criterion. That is, the minimum time between two impulses that are still discernible is* $\Delta t = 1/2f_c$*, where* f_c *is the cut-off frequency of the video bandwidth. With an active line time of 52.455 μs and the nominal bandwidth of 4.2 MHz, we obtain* $2 \cdot 4.2 \cdot 52.455 = 440$ *pixels per scan line. If the total bandwidth is used for imagery, 625 pixels can be resolved.*

7.3.3 Line Cameras

Classical film-based cameras used in photogrammetry are often divided into aerial and terrestrial (close-range) cameras. The same principle can be applied for solid-state cameras. A solid-state camera with a resolution comparable to a classical frame camera must have on the order of

[5]Includes non pictorial information, e.g. sound, chrominance and luminance.

$20,000 \times 20,000$ sensing elements. Such image sensors do not exist and will not likely become available commercially in the near future. However, it is possible to use several sensors in the image plane. An alternative solution to frame cameras is the line cameras. They use linear arrays, which are considerably cheaper than large staring arrays. Moreover, they can be butted together more easily to increase the number of sensing elements, and the read-out time is much shorter.

The principle of acquiring data with a line camera is sketched in Fig. 7.12(a). A linear CCD array is placed in the camera's focal plane. As discussed in Sec. 7.2 it accumulates charges during the integration (exposure) time from a narrow strip on the ground referred to as a line in this text. The charge is transferred to adjacent registers and read out serially (compare Fig. 7.6 on page 163). The projection of an array pixel on the ground is called IFOV (Instantaneous Field Of View). Its dimensions are $s_1 \times s_2$.

$$s_1 = \frac{H}{f} p_1 \tag{7.3}$$

$$s_2 = \frac{H}{f} p_2 \tag{7.4}$$

where p_1 and p_2 are the pixel dimensions of the CCD array, H is the flying height, and f the focal length. With this and the number of pixels, n, of the array, we obtain the width, L_w, and the length, L_l, of the line as follows

$$L_w = s_1 \tag{7.5}$$

$$L_l = s_2 \cdot n \tag{7.6}$$

During integration time, t_i, the platform moves by

$$\boxed{\Delta s = v \cdot t_i} \tag{7.7}$$

Fig. 7.12(b) illustrates how the line moves from the initial position to the final position of an integration cycle. Consequently, an array pixel collects photons from an area on the ground with dimensions of $(s_1 + \Delta s) \times s_2$. The diagram in the center of Fig. 7.12(b) shows the relationship between irradiance and time. We realize that along the edges of the line only a minuscule amount of energy contributes to the recorded charge. Therefore, the effective IFOV is smaller. For all practical purposes one assumes a line width of $L_w = \Delta s$.

Since the charge accumulated during the integration time is transferred very quickly to the shift registers, the array is immediately ready for a new integration cycle. This important property of CCD linear arrays ensures continuous ground coverage.

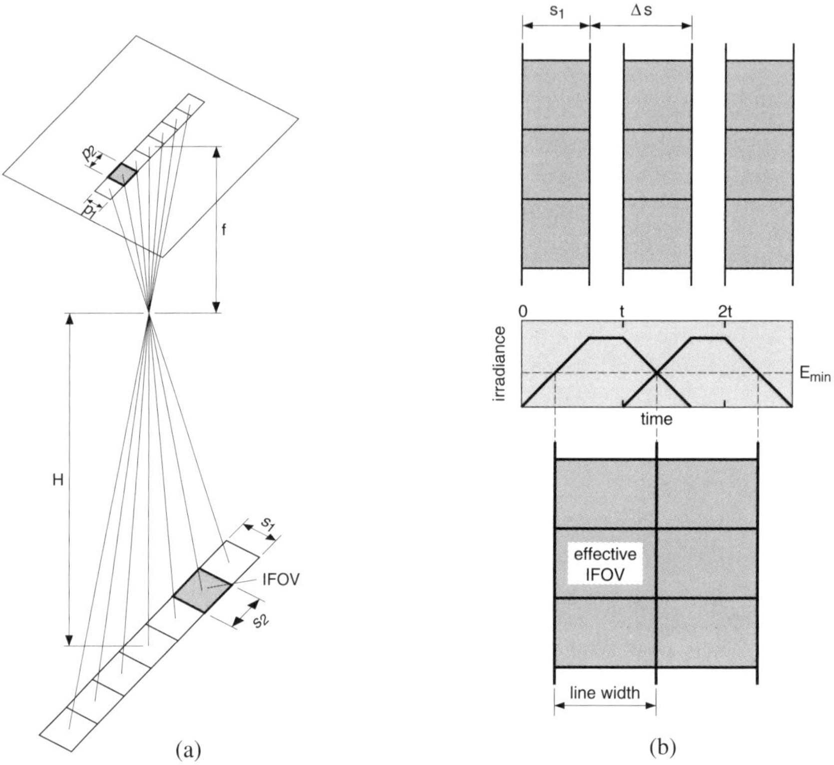

Figure 7.12: Principle of data acquisition with a line camera (a). The pixel dimensions of the sensor, mounted in the focal plane, are p_1 and p_2. The linear array projects a line on the ground whose width and length are L_w, L_l, respectively. The size of a pixel on the ground is $s_1 \times s_2$, referred to as IFOV (Instantaneous Field Of View). In (b) the effect of the moving platform is shown. During integration time, the linear array moves by Δs. Theoretically, charge is accumulated from an area with a width of $s_1 + \Delta s$. That is, the ground pixel is now larger in the flight direction. However, the effective size is smaller because the small area at the beginning and end of the integration cycle does not contribute significantly to the total charge.

For obtaining ground coverage of equal scale in the direction of flight and perpendicular to it, we require $\Delta s = s_2$, or

$$\boxed{v = \frac{H}{f \cdot t_i} p_2} \tag{7.8}$$

If the condition for the velocity expressed in Eq. 7.8 is not satisfied, then unequal ground coverage results. Fig. 7.13 depicts several scenarios. The top row shows the lines on the ground while the bottom row is a sequential representation of the recorded irradiance by the CCD array. This is what you would see on a monitor where every line would be displayed at equal intervals, for example.

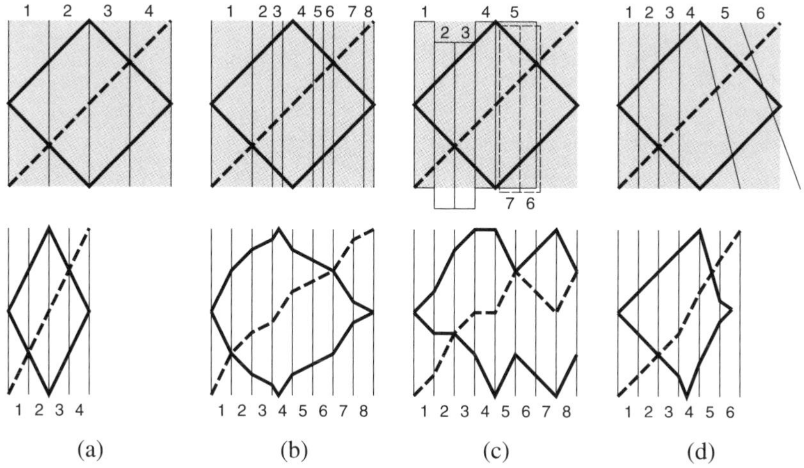

Figure 7.13: Illustration of unequal ground coverage by a linear array camera. The top row shows the lines on the ground and the bottom row is a time sequential representation of the sensor's response to incident radiant flux. The ideal case of (a) is reached if Eq. 7.8 is satisfied and the camera axis remains parallel for all lines. Less ideal cases are sketched in (b) and (c), where the velocity is either too slow or too fast. An awkward situation occurs when the camera axis quickly rotates between lines due to sudden platform movements, such as carrier vibration (d).

The ideal situation is reached if the velocity condition of Eq. 7.8 is satisfied and if the camera axis remains parallel from line to line. In that case, the ground pattern should be covered by 8 lines in our example. Fig. 7.13(a) demonstrates the case of $v_a = 2 \cdot v$, that is, the ground pattern is covered in half the time, or by four lines. Fig. 7.13(b) is an example of irregular velocity across our ground pattern. The result is a distorted image. If you think that such an erratic change in velocity is unrealistic then you are right. However, an apparent change in velocity may be caused by a rotation of the camera axis, due to pitch and roll, for example. Imagine a rotation about the longitudinal axis of the platform carrier, known as roll in case of airplanes, or as ω -rotation to photogrammetrists. The consequence is a lateral displacement of the lines, as illustrated in Fig. 7.13(c) (lines 2 and 3). An interesting situation occurs when a rotation perpendicular to the carrier is present (pitch, or φ-rotation). Now, the apparent velocity may even become negative (lines 6 and 7 in Fig. 7.13(c)). The resulting image is highly distorted and does not resemble the ground pattern any more. Finally, Fig. 7.13(d) shows the effect of crab (κ-rotation).

Clearly, the examples shown in Figs. 7.13(a)—(d) are highly exaggerated. They serve the purpose of pointing to a fundamental problem of line cameras, namely the determination of the exterior orientation of every line. With an integration time of, say, 3 ms, we need over 300 exterior orientations per second. This is obviously far beyond the capacity of any GPS/INS platform orientation system.

Example 7.4:

> *Suppose the following parameters for a data acquisition mission with a linear array camera: wide-angle camera, flying height $H = 6000$ m above ground, square pixel size of 10 μm, exposure time 1/300 sec, angular motion[6] of 1^0/sec. The ground pixel size is $s_1 = s_2 = 6000/0.15 \cdot 10^5 \approx 0.4$ m, the required velocity for a line spacing equal to the pixel size amounts to $v = 0.4/0.33$ m/ms $= 120$ m/sec. The angular motion caused by pitch (ϕ) during exposure time is equivalent to a movement along the flight path of $\Delta p = \tan(1^0) \cdot 6000 \cdot 1/300 = 0.35$ m, which is almost the line spacing. That is, the pitch offsets the line spacing and the new line will image the same ground as the previous line.*

The advantage of line cameras is the relatively simple design and low cost for the linear sensor array. By adding additional linear arrays with different spectral responsivity, color imagery can be acquired quite effectively. A well-known example of a one line camera is SPOT. The disadvantage of monoscopic ground coverage is circumvented by overlapping strips obtained from adjacent orbits.

Three-line cameras employ three linear arrays, which are mounted in the image plane in fore, center, and aft position. With this configuration, triple coverage of the surface is obtained. The principle is sketched in Fig. 7.14. Here, the challenge is to register the three individual strips, for

[6] *Light* (1996) reports about angular motions of 1^0/sec for ω and ϕ, and 0.5^0/sec for κ for typical flights undertaken for the National Aerial Photography Program.

example, by matching them, using common features. For this to succeed, good approximations are required, provided by GPS/INS on board.

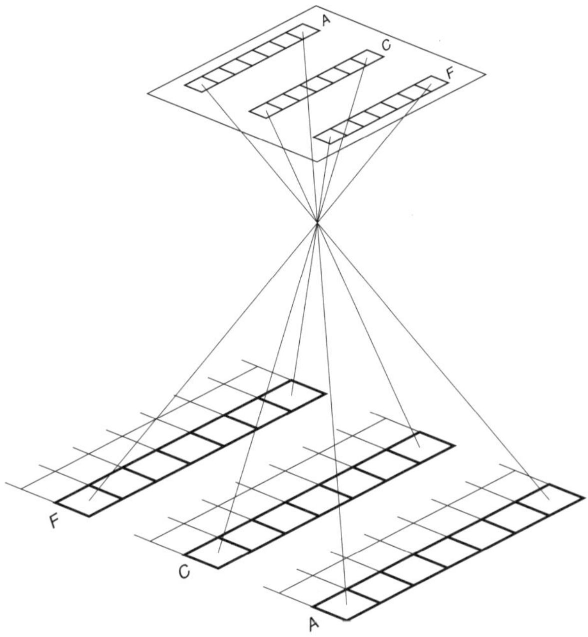

Figure 7.14: Principle of a three-line camera. Three linear arrays are mounted in the focal plane in fore, center and aft position. Each linear array provides continuous ground coverage (see also Fig. 7.12(a)). In this fashion, three independent strips are obtained. The challenge is to associate (register) the three independent strips, for example by matching them.

Examples of three-line cameras include MOMS and its variants—cameras developed in Germany and used in the German space program, see, e.g. *Hofmann et al.* (1993), *Seige and Meissner* (1993), *Sandau and Eckhardt* (1996). In *Kornus* (1999), the author describes the 3-D reconstruction of points from the MOMS camera.

7.3.4 Comparison Between Analog and Digital Cameras

For those of you interested in comparing solid-state cameras with traditional photogrammetric cameras[7], this section highlights similarities and

[7] A photogrammetric camera is one whose metric properties, in essence the interior orientation, remain constant between two consecutive calibrations.

differences in functionality, operation and performance. However, to compare apples to apples, it is necessary to consider film and photographic processes together with the camera. Taking this view to the extreme we should actually compare the entire data acquisition process. For aerial applications this would include the platform (e.g. motion, vibration) and the atmosphere (e.g. haze, turbulence) as additional factors influencing the overall performance.

An essential assumption in photogrammetry is that the data acquisition is performed with cameras whose metric properties remain constant within a wide range of operational conditions (temperature, pressure, etc.). The metric properties comprise the position of the principal point (projection center), calibrated focal length, including radial and tangential distortion, position of the fiducial marks, flatness of focal plane, and the lens' resolving power. As a consequence, metric cameras must be constructed in such a way that the relative positions of lens assembly, optical axis, and focal plane with fiducial marks—collectively called interior orientation—remain constant during the period between two calibration procedures. One way to ensure high stability is to combine the lens and the focal plane in one construction element known as the inner cone, and to separate it from the other camera parts, such as magazine, drive mechanism and camera body. Quite often, solid-state cameras do not meet these stringent metric characteristics.

Another problem is apparent from Fig. 7.15, namely the resolution vs. angular coverage dilemma with small sensor arrays. The top figure depicts a film-based camera. Its angular coverage is defined by the focal length. The maximum value of over 100^0 is obtained by superwide angle cameras. The angular resolution can be expressed by the ratio lp/mm to focal length.

Example 7.5:

> *For aerial films with over 100 lp/mm and a superwide angle camera with a focal length of 88 mm the angular resolution amounts to* $1/(200 \cdot 88) \approx 5.7 \cdot 10^{-5}$. *A CCD camera with a high resolution sensor (array size 2.5 in) and focal length of 60 mm has an angular coverage of about* 72^0 *and an angular resolution of* $2 \times 15\,\mu m/60\,mm = 5 \cdot 10^{-4}$.

To improve the resolution, a longer focal length should be chosen at the cost of decreasing angular coverage. The same "back of the envelope calculation" can be repeated with smaller sensor arrays (e.g. one inch or less) and cameras with C-mounts (f = 17.526 mm) to demonstrate the dilemma of angular coverage vs. resolution.

Let us conclude the comparison of film-based and solid-state cameras with a more intuitively motivated approach towards resolution and pixel

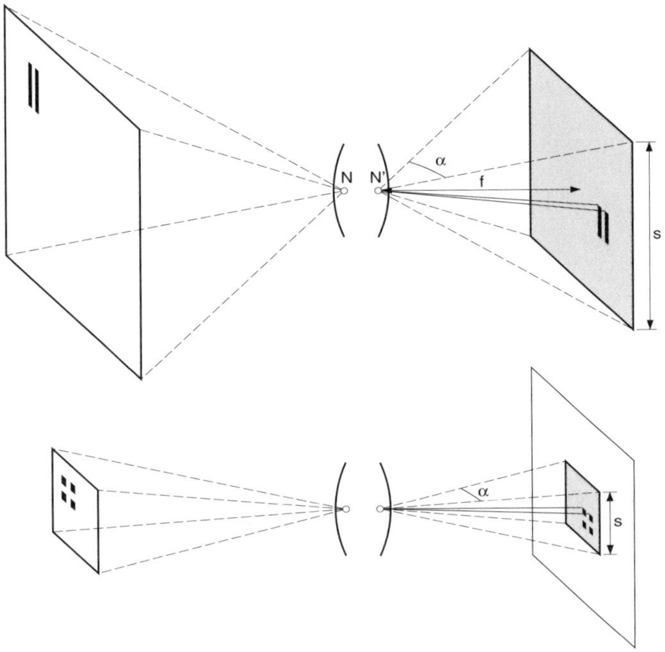

Figure 7.15: The top figure is a diagram of a film-based camera. The object space is imaged by the optical system on the focal plane where the film is exposed. The angular coverage, α, is determined by the focal length, f, and the film format, s. The resolution is often expressed in line pairs per mm (lp/mm). The angular resolution then, is $\frac{1}{f \cdot n}$, with n the number of lp/mm. The bottom figure illustrates the solid-state camera. Note the difference in the image format; except for very high resolution sensors, the size of the detector chip is usually less than one inch, e.g. $s = 2/3''$. Assuming the same focal length, a smaller angular coverage results. The angular resolution is obtained by dividing the pixel spacing by the focal length. To increase the angular coverage, a shorter focal length can be chosen, but then the angular resolution may drop below acceptable values. On the other hand, a long focal length results in a better angular resolution, but at the cost of a smaller angular coverage. This is the dilemma of using sensors with array sizes equal or smaller than one inch.

size. Suspended in the film emulsion are microscopic crystals of silver halide whose size ranges from a few tenths of a micrometer (fine grained film) to a few micrometers (coarse grained film). The crystals may have as many as 10^{10} silver and halide ions, for example bromide. During the time of exposure, photons may be absorbed by a crystal, thereby liberating an electron from a bound state. The mobile electrons may combine with mobile silver ions, forming a silver atom. A few silver atoms are sufficient to act as a catalyst during the development, where crystals with a silver speck are completely reduced to silver. During the same time, unexposed crystals are washed out. The net result is a negative that ultimately carries binary information: an original crystal in the emulsion is either present in the form of an aggregate of silver (also called corn or grain), or it is absent (washed out).

How do we perceive gray shades of a black and white film which is binary at the corn level? Let us consider a tiny region that contains n silver halide crystals. With the binary state of the crystal (either developable or not), we can theoretically represent 2^n states. Suppose a square-shaped region for a moment, side $s = r\sqrt{n}$, with r the average spacing of crystals. Then, as an absolute minimum, we can represent n different gray levels. Since the distribution of developed crystals (corns) within our imaginary region matters, more than the minimum number is perceived as gray shades. By the same token, fewer than the maximum of 2^n are obtained. In conclusion, our tiny region is the smallest unit that carries gray level information. The number of gray levels is between n and 2^n. We have established an informal relationship between the size and distribution of corns and a pixel.

Example 7.6:

> *Suppose a fine-grained emulsion has a crystal size of 1 μm. Lets further assume the crystals are densely packed (spacing equals crystal size). An area of 5 μm× 5 μm contains 25 crystals. Thus, it can represent between 25 and 2^{25} gray levels. The upper limit is the theoretical maximum. Since the position of the developed crystal within the area under consideration matters, more than the minimum of 25 gray shades are possible, say 100. Then it follows that the pixel size should be about 5 μm. How does this compare to the resolution of a film, for example expressed in lp/mm? The fine-grained emulsion in our example has a resolution of 100 lp/mm. From this, a pixel size of $1000/(2\cdot 100) = 5$ μm is obtained. Surprised?*

Fig. 7.16 on page 182 illustrates the process of obtaining a latent image and a negative. The latent image may be compared with the charge accumulated by a CCD sensor. The developed film, on the other hand, may be compared to the measured signal at the sense node. The analogy goes further; a few photons may form a developable silver speck or a measurable charge. Neither the latent image nor the accumulated charge can be viewed. The process of developing the latent image compares to transferring and measuring the charge. Moreover, both processes are subject to large amplifications.

Table 7.1: Comparison of film-based and CCD Cameras.

	Film-Based Camera	*CCD Camera*
focal lengths	3.5″,6″,8.25″, 12″ cones available	50 to 80 mm for large frame arrays, C - mount for smaller arrays 17.526 mm
format	$9'' \times 9''$	$2.5'' \times 2.5''$ for high resolution sensors, less than one inch for low resolution
angular coverage	60^0 to 125^0	similar values for high resolution sensors, 25^0 for chip sizes $< 1''$ and C - mount
photo-sensitive material	emulsion with silver halide	CCD sensor
motion compensation	modern aerial cameras have mechanical and/or optical FMC; the effect of angular motions is reduced by platform stabilizers	FMC is realized by CCDs with TDI architecture
exposure	photons liberate e^- that combine with Ag^+ to form a silver speck on the crystal (latent image)	photons generate electron/hole pairs, e^- are collected at capacitor (pixel site)
actual image	obtained from latent image as negative or positive during development process	obtained by transferring charge to the sense node where it is measured, amplified and output
spectral sensitivity	optical dyes added to inherently UV sensitive emulsion to extend spectral range to near IR. Different emulsions necessary for different spectral ranges	silicon inherently sensitive to photons with $\lambda < 1.1\ \mu m$
dynamic range	6 bits for b/w	10—12 bits
resolution	lens: AWAR > 100 lp/mm film: > 100 lp/mm image motion and atmosphere limiting factors of system resolution (50 to 60 lp/mm)	lens: probably lower unless same high quality aerial camera lens is used CCD: pixel pitch of 15 μm gives 30—40 lp/mm
advantages	proven, established technology very high resolution at large angular coverage film is large, inexpensive storage medium	large spectral sensitivity range high dynamic range digital image immediately available feedback from real-time processing increases flexibility of data acquisition
drawbacks	long processing time image only available as hard copy	high resolution sensors are very expensive and still inferior to film long read-out times for high resolution sensors high data transfer rates, huge storage capacity needed

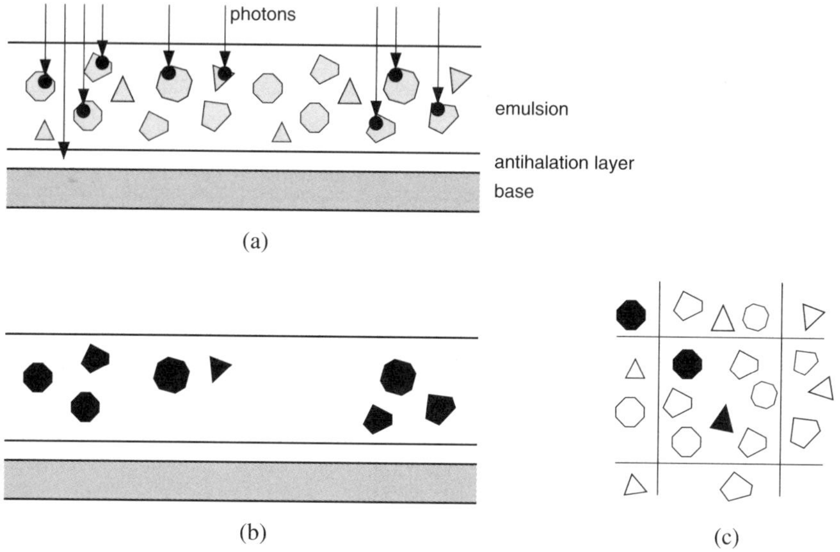

Figure 7.16: Cross section of film. In (a) irradiance is incident on the emulsion. Photons of appropriate wavelengths interact with the silver halide crystals by liberating electrons that combine with silver ions. A few silver atoms form the silver speck, indicated by a solid circle in (a). Crystals with a speck are entirely reduced to silver during the development process. Unexposed crystals are washed out. The resulting negative is shown in (b). A top view of the negative is depicted in (c) where the black crystals symbolize the corn. The white figures indicate the original locations of the undeveloped crystals. These crystals are washed out during the development process and are, in fact, no longer present. The presence or absence of a silver halide crystal constitutes a binary state. Gray levels are only perceived within regions that contain several original crystal locations. See text for more information.

7.4 Frame Grabbers

Frame grabbers capture images—up to 30 frames per second. Fig. 7.17 shows the major building blocks of a typical frame grabber. Some of the functionality may no longer be necessary or is moved into the camera. For example, the A/D converter may now be part of the digital camera. Moreover, on board storage and processing may no longer be necessary due to the availability of fast data transfer and increased processing power on the host computer.

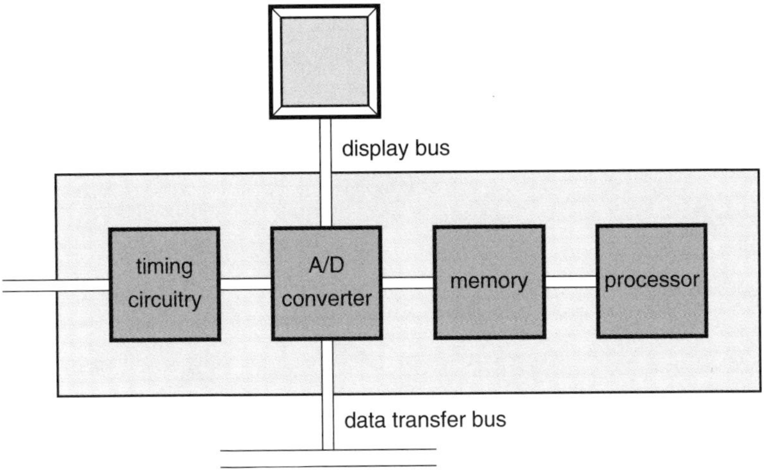

Figure 7.17: Schematic diagram of a frame grabber.

The frame grabber receives either an analog video signal or a digital signal from the camera.

7.4.1 Timing Circuitry

The timing circuitry senses the timing signal of the video wave, known as horizontal sync or h-sync (see Fig. 7.18 on the following page). This is critical for accurate image capture for it determines when to start and stop the sampling of the video signal.

The traditional approach is to create a clock from a reference frequency (Phase Locked Loop, PLL). The accuracy of this clock, for example expressed as standard deviation in units of nanoseconds (ns), determines the geometric accuracy of the pixels, called *pixel jitter*. Highly accurate frame grabbers have pixel jitters of about 2 nanoseconds.

7.4.2 A/D Converter

The principal task of the frame grabber is to convert the analog video signal into digital form. This is accomplished by sampling the signal in n^2 equal intervals, where $n \times n$ is the resolution of the camera, that is, the number of pixels. Fig. 7.18 on the next page shows a video signal. The h-sync signal marks the beginning of a frame. The time interval, ΔT, for one frame is about 33 milliseconds (ms). Consequently, one pixel has a time interval of $\Delta t_p = 33\ \text{ms}/n^2$. We realize that for high resolution images the clock that generates the sampling intervals must be very precise. Also, the synchronization with the video signal becomes crucial.

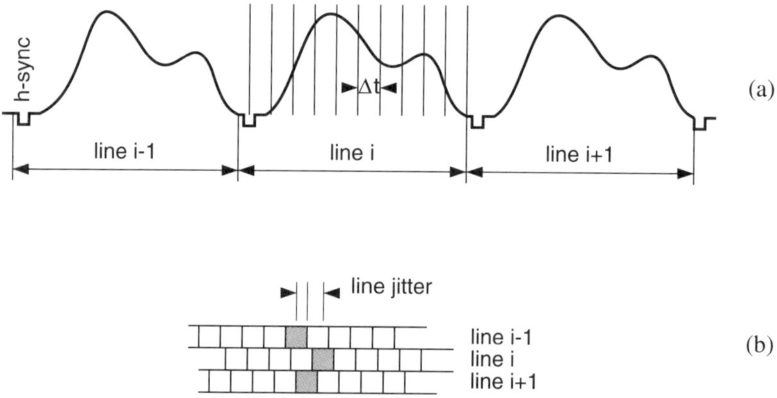

Figure 7.18: The video signal in (a) is digitized by dividing the time interval between two h-sync signals (one line) into equal intervals Δt. As shown in (b) identical columns in neighboring lines may be off (line jitter) due to errors in Δt and in locating the h-sync signal.

Example 7.7:

A video camera with a resolution of 2 K × 2 K pixels has a time interval $\Delta t = \frac{33\ ms}{4.146 \cdot 10^6} \approx 8\ ns$. A frame grabber with a pixel jitter of 4 ns can produce significant errors in the digital image as demonstrated in Fig. 7.18(b).

The digital signal, generated by the A/D converter, goes via system bus to the host computer for further processing and permanent storage. Now, the system bus may not be fast enough to process the 30 frames per second of a high resolution camera. To continue with the above example: the A/D converter produces a steady data stream of $30 \times 2048 \times 2048 = 120$ MB/sec. This is just about the maximum bandwidth of the PCI bus. This transfer rate may exceed the capacity of other bus systems, however. Therefore, the data must be temporarily stored on the frame grabber.

Table 7.2 on the facing page contains maximum transfer rates and typical throughput of some bus systems. The ISA bus was introduced in 1985

Table 7.2: Typical transfer rates of bus systems.

	ISA Bus	*EISA Bus*	*PCI bus*
introduced	1985	1988	1993
max transfer [MB/sec]	16	33	132
typical throughput [MB/sec]	2	8	65

with the personal computer. The VME bus is a dedicated I/O interface that can pass up to 40 MB/s. The non-proprietary PCI bus (Peripheral Component Interconnect) was developed so that 32-bit microprocessors, operating at speeds of 33 MHz and beyond, do not have to wait until data from hard disks and video boards arrive.

Problems

1. What, in your view, are the major advantages of CCD cameras for photogrammetric applications?

2. What are some of the obstacles for CCD cameras to overcome before they can replace aerial film-based cameras completely?

3. What is the relationship between the FMC of an aerial camera and the TDI architecture of a CCD sensor?

4. Explain why continuous ground coverage is obtained with line cameras that operate with open shutter.

5. With a three line camera, three continuous strips are obtained. Elaborate on the differences between the strips. Specifically, what radiometric differences do you expect between the forward looking and after looking strips?

References

Beyer, H. A. (1992). *Geometric and radiometric analysis of a CCD-Camera based on a photogrammetric close-range system.* Tech. Report, No. 51, ETHZ.

Budde, W. (1983). *Physical Detectors of Optical Radiation.* Academic Press, New York, NY.

Chen, Y. and T. Schenk (1992). A rigorous calibration method for digital cameras. In *International Archives of Photogrammetry and Remote Sensing*, **29**(B1), 199–205.

Dereniak, E. L. and D. G. Crowe (1984). *Optical Radiation Detectors.* Wiley & Sons, New York, NY.

El-Habrouk, H., X. Li and W. Faig (1996). Determination of geometric characteristics of a digital camera by self-calibration. In *International Archives of Photogrammetry and Remote Sensing*, **31**(B1), 60–64.

Hofmann, O., A. Kaltenecker and F. Müller (1993). Das flugzeuggestützte digitale Dreizeilenaufnahme- und Auswertesystem DPA—erste Erprobungsberichte. In Ackermann/Hobbie (eds.) *Photogrammetric Week '93*, 97–107, Wichmann Verlag.

Holst, G. C. (1996). *CCD Arrays, Cameras, and Displays.* JCD Publishing and SPIE Optical Engineering Press, 2932 Cove Trail, Winter Park, FL 32798.

Kornus, W., M. Lehner, F. Blechinger and E. Putz (1996). Geometric calibration of the stereoscopic CCD-Linescanner MOMS-2P. In *International Archives of Photogrammetry and Remote Sensing*, **31**(B1), 90–98.

Kornus, W. (1999). *Dreidimensionale Objektrekonstruktion mit digitalen Dreizeilenscannerdaten des Weltraumprojekts MOMS-02/D2.* Dissertation, DGK-C, Heft 496, Munich.

Lenz, R. and D. Fritsch (1988). On the accuracy of videometry. In *International Archives of Photogrammetry and Remote Sensing*, **27**(B5), 335–345.

Light, D. (1996). Film cameras or digital sensors? the challenge ahead for aerial imaging. *Photogrammetric Engineering and Remote Sensing, PERS,* **62**(3), 285–291.

Sandau, R. and A. Eckhardt (1996). The stereocamera family WAOSS/WAAG for spaceborne/airborne applications. In *International Archives of Photogrammetry and Remote Sensing*, **31**(B1), 170–175.

Seige, P. and D. Meissner (1993). MOMS-02: An advanced high resolution multispectral stereo scanner for Earth observation. *Geo-Informations-Systeme*, **6**(1), 4–11.

Seedahmed, G., D.C. Merchant and T. Schenk (1998). Experimental results of digital camera calibration. In *International Archives of Photogrammetry and Remote Sensing*, **32**(3/1), 91-96.

Theuwissen, A. J. P. (1995). *Solid-State Imaging with Charge-Coupled Devices.* Kluwer Academic Publishers, Dordrecht, The Netherlands.

Yang, E. S. (1988). *Microelectronic Devices.* McGraw-Hill, New York, NY.

Chapter 8

Scanners

Despite the astounding progress in solid-state cameras and their increased use in photogrammetry, film-based cameras remain the main data acquisition source for aerial applications. Modern aerial cameras are an engineering feat. They provide, together with the film's impressive geometric and radiometric resolution, high quality photographs.

Scanners, in general, convert hard copy documents into digital form. In addition to digitizing (scanning) aerial and satellite photography, scanners are also employed to convert maps of all sorts. This chapter describes photogrammetric scanners. The first section summarizes typical requirements that explain the need for developing special scanners for photogrammetric applications. The following sections deal with the basic components and the working principle of flatbed scanners, including a short account of potential errors.

A number of scanners are commercially available. No attempt is made to describe or compare these systems; rather, emphasis is placed on general principles.

8.1 Introduction

The requirements for converting documents in photogrammetry and mapping, such as aerial and satellite photographs, can be stated as follows:

1. Format: size of aerial photographs $9'' \times 9''$, size of satellite photography may be as large as $12'' \times 18''$. Often, entire film rolls must be scanned.

2. Geometric resolution and accuracy: $15\mu m$ pixel size and $2\mu m$ pixel accuracy for applications such as aerial triangulation.

3. Radiometric resolution: the density range is 2.5 D for panchromatic and up to 3.5 D for color photography. This requires 10 to 12 bit resolution.

Considering the increasing trend toward information systems and the large volume of analog documents that need to be converted, the importance of scanners is stressed. *Desk top publishing scanners (DTP)*, suitable for digitizing office documents, are now standard peripheral devices of desk top publishing systems. The requirements listed above exceed the specifications of DTP scanners, however. Consequently, special scanners have been developed, tailored to the needs and peculiarities of the mapping industry.

Increased performance does not come for free, however. Generally, *photogrammetric scanners* are at least one order of magnitude more expensive than DTP scanners. This is the chief reason why DTP scanners are occasionally used in photogrammetry, especially when the geometric accuracy is not of utmost concern. *Baltsavias and Waegli* (1996) present interesting results concerning the quality of DTP scanners.

Considering the rapid progress in the development of high quality electronic cameras and their increased use in photogrammetry, one may wonder when all imagery will be acquired electronically, thereby eliminating the process of scanning films. Modern aerial data acquisition systems, comprising distortion-free cameras with motion compensation, films with great geometric stability, high resolution and impressive dynamic range, remain a challenge that digital data acquisition systems are not likely to meet in the near future.

8.2 Drum Scanners vs. Flatbed Scanners

Two radically different design principles are applied in building scanners. Fig. 8.1(a) shows the principle of a *drum scanner*. Here, the film (or map), is clamped on a drum. The photosensitive element is usually a photodiode. Due to the high sensitivity of photodiodes, drum scanners offer a large density range; that is, they render a faithful tonal representation. The photodiode moves along an axis parallel to the drum, pixel by pixel. In this fashion, one row of the digital image is obtained. The rotation of the drum provides the columns. As impressive as the radiometric resolution fidelity of drum scanners is, their use is hampered by geometric problems. Uneven drum surface, difficulty in pressing the film firmly against the drum, and differences in rotational speed are just some of the geometric problems. Moreover, calibrating drum scanners poses an additional problem, because high precision glass plates cannot be mounted.

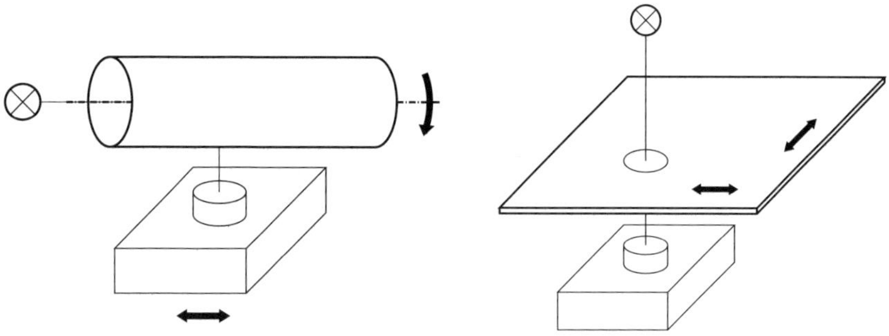

(a) (b)

Figure 8.1: Two different scanner design principles. In (a) a drum scanner is shown and in (b) a flatbed scanner.

Fig. 8.1(b) depicts the principle of a *flatbed scanner*. The film is placed on a photo carrier and illuminated from above. Solid-state sensors detect the amount of energy transmitted through the film. Usually, line sensors are employed, consisting of several thousand sensor elements. The photo carrier with the film is moved with respect to the fixed position of the sensor, the optics, and light source.

Today's scanners are predominantly flatbed constructions. The advances in solid-state sensors, such as increased number of sensing elements per chip and the decreasing size of the elements, account for the widespread use of flatbed scanners. Fig. 8.2 shows an example of a flatbed scanner that is equipped with a motorized film drive to scan film rolls.

8.3 Principal Components of Flatbed Scanners

Fig. 8.3 on page 193 provides a schematic diagram of a flatbed scanner with its principal components.

8.3.1 Illumination and Optic System

There are a number of challenging and partly contradictory requirements that the illumination system of a scanner must meet. For one, the light source must be stable, so that the flux incident on the line sensor is solely a function of the film density. To compensate for variations in the light

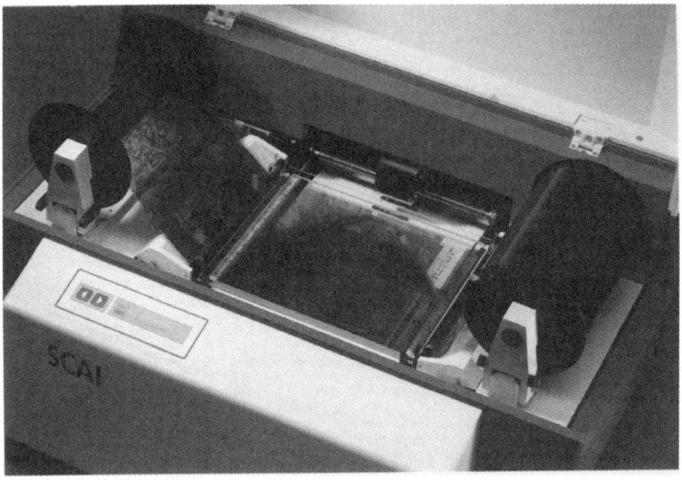

Figure 8.2: Zeiss' SCAI precision scanner equipped with a film drive device to scan entire film rolls. Two motorized film reels allow rewinding the film in both directions. In order to prevent scratches during film transport, a lifting device presses the film against deflection rollers. An electronic frame counter enables automatic positioning to preselected photos. Courtesy Intergraph Corporation, Huntsville, AL.

source, its energy must be measured directly. Then, corrections to the recorded energy from the sensors can be applied accordingly.

Another requirement stems from the dynamic scanning process. The saturation energy of the sensor is the product of integration time and the incident flux. To keep the integration time as short as possible, the light source must radiate enough energy. Typical light sources are halogen lamps with a power of 100 to 300 Watts, or fluorescent lamps. Such lamps may cause considerable heat diffusion, which, in turn, may affect the sensor stability and the film. To prevent these undesirable effects, the thermally insulated lamp is positioned as far from the sensor and film as possible. In this case, a fiber optic cable guides the light to the illumination position.

Another design aspect is the question of whether diffuse or direct light should be used. In the case of diffuse illumination, a light-scattering glass plate is placed on top of the film. Diffuse illumination offers the advantage that small blemishes on the base of the emulsion, such as dust particles or scratches, will not be imaged. Additionally, diffuse light produces a smaller image noise than direct light sources.

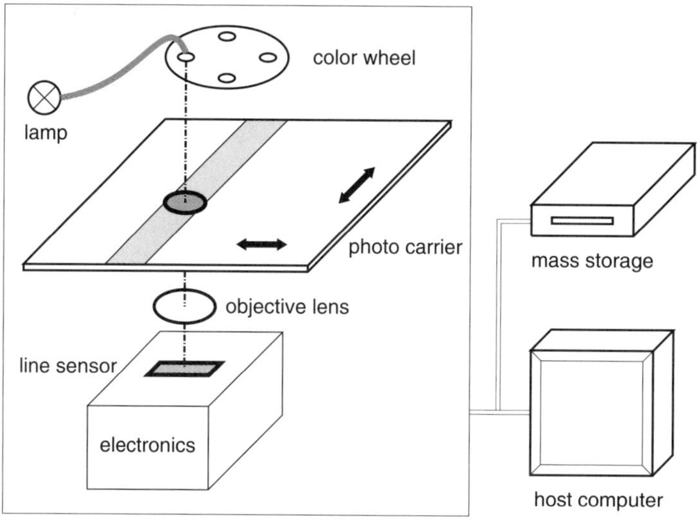

Figure 8.3: Schematic diagram of a flatbed scanner with its principal components, including host computer and mass storage device.

Direct illumination requires a condenser lens. Its function is to generate coherent light from a point light source. An advantage of direct illumination arises from the large depth of focus (due to the rather narrow bundle of light rays emanating from the light source), alleviating problems associated with defocalization.

Part of the illumination system is the color wheel. It contains the necessary filters for scanning color film. With this solution, the film is scanned three times. Another solution to scanning color images is using three separate line sensors, each sensitive to a primary color. In that case all three colors are scanned in one pass.

The lens system must be corrected for radial distortion and aberrations. A critical issue is to ensure that the film patch is precisely imaged on the plane containing the sensor elements. Moreover, the relative position of objective lens, film and line sensor must remain constant during the scanning process.

The pixel size of the scanned image in the direction of the line sensor depends on the sensor element size and the magnification of the optical system. Most sensor elements have a size of about 10 μm (see also Fig. 7.3 on page 160). The dimension of the pixel in the scan direction is determined by the scanning speed and integration time. Some scanners work with a fixed pixel size. Instead of optically changing the resolution, e.g. by zoom optics, software performs the task by resampling the scanned image to the desired resolution.

8.3.2 Photo Carrier

The photo carrier of photogrammetric scanners is usually a modified version of the analytical plotter carrier stage. Thus, high reliability and a geometric accuracy of about 2 μm can be expected. The photo carrier sits on a two-axis cross slide. Motion is controlled by a servo system through a friction drive, for example. Linear encoders with an accuracy of one micron or less provide the positional feedback. The maximum speed is typically between 10 and 20 mm/sec.

Some scanners provide an option for scanning film rolls. Conceptually, the solution is similar to that employed in aerial cameras, except that for scanners, no vacuum is applied to press the film firmly against the glass plate. With this option, the manual loading of film is circumvented and it is conceivable that the scanning of an entire film roll containing several hundred frames is carried out without assistance of an operator.

As the photo carrier moves the film across the line sensor, a small strip, called a *swath*, is digitized. Its width is determined by the physical dimension of the sensor, multiplied by the magnification of the optical system. At the end of a swath, the carrier first moves perpendicular to the scan direction by the swath width and only then resumes the scanning operation. To prevent overlap or gaps between adjacent swaths, the shift must be very accurate. Moreover, smooth movement at constant speed in the scanning direction is essential as this determines the pixel size in the scan direction.

8.3.3 Sensor

The sensor's function is to 'measure' the film density. The incident flux Φ_i interacts with the film, where it is either absorbed, reflected or transmitted. The ratio of transmitted flux to incident flux is the *transmittance*, $\tau = \Phi_t/\Phi_i$. The *density*, D, is defined as the logarithm of the reciprocal of the transmittance, or $D = \log(1/\tau)$. The human visual system has nearly a logarithmic luminous response; that is, it is proportional to the density of the film. A density scale of equal steps is perceived as linear. It follows that the density range of a film should be digitized in equal steps. The density of black and white aerial films ranges from of 0.2 to 2.5 D, and of color films from 0.3 to 3.5 D. *Ziemann* (1996), provides a detailed account of digitizing photographic images.

Almost every flatbed scanner employs a CCD line sensor[1]. One of the reasons for using CCD line sensors instead of 2-D arrays is the greater sensitivity of line sensors.

Photomultipliers are the most sensitive and accurate sensing elements—they virtually count single photons. It should not come as a surprise that drum scanner equipped with photomultipliers provide linearity across the full density range of photographic material—a notorious problem with CCD sensors. Fig. 8.4 illustrates the sensitivity behavior of CCD sensors. It decreases with increasing density. As a consequence, the density difference ΔD in dark areas is resolved in many fewer gray values than the same density difference in brighter areas.

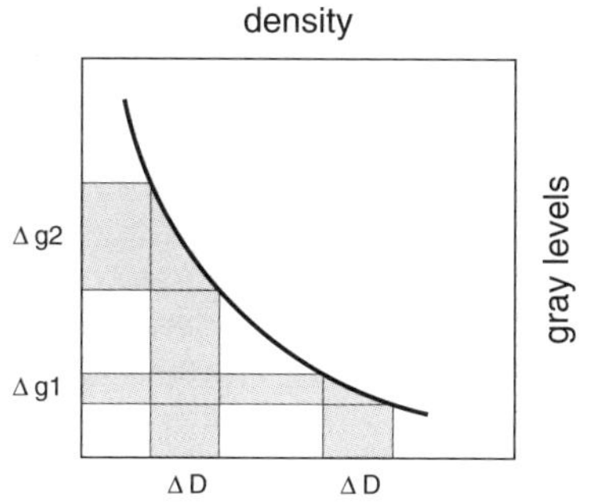

Figure 8.4: Sensitivity of CCD sensors as a function of density. CCD sensors respond linearly. The same density range ΔD is represented by fewer gray values (e.g. $\Delta g1$) at higher densities than at lower densities (e.g. $\Delta g2$). As a consequence, the tonal representation of dark and bright regions is not balanced.

8.3.4 Scanner Electronics

The scanner electronics consist of several components providing the following essential functionality.

A/D conversion: the charge accumulated in the CCD sensor elements is read out and converted to an analog video signal. The output of the scanner is a digital signal containing the gray values of all the pixels. The A/D conversion module quantizes the amplitude into 2^n equal

[1] DTP scanners use sensors with 5,000 to 7,500 elements; photogrammetric scanners usually use chips with fewer elements, e.g. 2,048.

steps (gray levels), where n is the number of bits required to represent the gray values. Although most scanners output 8 bits, the internal quantization is sometimes performed at a higher level, for example, with 10 to 12 bits. In that way, some of the problems with low sensitivity in dark areas are alleviated.

To accommodate fast conversion and individual corrections to every pixel, the technique of look-up-tables (LUT) is used. For every analog input value they contain the corresponding digital output. LUTs are generated during (internal) calibration procedures.

Sensor control: line sensors must be controlled; for example, the sensor plane must coincide with the image plane. If color sensitive elements are used, the control unit is responsible for ensuring that all three line sensors receive light from the same film area.

Carrier control: provides the necessary circuitry for the DC motors driving the cross-slide system.

Memory: the digital signal may be temporarily stored in internal memory. For example, the current swath is stored in memory while the previous swath is read out to disk or to the host computer.

Output interface: provides the link to the host computer or the mass storage device. Ideally, the data transfer rate should correspond to the scanning speed. For example, a scanning speed of 15 mm/sec generates a data stream of about 3 MB/sec (see Example 8.1). The PCI bus, discussed in 7.4, can easily handle such a data rate; other bus systems or direct connections to mass storage devices may slow down the scanning process.

Example 8.1:

Assume a pixel size of 10 μm, and a scanner with 2,048 sensor elements and 8 bits quantization. Suppose the scanner operates with a speed of 15 mm/sec. Then, the transfer rate r amounts to $\frac{15{,}000 \cdot 2{,}048}{10} = 3.076$ *MB/sec (Mega Bytes per second).*

8.3.5 Host Computer

The host computer plays an important role. First, it provides the graphical user interface for the operator to communicate with the scanner, for example, by setting parameters, entering file names, starting calibration procedures, or defining local areas to be scanned.

In addition to controlling and monitoring the scanning process, application software may analyze scanned images for completeness and consistency. For this purpose, the image must be displayed on a high quality monitor. Also, it may be desirable to perform the interior orientation right at the scanner workstation either semi-automatically or automatically.

8.4 Pixel Size

It is helpful to distinguish between the following types of pixels that exist for flatbed scanners (see also Fig. 8.5).

sensor pixel: refers to one sensor element. Its size is determined by the type of the line scanner. Typical sizes are between 10 and 15 μm.

scanner pixel: refers to the projected area of the sensor pixel onto the film. It is what the scanner outputs to the interface.

refined pixel: is the final result of the scanning and post processing procedure. This pixel is stored and eventually processed further by digital photogrammetry methods. If no post processing is performed on the host computer then the refined pixel is identical to the scanner pixel.

photo pixel: refers to the film resolution. This is not a physical pixel, more a conceptual realization.

Let us now examine the relationship between the different pixels (see also Fig. 8.5).

8.4.1 Relationship Between Sensor Pixel and Scanner Pixel

Suppose for a moment that the photo carrier does not move. The sensor elements are projected through the lens onto the film. In that case, the size of the scanner pixel would be equal the size of the sensor pixel, multiplied by the magnification of the optical system.

Now let the photo carrier move. The sensor pixel is projected onto the moving film during the exposure time. The pixel size in the scan direction is now a function of scanning speed and exposure time. With Δt being the integration time and v the scan velocity, the pixel size s_y in the scan direction becomes $s_y = v \cdot \Delta t$. The pixel size along the line sensor is $s_x = s_s \cdot m$ with s_s the sensor pixel size and m the magnification of the optical system.

We conclude that the size of the scanner pixel is controlled by the optical magnification of the sensor pixel in one direction, and by the velocity of

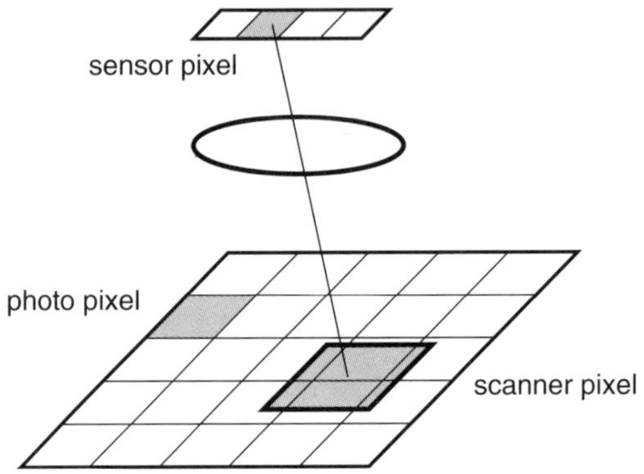

Figure 8.5: The notion of different pixels in the scanning process. The sensor pixel is a physical quantity referring to pixel pitch of the sensor. The photo pixel is an abstract quantity that relates to the "resolution" of the film document to be scanned. The scanner pixel is the projected sensor pixel—projected from the sensor through the sensor objective onto the film.

the photo carrier and integration time of the sensor elements in the other direction.

Example 8.2:

Assume a sensor pixel size of 13 μm, a desired scan pixel size of $10 \times 10\ \mu m$, *and a fixed integration time of 0.2 ms. How fast must the photo carrier move? We have from the above equation for the velocity* $v = \frac{s_y}{\Delta t} = \frac{10\ \mu m}{0.2\ ms} = 20\ mm/sec.$

8.4.2 Relationship Between Scanner Pixel and Photo Pixel

The central question of scanning is how large should the scanner pixel be so that the original document (film) is represented as faithfully as possible. This is equivalent to asking for the minimum size of the photo pixel, or the resolution. Film resolution is often expressed by the number of line pairs per millimeter that still can be discerned. This popular measure is quite subjective as it depends on the contrast of the film and human judgment. Modulation Transfer Functions (MTF) are better suited to express resolution.

The image formation process can be described by a linear system, at least as a first approximation. Referring to Chapter 2, the object to be imaged is the input signal, the lens system functions as a linear system, and the image constitutes the output. The system response is entirely characterized by the impulse response. An impulse can be thought of as a point source in object space. Then, the object is a linear combination of scaled and shifted impulses. How does the lens transform an impulse? The image of a point through an aperture is the point spread function. For example, the image of a point obtained by a pinhole camera is the diffraction pattern. An image, then, is obtained by convolving the light intensity function of the object by the point spread function. The same operation can be performed in the frequency domain. The Fourier transform of the point spread function is the optical transfer function and its magnitude (modulus) is the modulation transfer function.

The point spread function, characterized by the width of the Gaussian, σ, offers a measure for the resolution. For black and white films, σ varies between 20 μm and 30 μm; color emulsions have a slightly higher spread. From these values we conclude that the size of photo pixels is 10 μm to 15 μm for b/w, and 15 μm to 25 μm for color emulsions[2].

8.5 Potential Error Sources

Several researchers have investigated the quality of various scanners and reported performance characteristics and potential error sources. Baltsavias tested several scanners, including DTP scanners (*Baltsavias and Bill* (1994); *Baltsavias* (1994a); *Baltsavias and Waegli* (1996)). Kölbl examined the tonal representation of scanners in *Kölbl and Bach* (1996) and *Kölbl* (1994). *Bolte* (1996) investigated the geometric and radiometric behavior of a photogrammetric scanner. *Seywald et al.* (1994) propose specifications and procedures for how scanners should be tested. In *Bethel* (1994), the author investigates the MTF of a photogrammetric scanner.

We summarize in this section the most important aspects of quality control and the major error sources.

8.5.1 Geometry: Position and Resolution

High quality photogrammetric scanners should have a repeatable accuracy of 2 μm over the entire format. Even if this accuracy is maintained, potential problems may arise from misaligned line sensors, a situation illustrated

[2]According to the sampling theorem, the pixel size should be half of the highest frequency, or half of the spread function width in our case.

in Fig. 8.6. The result is a skewed image coordinate system. Moreover, adjacent swaths are separated by a small gap.

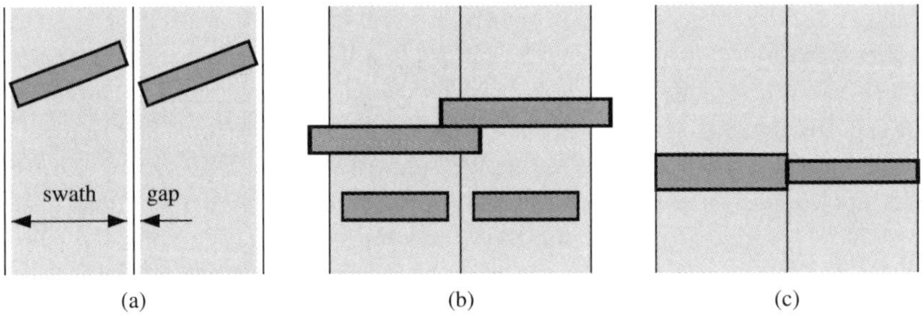

Figure 8.6: Illustration of some geometric errors. Misaligned line sensors cause a displacement of scan lines across swaths (a). In (b) the effect of a wrong scale perpendicular to the scan direction is shown, resulting in either an overlap or gap between adjacent swaths. Finally, a scale difference in the scan direction causes a rectangular pixel size (c).

A similar problem may occur from a wrong scale. A wrong scale perpendicular to the scan direction that is caused by a wrong optical magnification, would either cause an overlap between swaths or a gap as shown in Fig. 8.6(b). A scale error along the scan direction, caused by a wrong or uneven speed or poor synchronization between the sensor and the photo carrier, would result in a rectangular pixel shape. Also, pixels of the same row in adjacent swaths would not match exactly.

8.5.2 Radiometric Representation

Emulsions of aerial films have a large dynamic range and a logarithmic relationship between density and exposure. To faithfully represent these properties by the digital image, CCD sensors are confronted with a major challenge. As mentioned earlier, CCD sensors are linear.

Other radiometric problems may arise from color misregistration due to chromatic aberrations or positional offsets of the image patch projected to the color sensitive line sensors. Proper color balance may be difficult to achieve because the CCD sensor sensitivity is quite different in the visible portion of the spectrum. For example, blue has a much lower sensitivity than red and green.

The radiometric resolution is further reduced by electronic noise, such as thermal noise and blooming. A measure to reduce electronic noise is to cool the sensors, and to average multiple measurements. Also, a slower integration and read-out time helps to reduce the noise.

Problems

1. How would you determine an optimal pixel size when scanning a high quality aerial film?

2. Most flatbed scanners employ CCD sensors. CCD detectors are inherently linear with respect to illumination. Film material has a logarithmic response, however. What is the consequence when scanning films?

3. What is the relationship between the dynamic range of scanners and the density range of films?

4. Explain the difference between geometric resolution and geometric accuracy.

5. Determine the data transfer rate between scanner, computer, and hard disk of a flatbed scanner with a line array of 1K pixels (detector elements), pixel size 12μm. How long does it take to scan an aerial photograph?

6. An exposed and developed film is ultimately a binary representation; a silver crystal is either entirely reduced or not (black or white rendering). Why do we see and deal with gray shades?

References

Baltsavias, E. P. (1994a). Test and calibration procedures for image scanners. In *International Archives of Photogrammetry and Remote Sensing*, **30**(1), 163–170.

Baltsavias, E. P. (1994b). The Agfa Horizon Scanner—Characteristics, Testing, and Evaluation, In *International Archives of Photogrammetry and Remote Sensing*, **30**(1), 171–179.

Baltsavias, E. P. and R. Bill (1994). Scanners—A survey of current technology and future needs. In *International Archives of Photogrammetry and Remote Sensing*, **30**(1), 130–143.

Baltsavias, E. P. and B. Waegli (1996). Quality analysis and calibration of DTP scanners. In *International Archives of Photogrammetry and Remote Sensing*, **31**(B1), 13–19.

Bethel, J. (1994). Calibration of a photogrammetric image scanner. In *Proc. ACSM/ASPRS Ann. Convention*, **1**, 81–88.

Bolte, U., K. Jacobsen and H. Wehrmann (1996). Geometric and Radiometric Analysis of a Photogrammetric Image Scanner. In *International Archives of Photogrammetry and Remote Sensing*, **31**(B1), 72–77.

Kölbl, O. (1994). Survey of High-Quality Photographic Scanners. In *Mapping and Remote Sensing Tools for the 21st Century*, American Society for Photogrammetry and Remote Sensing, 7–14.

Kölbl, O. and U. Bach (1996). Tone Reproduction of Photogrammetric Scanners. *Photogrammetric Engineering and Remote Sensing, PERS*, **62**(6), 687–694.

Seywald, R., F. Leberl and W. Kellerer (1994). Requirements of a system to analyze film scanners. In *International Archives of Photogrammetry and Remote Sensing*, **30**(1), 144–149.

Ziemann, H. (1996). On the digitization of photographic images. In *International Archives of Photogrammetry and Remote Sensing*, **31**(B1), 125–129.

Chapter 9

Digital Photogrammetric Workstations

Probably the single most significant product of digital photogrammetry is the *digital photogrammetric workstation (DPW)*, also called a *softcopy workstation.* The role of DPWs in digital photogrammetry is equivalent to that of analytical plotters in analytical photogrammetry.

The development of DPWs is greatly influenced by computer technology. Considering the dynamic nature of this field, it is not surprising that digital photogrammetric workstations are undergoing constant changes, particularly in terms of performance, comfort level, components, costs, and vendors. It would be nearly impossible to provide a comprehensive list of the current products which are commercially available, much less describe them in some detail. Rather, the common aspects, such as architecture and functionality, are emphasized.

The next section provides some background information, including a few historical remarks and an attempt to classify the systems. This is followed by a description of the basic system architecture and functionality. Finally, the most important applications are briefly discussed.

To build on common ground, I frequently compare the performance and functionality of DPWs with those of analytical plotters. Sec. 9.5 summarizes the advantages and the shortcomings of DPWs relative to analytical plotters.

9.1 Background

Great strides have been made in digital photogrammetry during the past few years due to the availability of new hardware and software, such as powerful image processing workstations and vastly increased storage capacity. Research and development efforts have resulted in operational products that are increasingly being used by government organizations and private companies to solve practical photogrammetric problems. We are witnessing the transition from conventional to digital photogrammetry. DPWs play a key role in this transition.

9.1.1 Digital Photogrammetric Workstation and Digital Photogrammetry Environment

Fig. 9.1 depicts a schematic diagram of a digital photogrammetry environment. On the input side we have a digital camera or a scanner with which existing aerial photographs are digitized. At the heart of the processing side is the DPW. The output side may comprise a film recorder to produce hard copies in raster format and a plotter for providing hard copies in vector format. Some authors include the scanner and film recorder as components of the softcopy workstation. The view presented here is that a DPW is a separate, unique part of a digital photogrammetric system.

As discussed in the previous chapters, digital images are obtained directly by using electronic cameras or indirectly by scanning existing photographs. The accuracy of digital photogrammetry products depends largely on the accuracy of electronic cameras or scanners, and on the algorithms used. In contrast to analytical plotters (and even more so to analog stereoplotters), the hardware of DPWs has no noticeable effect on the accuracy.

Figs. 9.2 and 9.3 show typical digital photogrammetric workstations. At first sight they look much like ordinary graphics workstations. The major differences are the stereo display, 3-D measuring system, and increased storage capacity to hold all the digital images of an entire project. Sec. 9.2 elaborates further on these aspects.

The station shown in Fig. 9.2 features two separate monitors. In this fashion the stereo monitor is entirely dedicated to displaying imagery only. Additional information, such as the graphical user interface, is displayed on the second monitor. As an option to the 3-D pointing device (trackball), the system can be equipped with handwheels to more closely simulate the operation on a classical instrument.

The main characteristic of Intergraph's ImageStation Z is the 28-inch panoramic monitor that provides a large field of view for stereo display (see Fig. 9.3, label 1). Liquid crystal glasses (label 3) ensure high-quality stereo viewing. The infrared emitter on top of the monitor (label 4) provides synchronization of the glasses and allows group viewing. The 3-D pointing device (label 6) allows freehand digitizing and the 10 buttons facilitate easy menu selection.

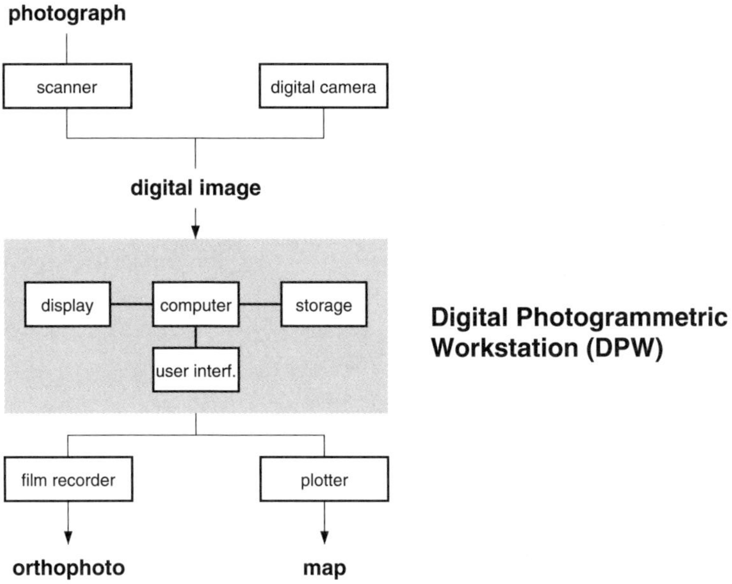

Figure 9.1: Schematic diagram of digital photogrammetry environment with the digital photogrammetric workstation (softcopy workstation) as the major component.

9.1.2 *Historical Perspective*

Gülch (1994) presents an interesting view on the development of digital photogrammetric workstations. He distinguishes three phases. The first phase begins in 1955 and ends in 1981. This period is characterized by conceptual ideas and scenario descriptions of an all digital photogrammetry environment. Because of inappropriate hardware and software, and lack of digital imagery, no real systems could be developed; at least not in the commercial sector. It is interesting to note that shortly after the invention of the analytical plotter, its inventor, Helava, mentioned the possibility of replacing the human operator by an automatic correlator—a typical consideration of digital photogrammetry.

The first reasonably detailed concept of a DPW was described in *Sarjakoski* (1981). Referred to as a fully digital stereoplotter, the functionality strongly resembles that of an analytical plotter, the major difference being the fact that the photographs are replaced by digital images. Sarjakoski proposed building the digital stereoplotter using image processing

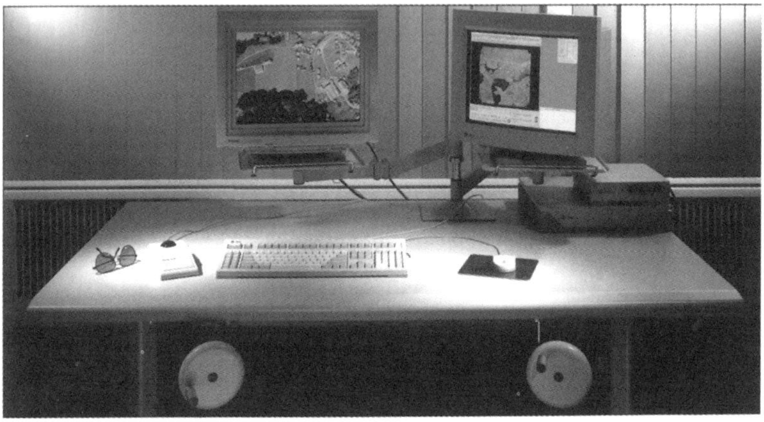

Figure 9.2: Typical digital photogrammetric workstation. The system shown here offers optional handwheels to emulate operation on classical photogrammetric plotters. Courtesy LH Systems, Inc., San Diego, CA.

systems and analytical plotter software as a basis. *Case* (1982) provided another fundamental concept of a digital image exploitation system. Again, the proposed system had the functionality of an analytical plotter with the potential of automating photogrammetric tasks such as DEM generation.

These two design concepts paved the way for the development of DPWs from 1982—1988, which Gülch considers the second phase in the rather short history of DPWs. Additional concepts were presented and experiments began with prototypes in different research institutions. Finally, at the occasion of the ISPRS Congress in Kyoto in 1988, Kern, then a leading manufacturer of photogrammetric instruments, introduced the DPS1, the first commercially available digital photogrammetric workstation, described in *Cogan* (1988). This also signaled the confluence of system development in remote sensing and digital photogrammetry—a rather remarkable event. Major efforts were directed towards implementing hardware components and developing low-level system software. The application software was "borrowed" from analytical plotters. The systems of that period were very much characterized by minimal functionality and performance; it is not surprising that the photogrammetric community met these first softcopy workstations with a healthy dose of skepticism. No first generation system survived!

Finally, the third phase of DPW development began with increased activities by researchers, developers and photogrammetric societies. For ex-

Figure 9.3: Digital photogrammetric workstation. Shown is Intergraph's ImageStation Z. Main characteristic is the large stereo display of the 28-inch panoramic monitor. Courtesy Intergraph Corporation, Huntsville, AL.

ample, an ISPRS working group was charged with the task of defining the functionality and performance of digital photogrammetric systems, including a critical evaluation of existing systems. The working group provided the following broad definition: "A digital photogrammetric system is defined as hardware and software to derive photogrammetric products from digital imagery using manual and automated techniques."

9.1.3 Current Status

The migration of digital photogrammetric workstations from military applications into the civilian market is complete. DPWs are currently sold in numbers exceeding those of analytical plotters. It appears that by the turn of the century analytical plotters will be superseded by digital photogrammetric workstations.

Today it is feasible to build DPWs from commercially available off-the-shelf products. This offers several advantages, such as:

- reduction in development time and cost;
- open architecture, platform independence;
- portability of application software;
- more frequent system updates.

At the time of the writing of this book, more than a dozen vendors offer DPWs. When comparing surveys of DPWs, a heterogeneous picture emerges; see, for example, *Heipke* (1996); *Grün* (1996). New vendors compete for a share in the softcopy workstation market while others quit the race.

In contrast to the first generation systems, the recently developed DPWs meet the performance criteria one can reasonably expect for solving photogrammetric tasks efficiently and reliably. As a consequence, more and more photogrammetric organizations—government agencies and private companies alike—are introducing DPWs into their production environment.

In a recent survey, *Boniface* (1994) concludes that users enthusiastically embrace the new technology. DPWs are being used for digital orthophoto production, aerial triangulation, DEM generation, and, though to a lesser extent, stereocompilation.

9.1.4 Classification of Digital Photogrammetric Workstations

Over the years different classification schemes have been proposed. *Walker* (1996) argues for a practical taxonomy comprising five categories with functionality being the prime classification criterion. Systems with complete photogrammetric functions are ranked in the top category while systems with limited functionality and performance are at the lower end.

Dowman (1991) proposes the following four categories:

1. performance and function of an analytical plotter with automatic feature extraction;
2. performance and function of an analytical plotter with computer assisted feature extraction;
3. systems designed for specific applications with high performance but limited functionality;
4. limited performance and functionality but low cost.

The products that can be derived from softcopy workstations is yet another way to classify the systems and was proposed in *Heipke* (1995). The classes include:

Stereo DPW: primarily used for interactive stereo plotting (compilation). They include high end and low end systems.

Mono DPW: used for planimetric plotting, for example, by digitizing orthophotos. Height information may come from a DEM.

Aerial triangulation DPW: has specific functionality to perform point transfer and the measurement of features in multiple images as automatically as possible.

DTM DPW: provides for automatically derived DTMs with the possibility of interactive editing and quality control.

Orthoimage DPW: has a special module for orthoprojection and mosaicking. Such modules are usually added to systems developed for remote sensing applications.

9.2 Basic System Components

Fig. 9.4 depicts the basic system components of a digital photogrammetric workstation.

CPU the central processing unit should be reasonably fast considering the amount of computations to be performed. Many processes lend themselves to parallel processing. Parallel processing machines are available at reasonable prices. However, programming that takes advantage of them is still a rare commodity and prevents a more wide spread use of the workstations.

OS the operating system should be 32 bit based and suitable for real-time processing. UNIX satisfies these needs; in fact, UNIX based workstations were the systems of choice for DPWs until the emergence of Windows and NT that make PCs a serious competitor of UNIX based workstations.

main memory due to the large amount of data to be processed, sufficient memory should be available. Typical DPW configurations have 64 MB, or more, of RAM.

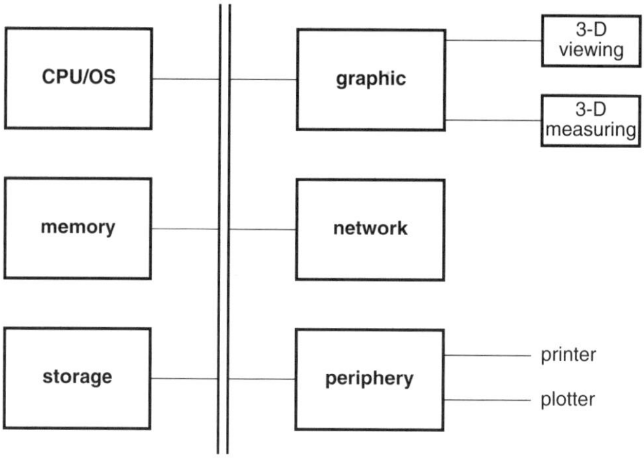

Figure 9.4: Basic system components of a digital photogrammetric workstation.

storage system must accommodate the efficient storage of several images. It usually consists of a fast access storage device, e.g. hard disks, and mass storage media with slower access times. Sec. 9.3.1 discusses the storage system in more detail.

graphics system the graphics display system is another crucial component of the DPW. The purpose of the display processor is to fetch data, such as raster (images) or vector data (GIS); process and store it in the display memory; and update the monitor. The display system also handles the mouse input and the cursor.

3-D viewing system is a distinct component of a DPWs usually not found in other workstations. It should allow viewing a photogrammetric model comfortably and possibly in color. For a human operator to see stereoscopically, the left and right image must be separated. Sec. 9.3.3 discusses the principles of stereo viewing.

3-D measuring device is used for stereo measurements by the operator. The solution may range from a combination of a 2-D mouse and trackball to an elaborate device with several programmable function buttons.

network a modern DPW hardly works in isolation. It is often connected to the scanning system and to other workstations, such as a geographic information system. The client/server concept provides an adequate solution in this scenario of multiple workstations and shared resources (e.g. printers, plotters).

user interface may consist of hardware components such as keyboard, mouse, and auxiliary devices like handwheels and footwheels (to emulate an analytical plotter environment). A crucial component is the graphical user interface (GUI).

9.3 Basic System Functionality

The basic system functionality can be divided into the following categories:

1. *Archiving:* store and access images, including image compression and decompression.
2. *Processing:* basic image processing tasks, such as enhancement and resampling.
3. *Display and Roam:* display images or sub-images, zoom in and out, roam within a model or an entire project.
4. *3-D Measurement:* interactively measure points and features to subpixel accuracy.
5. *Superpositioning:* measured data or existing digital maps must be superimposed on the displayed images.

A detailed discussion about the entire system functionality is beyond the scope of this book. We will focus on the storage system, on the display and measuring system, and on roaming.

9.3.1 Storage System

A medium size project in photogrammetric mapping contains hundreds of photographs. It is not uncommon to deal with thousands of photographs in large projects. Assuming digital images with 16 K $\times$ 16 K resolution (pixel size approx. 13 μm), a storage capacity of 256 MB per uncompressed black and white image is required. Consider a compression rate of three and we arrive at the typical number of 80 MB per image. To store a medium size project on-line imposes heavy demands on storage.

Photogrammetry is not the only imaging application with high demands on storage, however. In medical imaging, for example, imaging libraries in the terabyte size are typical. Other examples of high storage demand applications include weather tracking and monitoring, compound document management, and interactive video. These applications have a much higher market volume than photogrammetry; therefore, it is appealing for companies to further develop storage technologies.

The storage requirements in digital photogrammetry can be met through a carefully selected combination of available storage technologies. The options include:

hard disks: are an obvious choice, because of fast access and high performance capabilities. However, the high cost of disk space[1] would make it economically unfeasible to store entire projects on disk drives. Therefore, hard disk drives are typically used for interactive and real-time applications, such as roaming or displaying spatially related images.

optical disks: have slower access times and lower data transfer rates but at lower cost (e.g. $10 to $15 per GB, depending on technology). The classical CD ROM and CD-R (writable) with a capacity of approximately 0.65 GB can hold only one stereomodel. A major effort is being devoted to increasing this capacity by an order of magnitude and make the medium a rewritable one. Until such systems become commercially available (including accepted standards), CDs are used mostly as a distribution medium.

magnetic tape: offers the lowest media cost per GB (up to two orders of magnitude less than hard disk drives). Because of its slow performance (due to sequential access), magnetic tapes are primarily used as backup devices. Recent advances in tape technology, however, make this device a viable option for on-line imaging applications. Juke boxes with Exabyte or DLT (digital linear tape) cartridges (capacity of 20 to 40 GB per medium) lend themselves to use in on-line image libraries with capacities of hundreds of gigabytes.

When designing a hierarchical storage system, factors such as storage capacity, access time, and transfer rates must be considered. Moreover, the way data is accessed, for example randomly or sequentially, is important. Imagery inherently requires random access; think of roaming within a stereomodel. This seems to preclude the use of magnetic tapes for on-line applications. Clearly, one would not want to roam within a model stored on tape. However, if entire models are loaded from tape to hard disk, the access mode is not important, only the sustained transfer rate.

[1] Every 18 months the storage capacity doubles, while the price per bit halves. As this book is written, hard disk drives sold for less than $100 per GB.

Table 9.1: Magnification and size of field of view of analytical plotters.

Magnification	*Field of View, [mm]*		
	BC 1	*C120*	*P 1*
5 ×	30	29	40
10 ×	20	21	23
15 ×	15	14	16
20 ×	9	10	10

9.3.2 Viewing and Measuring System

An important aspect of any photogrammetric measuring system, be it analog or digital, is the viewing component. Viewing and measuring is typically performed stereoscopically, although certain operations do not require stereo capability.

As discussed in Chapter 4, humans can discern 7 to 8 lp/mm at a normal viewing distance of 25 cm. To exploit the resolution of aerial films, say 70 lp/mm, it must be viewed under magnification. The oculars of analytical plotters have zoom optics that allow viewing the model at different magnifications[2]. Obviously, the larger the magnification the smaller the field of view. Table 9.1 lists zoom values and the size of the corresponding film area that appears in the oculars. Feature extraction (compilation) is usually performed with a magnification of 8 to 10 times. With higher magnification, the graininess of the film reduces the quality of stereoviewing. It is also worth pointing out that stereoscopic viewing requires a minimum field of view. Operators consider viewing as comfortable if the field of view (size of image patch you look at) assumes values that are listed in Table 9.1.

Let us now compare the viewing capabilities of analytical plotters with those of DPWs. First, we realize that this function is performed by the graphics subsystem, that is, by the monitor(s). To continue with the previous example of a film with 70 lp/mm resolution, viewed 10 × magnified, we read from Table 9.1 that the corresponding area on the film has a diameter of 20 mm. To preserve the high film resolution it ought to be digitized with a pixel size of approximately 6 μm $1000/(2 \times 70)$. It follows that the monitor should display more than $3K \times 3K$ pixels. Monitors with this sort of resolution do not exist or are prohibitively expensive, particularly when considering color imagery and true color rendition (24+ bit planes).

[2]Typical magnification values range from 5 to 20 times.

If we relax the high resolution requirements and assume that images are digitized with a pixel size of 15 μm, then a monitor with the popular resolution of 1280 $\times$ 1024 would display an area that is quite comparable to that of analytical plotters.

Magnification, known under the more popular terms *zooming in/out*, is achieved by changing the ratio of the number of image pixels displayed to the number of monitor pixels. To zoom in, more monitor pixels are used than image pixels. As a consequence, the size of the image viewed decreases and stereoscopic viewing may be affected.

The analog to the floating point mark of analytical plotters is the three dimensional cursor that is created by using a pattern of pixels, such as a cross or a circle. The cursor must be generated by bitplane(s) that are not used for displaying the image. The cursor moves in increments of pixels, which may appear jerky compared to the smooth motion of analytical plotters. One advantage of cursors, however, is that they can be represented in any desirable shape and color.

The accuracy of interactive measurements depends on how well one can identify a feature, on the resolution, and on the cursor size. Ultimately, the pixel size sets the lower limit. Assuming that the maximum error is 2 pixels, the standard deviation is approximately 0.5 pixel. A better sub-pixel accuracy can be obtained in two ways. A straightforward solution is to use more monitor pixels than image pixels. Fig. 9.5(a) exemplifies the situation. Suppose we use 3 $\times$ 3 monitor pixels to display one image pixel. The standard deviation of a measurement is now 0.15 image pixels[3]. As pointed out earlier, using more monitor pixels for displaying an image pixel reduces the size of the field of view. In the example above, an area of only 6 mm diameter would be seen—hardly enough to support stereopsis.

To circumvent the problem of reducing the field of view, an alternative approach to sub-pixel measuring accuracy is often preferred. Here, the cursor is fixed in the monitor's center and the image is moved instead. Now the image does not need to move in increments of pixels. Resampling at sub-pixel locations allows smaller movements. This solution requires resampling in real time to assure smooth movement.

Yet another aspect is the illumination of the viewing system; so crucial when it comes to interpreting imagery. The brightness of the screen drops to 25% of the original brightness when polarization techniques are used[4]. Moreover, the phosphor latency causes ghost images. All these factors reduce the image quality.

[3]As before, the standard deviation is assumed to be 0.5 monitor pixel. We then obtain an accuracy of 0.5 $\times$ 1/3 image pixel in the image domain.

[4]Polarization absorbs half of the light. Another half is lost because the image is only viewed during half of the time usually available when viewing in monoscopic mode.

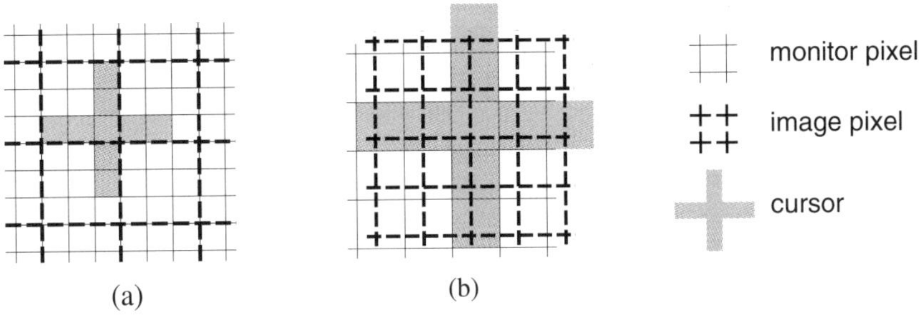

Figure 9.5: Two solutions to sub-pixel accuracy measurements. In (a), an image pixel is displayed to m monitor pixels, $m > 1$. The cursor moves in increments of monitor pixels, corresponding to $1/m$ image pixels. In (b) the image is moved under the fixed cursor position in increments smaller than an image pixel. This requires resampling the image at sub-pixel locations.

In conclusion, we realize that viewing on a DPW is hampered in several ways and is far inferior to viewing the same scene on an analytical plotter. To alleviate the problems, high resolution monitors should be used.

9.3.3 Stereoscopic Viewing

An essential component of a DPW is the stereoscopic viewing system (even though a number of photogrammetric operations can be performed monoscopically). For a human operator to see stereoscopically, the left and right image must be separated. This separation is accomplished in different ways; for example, spatially, spectrally, or temporally (Table 9.2).

One may argue that the simplest way to achieve stereoscopic viewing is by displaying the two images of a stereopair on two separate monitors. Viewing is achieved by means of optical trains, e.g. a stereoscope, or by polarization. Matra adopted this principle by arranging the two monitors at right angles, with horizontal and vertical polarization sheets in front of them.

An example of the split-screen solution is shown in Fig. 9.6. Here, the left and right images are displayed on the left and right half of the monitor, respectively. A stereoscope, mounted in front of the monitor, provides viewing. Obviously, this solution permits only one person to view the model. A possible disadvantage is the resolution, because only half of the screen resolution[5] is available for displaying the model.

[5]Actually, only the horizontal resolution is halved while the vertical resolution remains

Table 9.2: Separation of images for stereoscopic viewing. (Adapted from *Heipke* (1995)).

Separation	*Implementation*
spatial	2 monitors + stereoscope 1 monitor + stereoscope (split screen) 2 monitors + polarization
spectral	anaglyphic polarization
temporal	alternate display of left and right image synchronized by polarization

Figure 9.6: Example of a split-screen viewing system. Shown is the DVP digital photogrammetric workstation. Courtesy of DVP Geomatics, Inc., Quebec.

The most popular realization of spectral separation is by anaglyphs. The restriction to monochromatic imagery and the reduced resolution outweigh the advantage of simplicity and low cost. Most systems today use

the same as in dual monitor systems.

temporal separation in conjunction with polarized light. The left and right image is displayed in quick succession on the same screen. In order to achieve a flicker-free display, the images must be refreshed at a rate of 60 Hz per image, requiring a 120 Hz monitor.

Two solutions are available for viewing the stereo model. As illustrated in Fig. 9.7(a), a polarization screen is mounted in front of the display unit. It polarizes the light emitted from the display in synchronization with the monitor. An operator wearing polarized glasses will only see the left image with the left eye as the polarization blocks any visual input to the right eye. During the next display cycle, the situation is reversed and the left eye is prevented from seeing the right image. The system depicted in Fig. 9.2 on page 206 employs the polarization solution.

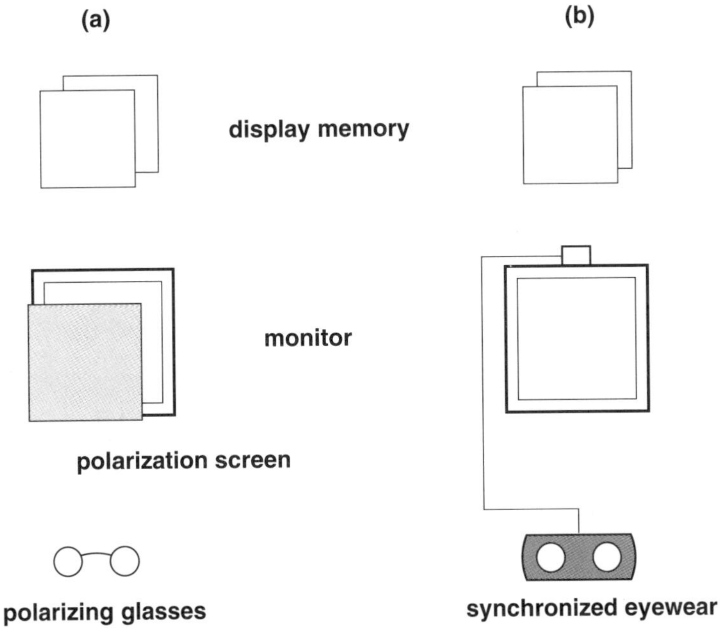

Figure 9.7: Schematic diagram of the temporal separation of the left and right image of a stereopair for stereoscopic viewing. In (a), a polarizing screen is mounted in front of the display. Another solution is sketched in (b). The screen is viewed through eyewear with alternating shutters. See text for detailed explanations.

The second solution, depicted in Fig. 9.7(b), is more popular and less expensive to realize. It is based on active eyewear containing alternating shutters, realized, for example, by *liquid crystal displays* (LCD). The synchronization with the screen is achieved by an infrared emitter, usually mounted on top of the monitor (Fig. 9.3 on page 207 shows an example). Understandably, the goggles are heavier and more expensive compared to the simple polarizing glasses of the first solution. On the other hand, the polarizing screen and the monitor are a tightly coupled unit, offering less flexibility in the selection of monitors.

9.3.4 Roaming

Roaming refers to moving the 3-D pointing device. This can be accomplished in two ways. In the simpler solution, the cursor moves on the screen according to the movements of the pointing device (e.g. mouse) by the operator. The preferred solution, however, is to keep the cursor locked in the screen center, which requires redisplaying the images. This is similar to the operation of analytical plotters where the floating point mark is always in the center of the field of view.

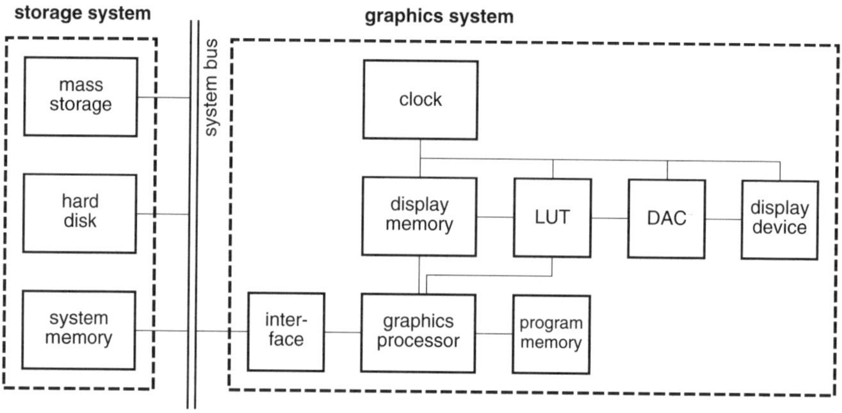

Figure 9.8: Schematic diagram of storage system, graphic system and display.

The following discussion refers to the second solution. Suppose we have a stereo DPW with a 1280×1024 resolution, true color monitor, and imagery digitized to 15 μm pixel size (or approximately $16K \times 16K$ pixels). Let us now freely roam within a stereomodel, much as we would do it on an analytical plotter and analyze the consequences in terms of transfer rates and memory size.

Fig. 9.8 schematically depicts the storage and graphic systems. The essential components of the graphic system include the graphics processor, the display memory, the digital-to-analog converter (DAC), and the display device (CRT monitor in our case). The display memory contains the portion of the image that is displayed on the monitor. Usually, the display memory is larger than the screen resolution to allow roaming in real-time. As soon as we roam out of the display memory, new image data must be fetched from disk and transmitted to the graphics system.

Graphics systems come in the form of high-performance graphics boards, such as RealiZm or Vitec boards. These state-of-the-art graphics systems are as complex as the system CPU. The interaction of the graphics system with the entire DPW, e.g. requesting new image data, is a critical measure of system performance.

Factors such as storage organization, bandwidths, and additional processing cause delays in the stereo display. Let us further reflect on these issues.

With an image compression rate of three, approximately 240 MB are required to store one color image. Consequently, a 24 GB mass storage system could store 100 images on-line. By the same token, a hard disk with 2.4 GB capacity could hold 10 compressed color images.

Since we want true color display, approximately 2×4 MB are required to hold the two images of the stereomodel[6]. As discussed in the previous section, the left and right image must be displayed alternately at a frequency of 120 Hz to obtain an acceptable model[7]. The bandwidth of the display memory amounts to $1280 \times 1024 \times 3 \times 120 = 472$ MB/sec. Only high speed, dual port memory, such as VRAM (video RAM) satisfies such high transfer rates. For less demanding operations, such as storing programs or fonts, less expensive memory is used in high performance graphics workstations.

At what rate should one be able to roam? Skilled operators can trace contour lines at a speed of 20 mm/sec. *Case* (1982) requests that the display on the monitor should be "crossed" within 2 seconds, in any direction. This translates to $1280 \times 0.015/2 \approx 10$ mm/sec in our example. In *Graham et al.* (1997), the authors state a maximum roam rate of 200 pixels/sec on Intergraph's ImageStation Z softcopy workstation. As soon as we begin to move the pointing device, new portions of the model must be displayed. To avoid immediate disk transfer, the display memory is larger than the monitor, usually four times. Thus, we can roam without problems within a

---

[6] $1280 \times 1024 \times 3$ Bytes = 3,932,160 Bytes.

[7] Screen flicker is most noticeable far out in one's vision periphery. Therefore, large screen sizes require higher refresh rates. Studies indicate that for 17-inch screens refresh rates of 75 Hz are acceptable. For DPWs larger monitors are required; therefore with a refresh rate of 60 Hz for one image we still experience annoying flicker at the edges.

distance twice as long as the screen window at the cost of increased display memory size (32 MB of VRAM in our example).

Suppose we move the cursor with a speed of 10 mm/sec toward one edge. When will we hit the edge of the display memory? Assuming we begin at the center, after one second the edge is reached and the display memory must be updated with new data. To assure continuous roaming, at least within one stereomodel, the display memory must be updated before the screen window reaches the limit. The new position of the window is predicted by analyzing the roaming trajectory. A look-ahead algorithm determines the most likely positions and triggers the loading of image data through the hierarchy of the storage system.

Referring again to our example, we have one second to completely update the display memory. Given its size of 32 MB, data must be transferred at a rate of 32 MB/sec from hard disk via system bus to the display memory. The bottle necks are the interfaces, particularly the hard disk interface. Today's systems do not offer such bandwidths, except perhaps SCSI-2 devices[8]. A PCI interface (peripheral component interface) on the graphics system will easily accommodate the required bandwidth.

A possible solution to get around the hard disk bottleneck is to dedicate system memory for storing an even larger portion of the stereomodel, serving as a sort of relay station between hard disk and display memory. This caching technique, widely used by operating systems to increase the efficiency of data transfer from disk to memory, offers additional flexibility to the roaming prediction scheme. It is quite unlikely that we will move the pointing device with a constant speed across the entire model (features to be digitized are usually confined to rather small areas). That is, the content of the system memory does not change rapidly.

Fig. 9.9 depicts the various windows related to the size of a digital image. In our example, the size of the display window is 19.2 mm $\times$ 15.4 mm, the display memory size is 4$\times$ larger, and the dedicated system memory again could be 4$\times$ larger. Finally, the hard disk holds more than one stereopair.

9.4 Application Functionality

The system functionality discussed in the previous section basically determines the comfort level of a DPW. For example, if the roaming is not optimally realized, then it may take seconds before the monitor displays new sub-images. Another important aspect of comparing different workstations is to examine their functionality for solving photogrammetric ap-

[8]Fast wide SCSI-2 devices, available as options, sustain transfer rates of 20 MB/sec. This would be sufficient for roaming within a b/w stereo model.

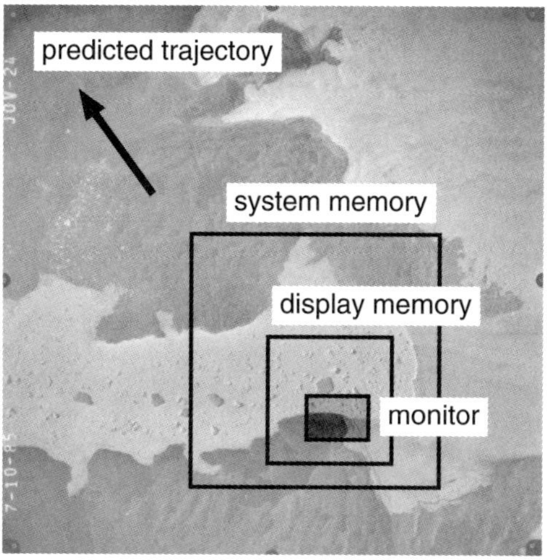

Figure 9.9: Schematic diagram of the various windows related to the size of an image. Real-time roaming is possible within the display memory. System memory holds a larger portion of the image. The location is predicted by analyzing the trajectory of recent cursor movements.

plications. In this section I briefly discuss the orientation procedures, aerial triangulation, DEM generation, and feature extraction, with an eye toward comparing the same procedures on analytical plotters. A more detailed account on these core applications is given in the following chapters.

9.4.1 Preparation

The preparation phase begins with the project definition, including information about project size, tolerances, average scale, camera calibration data, ground control points, and system parameters. For every model, additional information is usually required, such as the identification of the diapositives and their position on the stages, viewing mode (ortho/pseudo), and orientation procedures. By and large, this first phase of entering essential information is identical on analytical plotters and DPWs.

The next preparatory step involves the adjustment of the viewing conditions, accomplished on the analytical plotter with the help of various buttons and dials that are usually arranged in close proximity to the viewer's eye pieces. First, the illumination of the diapositives is adjusted until they appear equally bright. On DPWs, image enhancement techniques allow for more subtle adjustments. Basically, the whole range of photo lab processing techniques is available at the operator's fingertips!

9.4.2 *Orientation Procedures*

After the diapositives are placed on the stages and the viewing control is set the model is ready for orientation. The procedure is essentially the same on both systems. In the following I will mainly concentrate on the differences and advantages/disadvantages associated with DPWs. To keep the discussion to the essentials, a conventional aerial case is assumed.

Interior Orientation

The interior orientation process determines the relationship between the stage- and photo-coordinate systems. It determines the exact position of the diapositive on the carrier stage. Accordingly, the interior orientation on a DPW relates the digital image to the digitized diapositives. Therefore, the mathematical model must consider the scanner geometry and its systematic errors. The transformation parameters, then, relate pixel- to photo-coordinates.

As on analytical plotters, we expect that the DPW "drives" the cursor into the vicinity of the first fiducial mark. Precise measurement of the fiducial is left to the operator. However, one may expect that pattern recognition software would determine the precise location. Another advantage can be seen in the fact that all fiducial marks can be displayed simultaneously on the monitor, each in a separate window with a cursor placed on the predicted location.

Although the time required to measure the interior orientation interactively is very short, it may be justified to perform it autonomously, for example, immediately after scanning. In that case, the digital image would include the transformation parameters. Such an approach is discussed in Chapter 13.

Relative Orientation

The task of establishing a model is accomplished by measuring a number of conjugate points (parallax points), followed by computing the five orientation parameters. Again, we expect the analytical plotter to drive to a pre-specified pattern of parallax points, where the parallax is removed by moving one stage with respect to a locked position of the other stage. This same procedure can be accomplished on the DPW by keeping one of the stereo cursors in a fixed position while moving the other one. We may even expect from the DPW that conjugate points are found automatically, especially when the operator provides good approximations. For example, *Krupnik and Schenk* (1994) describe an automatic procedure that keeps the cursor on the ground.

After having measured enough points, the orientation parameters can be determined. This is the turning point on analytical plotters to switch from stereo-comparator mode to model mode. That is, the orientation elements are now used in the real-time loop to maintain the model. The transition between the two modes goes almost unnoticed; if everything went well with the first six parallax points then, all of a sudden, you have a model established. Now we wonder, again, how this all works on a softcopy workstation. The measurement of points and the computation is indeed very similar, but not the sudden switch to model mode, unless we have a system with centrally-fixed cursor and real-time roaming. The majority of existing DPWs must take a deep breath, however, to switch from comparator to model mode, because the images need to be resampled to epipolar geometry. By that we refer to the process of transforming the actual images into a normalized position such that the rows become parallel with the stereo-base (see Chapter 12). In that case, conjugate points lie on the same row. This dramatically reduces the search space, resulting in faster matching procedures and makes roaming easier (see 10.3.1).

By the way, the relative orientation can be automated to a degree where it runs as an autonomous batch process (see Chapter 14). This is particularly appealing because the resampling to epipolar geometry can immediately follow the orientation. Moreover, this batch process can run in a less costly environment. Given the scenario of autonomous interior and relative orientation, the first time the operator deals with a model on the DPW is after it has been oriented and resampled—the model is already established!

Absolute Orientation

During the process of absolute orientation, the relationship between the model and the object space is established. For aerial scenes this requires the measurement of control points. Quite often, the identification of control points poses significant problems because targets may only be partially visible, or are of low contrast compared to the background. If the identification is difficult for human operators, it is more difficult for the computer. As a consequence, automatic control point recognition is years away from becoming practical.

9.4.3 *Digital Aerial Triangulation*

Aerial triangulation is a good example to demonstrate the potential of DPWs to automate photogrammetric processes, thereby, substantially increasing the price performance ratio of photogrammetrically derived products. Even though today aerial triangulation is a well-established process, its performance and reliability can be considerably improved.

Traditionally, aerial triangulation begins with preparing and annotating photographs, whereby a suitable number of well distributed points are carefully selected such that they appear on as many photographs as possible. Once the preparation is completed, the points must be transferred to all photographs. This latter phase is quite crucial, particularly the transfer of strip tie points. In fact, the success of an aerial triangulation project depends largely on the quality of the point transfer. Only after the points are transferred or clearly identified can the measuring process begin.

It is here where automatic aerial triangulation offers significant improvements. The point transfer can be combined with simultaneous measurement of conjugate points on all images involved. Basically, aerial triangulation is well suited for automation. In the preparation phase, for example, block points can be automatically selected at suitable locations. A distinct advantage is the possibility to simultaneously determine conjugate points that appear on more than two photographs by a procedure known as *multiple image matching*. In traditional aerial triangulation, an operator can normally only view two images at the same time. Conjugate points in multiple images are determined with the help of a point transfer device.

Yet another advantage of automatic aerial triangulation is the fact that there is virtually no economic limit to the number of measured points per photo. If implemented as a batch process, the computation time does not really matter a great deal. Increasing the number of points from the typical 9 point pattern per image to, say, 50 or even 100 points considerably increases the reliability and the accuracy.

9.4.4 *Automatic DEM Generation*

One of the major goals of digital photogrammetry has always been, and will continue to be, the automatic generation of DEMs. Considerable progress has been made during the last few years, but many problems persist, preventing general, robust solutions.

DEMs for the purpose of orthophoto production are quite successful since utmost accuracy is not required. The automatic DEM generation is not bound to DPWs. Actually, it is more economical to run the programs in a batch mode. The DPW comes in handy for the subsequent quality control—a very important aspect of DEM generation. Unfortunately, problems are not always detected and subsequently reported by DEM programs. For many applications, a visual inspection is inevitable. The manual checking procedure and the interactive editing reduce the savings gained in the automatic DEM generation.

9.4.5 *Digital Orthophoto Production*

Recently, an increasing interest in orthophotos can be observed. This may be attributed to various factors, but most importantly it is due to digital photogrammetry. Automatically generated DEMs are often accurate enough for orthophotos, especially when derived from small scale imagery. DPWs play a central role in the production of orthophotos. The image processing capabilities allow for radiometric adjustments and the superpositioning of additional information, such as vector data or text. Moreover, the visualization facilities make DPWs an important component in the production of orthophotos.

Variations of orthophoto production include draping radiometric information on perspective views of surfaces. Another interesting use of DPWs is for the generation of animated sequences. An example is the animated view of a "ghost" driver through a city (analogous to flight simulation, but with increased demand for true 3-D scenes).

9.5 *Analytical Plotters vs. DPWs*

During the discussion of the basic system functionality and the orientation procedure, the advantages and disadvantages of DPWs became apparent. Probably the most severe shortcoming of today's DPWs is the viewing and roaming quality, which is far inferior to that of analytical plotters. Consider moving the floating point mark in an oriented model on the analytical plotter. Regardless on how quickly you move, the model is always there, viewed at superb quality. In contrast, DPWs behave poorly when an attempt is made to move the cursor from one model boundary to another.

Several factors influence the viewing quality. For one, the monitor resolution sets a limit to the size of the field of view. Small fields of view reduce the capability to view a model stereoscopically. This makes image interpretation more difficult. Flickering, particularly noticeable on large monitors, is still a nuisance, despite 120 Hz refresh rates. A third factor is the reduction in illumination due to polarizing and alternate image displays. Finally, the ease and simplicity of optical image manipulation, such as rotation, cannot be matched on DPWs. Resampling is a time consuming process and may even reduce the image quality.

The advantages of DPWs outweigh these shortcomings by far, however. The benefits are well documented, see, e.g., *Kaiser* (1991); *Helava* (1991); *Gagnon et al.* (1995); *Madani* (1993); *Miller and Walker* (1993), and *Walker* (1996). *Madani* (1999) provides a recent account on DPW products from Z/I Imaging. The following are a few noteworthy factors.

- image processing capabilities are available at the operator's fingertips. Enlargements, reductions, contrast enhancements, and dodging do not require a photo lab any more—DPWs have a built-in photo lab.
- traditional photogrammetric equipment, such as point transfer devices and comparators, are no longer required—their functionality is assumed by DPWs. Digital photogrammetric workstations are much more universal than analytical plotters.
- the absence of any moving mechanical-optical parts makes DPWs more reliable and potentially more accurate since no calibration procedures are necessary. *Madani* (1993) reports about accuracy comparisons between DPWs and analytical plotters.
- DPWs offer more flexibility in viewing and measuring several images simultaneously. This is a great advantage in identifying and measuring control points and tie points.
- several persons can view a model stereoscopically. This is interesting for applications where design data is superimposed on a model. Viewing a model stereoscopically directly, that is, without optical system, is also considered an advantage by many operators.
- DPWs are more user friendly than analytical plotters. As more photogrammetric procedures will be automated, the operation of a DPW requires less specialized operators.

Among the many potentials of DPWs is the possibility to increase the user base. To illustrate this point, compare the skill level of an operator working on an analog stereoplotter, analytical plotter, and digital photogrammetry workstation. There is clearly a trend away from very special photogrammetric know-how to more generally available know-how on how to use a computer. The fact that stereo models can be viewed without optical mechanical devices and the possibility to embed photogrammetric processes in user friendly graphical user interfaces raise the chances that non-photogrammetrists can successfully use photogrammetric techniques.

Schenk and Toth (1997) suggest a way to increase the roaming capabilities of DPWs beyond the stereomodel. Roaming should be performed in the entire project and it should include vector data, DEMs, databases, and design data. Such a generalized roaming scheme would further increase the efficiency and user friendliness of DPWs.

One of the biggest advantages, however, lies in the potential to automate photogrammetric applications, such as aerial triangulation, DEM generation, and orthophoto production.

9.6 Summary

In this chapter we have discussed some of the fundamental functionality and issues of digital photogrammetric workstations. DPWs are interactive workstations. Their role in digital photogrammetry is even more prominent than that of analytical plotters in classical photogrammetry.

Softcopy workstations have come a long way from expensive systems built with many proprietary components to affordable systems comprising mostly off-the-shelf components. During the last few years, DPWs have migrated from military applications into the commercial market place, see, e.g., *Walker* (1996). In the not too distant future, DPWs will supersede analytical plotters.

Mainly for didactic reasons, I have emphasized the comparison of DPWs with analytical plotters throughout this chapter. However, this does not suggest that we attempt to emulate the analytical plotter by a DPW. In order to exploit the full potential of digital photogrammetry one must take a fresh look at traditional procedures and solve them considering the new technology and tools—an aspect forcefully brought forward in *Schenk and Toth* (1997).

In today's DPWs a lot of effort has been spent on developing basic system functionality, such as storing images, displaying and viewing stereo images, providing image processing capabilities, and performing the necessary orientation procedures. As more DPWs are used in operational environments, the demand for more automation will increase. To meet this demand, the harder problems must be addressed on a much broader basis, as indicated in *Schenk and Toth* (1992).

Problems

1. What are the major advantages of digital photogrammetric workstations when compared to analytical plotters?

2. Monitors are not very precise devices. The images displayed are distorted, for example. Explain why the measurements are not affected by imperfections of the display system.

3. Suppose a DPW uses a monitor with a resolution of 1024×1024 pixels. The images have a pixel size of $15 \mu m$. What is the size of the image patch you are looking at if displayed at maximum resolution? Compare this size with the field of view of analytical plotters at different magnifications.

4. What is the magnification in the previous example if the entire image is displayed on a 21 inch monitor? How does the monitor size influence the magnification?

5. In the previous examples, we looked at monoscopic displays. What will change in a stereoscopic display?

6. List some distinct differences between a DPW and a high-end graphical workstation.

7. Comment on the pros and cons of using a DPW vs. an analytical plotter to solve the task of digitizing contour lines directly.

References

Boniface, P. (1994). State-of-the-art in soft copy photogrammetry. In *Mapping and Remote Sensing Tools for the 21st Century*, American Society of Photogrammetry and Remote Sensing, 205-210.

Case, J. B. (1982). The digital Stereo Comparator/Compiler (DSCC). In *International Archives of Photogrammetry and Remote Sensing*, **24**(2), 23–29.

Cogan, L., D. Gugan, D. Hunter, D. Lutz, and S. Peny (1988). KERN DSP1–Digital stereophotogrammetric system. In *International Archives of Photogrammetry and Remote Sensing*, **27**(B2), 71–83.

Dowman, I. (1991). Design of digital photogrammetric workstations. In Ebner/Fritsch/Heipke (eds.) *Digital Photogrammetric Systems*, 28-38, Wichmann, Karlsruhe.

Gagnon, P. A., M. Boulianne, J.-P. Agnard, C. Nolette, and J. Coulomb (1995). Present status of the DVP system. *Geomatica*, **49**(4), 479-488.

Graham, L., K. Ellison, and S. Riddell (1997). The architecture of a softcopy photogrammetry system. *Photogrammetric Engineering and Remote Sensing*, **63**(8), 1013–1020.

Grün, A. (1996). Digital photogrammetric workstations: revisited. In *International Archives of Photogrammetry and Remote Sensing*, **31**(B2), 127-144.

Gülch, E. (1994). Fundamentals of softcopy photogrammetric workstations. In *Mapping and Remote Sensing Tools for the 21st Century*, American Society of Photogrammetry and Remote Sensing, 193-204.

Heipke, C. (1995). State-of-the-Art of Digital Photogrammetric Workstations for Topographic Applications. *Photogrammetric Engineering and Remote Sensing*, **61**(1), 49–56.

Heipke, C. (1996). *Digitale photogrammetrische Arbeitsstation.* DGK-C 450, 111 pages.

Helava, U. (1988). On system concepts for digital automation. In *International Archives of Photogrammetry and Remote Sensing*, **27**(B2), 171–190.

Helava, U. (1991). State of the art in digital photogrammetric workstations. *Photogrammetric Journal of Finland*, **12**(2), 65-76.

Kaiser, R. (1991). ImageStation: Intergraph's digital photogrammetric workstation. In Ebner/Fritsch/Heipke (eds.) *Digital Photogrammetric Systems*, Wichmann, Karlsruhe, 188-197.

Krupnik, A. and T. Schenk (1994). Predicting the reliability of matched points. In *Proc. ACSM/ASPRS Annual Convention*, **1**, 337–343.

Madani, M. (1993). How a digital photogrammetric workstation is compared to an analytical plotter. In *Proc. of the International Society for Optical Engineering (SPIE)*, **1943**, 266-272.

Madani, M. (1999). A complete digital photogrammetric system from Z/I Imaging. *International Archives of Photogrammetry and Remote Sensing*, **32**(5W11), 243–251.

Miller, S. B. and K. DeVenecia (1992). Softcopy photogrammetric workstations. *Photogrammetric Engineering and Remote Sensing*, **58**(1), 77–84.

Miller, S. B. and S. Walker (1993). Further Developments of Leica Digital Photogrammetric Systems by Helava. In *Proc. ACSM/ASPRS Annual Convention*, **3**, 256-263.

Sarjakoski, T. (1981). Concept of a completely digital stereoplotter. *The Photogrammetric Journal of Finland*, **8**(2), 95–100.

Schenk, T. and C. Toth (1992). Conceptual Issues of Softcopy Photogrammetric Workstations. *Photogrammetric Engineering and Remote Sensing*, **58**(1), 101–110.

Schenk, T. and C. Toth (1997). Conceptual framework of a generalized roaming scheme. *Photogrammetric Engineering and Remote Sensing*, **63**(8), 959–964.

Walker, S. (1996). Digital Photogrammetric Workstations 1992-1996, In *International Archives of Photogrammetry and Remote Sensing*, **31**(B2), 384-395.

Chapter 10

Image Matching Fundamentals

One of the most fundamental process in photogrammetry is to identify and to measure conjugate points in two or more overlapping photographs. Stereo photogrammetry relies entirely on conjugate points. In analog and analytical photogrammetry the identification of conjugate points is performed by a human operator; in digital photogrammetry one attempts to solve the problem automatically—a process known as *image matching*.

Considering the importance of image matching in photogrammetry, the subject is treated in two chapters. This chapter begins with an introduction, followed by typical problems and common solutions. Following the spirit of this book, I emphasize principles and general solutions, rather than providing a detailed description of the many different algorithms that have been developed to solve the matching problem. The chapter concludes with area-based matching methods. Additional matching methods are treated in Chapter 11.

The following discussions have the general image matching case in mind. However, most examples used refer to the standard aerial case with central projection.

10.1 Introduction

10.1.1 Historical Remarks

Image matching, or finding conjugate points automatically, has quite a long history. First experiments started in the fifties, most notably by Hobrough (*Hobrough* , 1959). The solution was analog in nature; correlators which were realized in hardware compared (correlated) the gray levels of two images. Wild Heerbrugg, then the leading manufacturer of photogrammetric equipment, introduced a correlator at the occasion of the 1968 ISPRS congress. This innovative instrument did not enjoy much success, however. For one, the photogrammetric community was rather skeptical toward this new concept. Furthermore, the instrument was not flexible and reliable enough.

From the early seventies to the mid-eighties, research related to image matching focused on digital correlation techniques. *Kreiling* (1976) was among the early pioneers to apply digital correlation for the purpose of generating DEMs and digital orthophotos. Efforts continued to add correlation techniques to instruments, see, e.g. *Helava* (1978) and *Hobrough* (1978). Despite considerable effort, no general solution was found. Researchers were puzzled because humans find conjugate points easily, without conscious effort. We know now that the human stereo system does not employ the correlation of gray levels at all as pointed out by *Grimson* (1981) and *Horn* (1983), for example. This is just another example of underestimating the complexity of the visual tasks that humans solve with great ease.

10.1.2 Terminology, Working Definitions

As might be suspected, terminology is not standardized at all; at times it is actually rather confusing. Image matching is sometimes called *automatic stereo matching* or simply *correlation.* In computer vision the term *correspondence problem* is preferred.

Because of the lack of a unified terminology, the following working definitions are introduced in this book.

conjugate entity: is a more general term than conjugate point. Conjugate entities are the images of object space features, including points, lines and areas.

matching entity: is the primitive, which is compared with primitives in other images to find conjugate entities. Primitives include gray levels, extracted features, and symbolic descriptions.

similarity measure: is a quantitative measure of how well matching entities correspond to each other. Generally, the degree of similarity is measured by a cost function. In its simplest form, this may be the cross-correlation coefficient or the standard deviation in least-squares matching.

matching method: performs the similarity measure of matching entities. The methods are usually named after the matching entity, for example area-based matching, feature-based matching and relational (symbolic) matching.

Table 10.1: Relationship between matching methods and matching entities.

Matching Method	*Similarity Measure*	*Matching Entities*
area-based	correlation, least-squares	gray levels
feature-based	cost function	edges, regions
symbolic	cost function	symbolic description

matching strategy: refers to the concept or overall scheme of the solution of the image matching problem. It encompasses the analysis of the matching environment, the selection of the matching method, and quality control of the match.

Table 10.1 shows how these terms are related. The first column lists the three best known matching methods. *Area-based matching* is associated with matching gray levels. That is, the gray level distribution of small areas of two images, called *image patches*, is compared and the similarity is measured by correlation or least-squares techniques. Area-based matching using correlation is often simply called *correlation*. Likewise, area-based matching with a least-squares approach for measuring similarity is known as *least-squares matching (LSM)*. Section 10.4 deals with area-based matching.

Area-based matching is quite popular in photogrammetry—presumably because of historical reasons. *Feature-based matching* is predominantly used in computer vision. Here, edges or other features derived from the original images are compared to determine conjugate features. The similarity, for example the shape, sign and strength of edges, is measured by a cost function. Chapter 11 describes this approach in more detail.

The third method, *symbolic matching*, refers to methods which compare symbolic descriptions of images and measure the similarity by a cost function. The symbolic descriptions may refer to gray levels or to derived features. They can be implemented as graphs, trees, semantic nets—just to mention a few possibilities. In contrast to the other methods, symbolic matching is not strictly based on geometric similarity properties. Instead of using the shape or location as a similarity criterion, it compares topological properties. Chapter 11 briefly describes *relational matching* as an example of symbolic matching.

Although the majority of matching problems involves two images (typically a stereopair), it is worth pointing out that matching several images simultaneously is important in aerial triangulation or in matching video sequences. Usually, the term *multiple image matching* is used when more than two images are involved.

10.1.3 Problem Statement

The problem of image matching can be stated as follows:

1. Select a matching entity in one image;
2. Find its conjugate (corresponding) entity in the other image;
3. Compute the 3-D location of the matched entity in object space;
4. Assess the quality of the match.

Obviously, the second step is the most difficult to solve. Although the other steps appear trivial at first sight, there are still interesting issues involved. Take a typical stereopair as an example. In which of the two images should the matching entity be selected? What entity should be selected and how is it determined in the first place? Now imagine we have successfully matched conjugate edges. How should we compute their location in object space if no conjugate points[1] are available? The matching strategy must address these issues.

10.2 Fundamental Problems of Image Matching

The purpose of this section is to elaborate on some of the problems that must be addressed and, hopefully, solved by the different matching methods.

10.2.1 Search Space, Uniqueness of Matching Entity

Suppose we work with a stereopair, resolution $32K \times 32K$ (pixel size $\approx 7\,\mu$m). According to step 1 of the problem statement let us pick pixel P in one image at location i_p, j_p, gray level g_p. If we search for the conjugate pixel in the other image in a brute force manner then the entire overlapping area must be checked for occurrences of gray level g_p. This would lead to $0.6 \times 32K \times 32K \approx 6.4 \cdot 10^8$ operations. If every pixel in the reference image is matched, the total number of operations would be on the order of 10^{17}.

[1]Having conjugate edges does not imply that the individual edge pixels are conjugate. See Sec. 14.5 for more details.

That is, the computational complexity is $\mathcal{O}(n^4)$, with n^2 the resolution of the image. Apart from the untractable computational problem there is the additional problem of ambiguity. To demonstrate this point let us assume that every gray level occurs with the same frequency (e.g. after histogram equalization). Thus we find $(0.6 \times 32K \times 32K)/256 = 2.5 \cdot 10^6$ matches. Add a noise level of ± 5 gray levels and we end up with having 10^7 potential solutions of which only one is correct (assuming opaque surfaces). This example vividly illustrates two problems:

combinatorial explosion occurs if the similarity measure between the matching entities is computed over the entire image (model).

ambiguity occurs if the matching entity is not characteristic (unique) enough.

A solution to the first problem is to restrict the *search space* for finding conjugate entities (see Section 10.3.1). The second problem must be tackled by selecting more unique matching entities (see Section 10.3.2).

10.2.2 Approximations, Constraints and Assumptions

A problem is said to be well-posed if (i) a solution exists, (ii) the solution is unique, and (iii) the solution depends continuously on the initial data. Image matching is an ill-posed problem because it violates several if not all of the conditions a well-posed problem must meet. For example, no solution (occlusion) or no unique solution (ambiguity) may exist. Image matching belongs to the class of inverse problems which are known for their ill-posed nature. A crucial question is how image matching is made well-posed. The approaches range from utter ignorance to sophisticated regularization methods.

A straightforward way to make image matching well-posed is to restrict the space of possible solutions, for example by setting bounds. The underlying function of the similarity measure is non-monotonic and rather flat around peaks and troughs. As illustrated in Fig. 10.1 we must begin the image matching process rather close to the true solution; otherwise we may end up with a secondary minimum or require too many iterations to arrive at the global minimum. Consequently, we need very good approximations to assure convergence. The convergence radius is also called *pull-in range*. Area-based methods must begin the search process as close as a few pixels from the true conjugate location. Feature-based matching and symbolic matching have a much larger convergence radius. Approximations and search space are closely related; the better the approximations the smaller the search space.

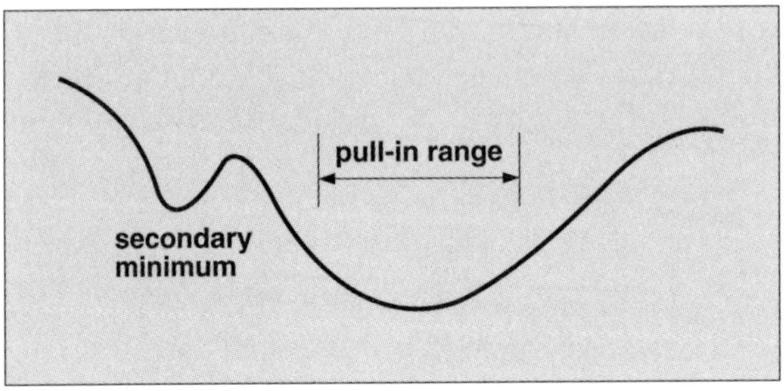

Figure 10.1: Illustration of the ill-posed nature of image matching. The optimization function is non-monotonic and the convergence radius (pull-in range) is rather small.

A bold step toward making image matching well-posed is to introduce constraints. We recall from adjustment theory that additional information about the parameters, e.g. in the form of constraints, may dramatically improve the solution. Such is the case in image matching. Similarity measures which are based on geometric properties of the matching entities can take advantage of geometric constraints. An example of a strong geometric constraint is the fact that the perspective centers, the conjugate points, and the corresponding object point all must lie in one plane—the epipolar plane.

Constraining the solution works well, but the dilemma is to find just the right balance between a general solution and a very restricted one. Using a least-squares approach for the similarity measure offers an elegant possibility to include various constraints and to treat them simultaneously with the matching.

Constraints can be viewed as additional information about the parameters. In that regard additional information is "knowledge" about the matching process. Knowledge is not restricted to geometric conditions, rather it encompasses all sorts of information useful to assure a plausible solution.

Assumptions about the matching problem are another way of restricting the solution space. A good example is aerial stereopairs where we safely assume that the forward overlap is about 60%. This is useful because it permits us to predict the conjugate location fairly well. As long as we work with aerial images there is nothing wrong with this kind of assumption. It is important to know, though, what kind of assumptions are built into a specific matching program to prevent it from being used in situations that violate these assumptions.

In summary, the problem is twofold: (i) to obtain good enough approximations and (ii) to strike a fine balance between assumptions made and constraints imposed on the solution, on the one hand, and to maintain generality on the other hand.

10.2.3 Geometric Distortions of Matching Entities

The compelling conclusion from Section 10.2.1 is to make the matching entities as unique as possible so that only one solution (match) exists. Instead of comparing the gray level of one pixel in area-based matching, several pixels must be considered in the similarity measure. Thus, the gray levels of an image patch, size $n \times m$ pixels, are compared with those of a patch in the other image. Suppose for a moment that both image patches are centered on their true conjugate locations. The similarity measure yields a maximum if the gray levels of every pixel compared are identical. This is an ideal situation which will never occur in reality. Noise, changing illumination and reflection properties between two consecutive images cause differences in the gray levels. Another source that causes two image patches to appear different, even in their true conjugate locations, are geometric distortions due to the central projection and due to relief. This section focuses on various geometric distortions.

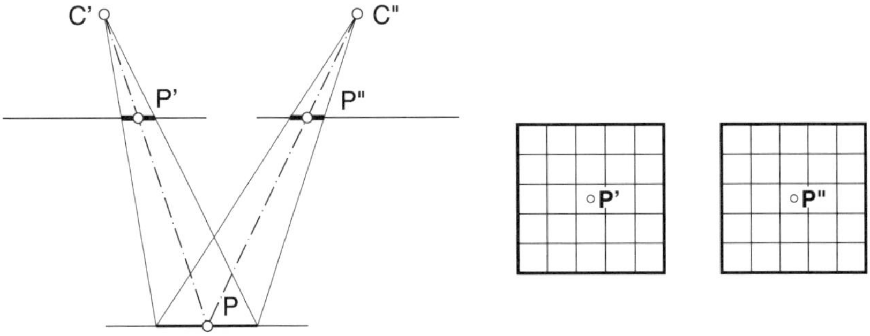

(a) (b)

Figure 10.2: A flat surface patch with point P in its center is imaged in two image patches, whose centers are in P' and P'' (a). As indicated in (b), the two image patches are identical in this ideal case.

We begin with Fig. 10.2 which depicts a stereopair and two image patches, each with a size of 5×5 pixels, centered at their conjugate locations. The ideal case refers to the situation where every pixel pair corresponds to the same spot in object space. That is, every pixel in the image patches must be conjugate. The geometric condition for this ideal case is that both images and the surface patch must be parallel to the base. Any departure from this condition causes geometric distortions. Let us examine a few cases.

Geometric Distortion Due to Orientation Parameters

Fig. 10.2 depicts the situation of a true vertical stereopair with a horizontal air base and a flat, horizontal surface. We now consider a real stereopair and a flat surface patch with its surface normal parallel to the camera axis of the left image. The impact of scale and rotations of the right image is of interest. This configuration corresponds to a dependent relative orientation.

It should be stressed that the cases examined here refer to unoriented stereopairs. If the orientation is known then the images can be transformed and resampled into a normalized position (see Chapter 12) analogous to the ideal situation of Fig. 10.2. Since in photogrammetry we usually begin with unoriented stereopairs it is important to be aware of the geometric distortions caused by the orientation parameters.

a) Scale Difference Between the two Images: Fig. 10.3(a) shows a stereopair with different scale caused by different flying heights. The projection of two conjugate image patches (Fig. 10.3(b)) onto the flat surface reveals that the pixels within the image patches are now no longer conjugate because they refer to different locations in object space. To demonstrate the case, the left-upper pixel of both patches is dark shaded. The further we move from the center, the larger the discrepancies. Consequently, the similarity measure will be affected.

b) Different Rotation Angles Between the two Images: Fig. 10.4 demonstrates the effect of differences in the rotation angles. Since we are considering the case of dependent relative orientation, the rotation angles of the right image are with respect to the fixed position of the left image. Fig. 10.4(a) depicts the effect of a rotation of the right image about the x-axis (ω). The right image patch appears as a rectangle in its projection on the flat surface. The effect of a φ rotation (about the y-axis) is shown in Fig. 10.4(b). Finally, the κ rotation (about the z-axis) has an effect as illustrated in Fig. 10.4(c).

It is evident from Fig. 10.4 that a pixel to pixel comparison is wrong because the same pixel in the two image patches refers to a different position in the object space. Compare, for example, where the pixels of the upper-left corners appear in object space.

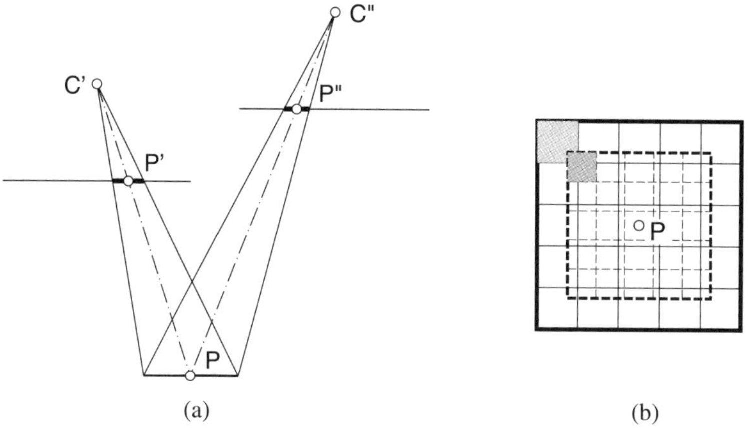

Figure 10.3: Stereopair with different image scale (a). In (b) the projection of two conjugate image patches onto a flat surface is shown. The dark shaded pixels of the left-upper corner of both image patches are assumed to be conjugate in area-based matching. Their different position on the surface illustrates the geometric distortion introduced by a scale difference.

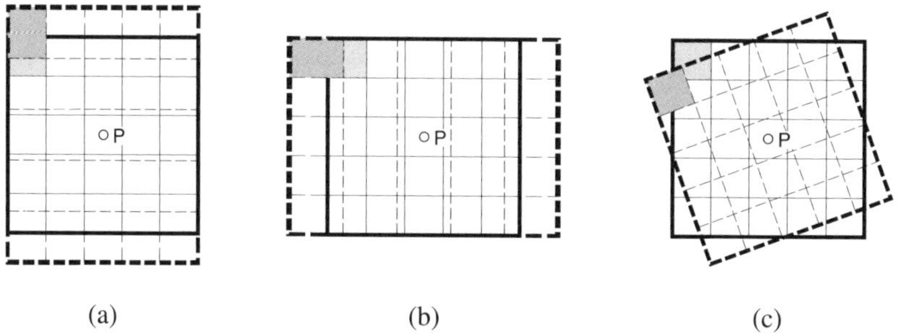

Figure 10.4: Effect of rotation differences between the left and right image of a stereopair. In (a) a rotation about the x-axis (flight direction) is shown; (b) and (c) demonstrate the effects of rotations about the y-axis and z-axis, respectively. In reality, all three rotations are combined.

Effect of Tilted Surface on Geometric Distortions

So far we have assumed that the flat surface patch in object space is parallel to the left image. This is why the left image patch always appeared as a perfect square in the preceding figures. Now we assume a true vertical stereopair but a tilted, flat surface. In Fig. 10.5 the surface is rotated about the *air base*[2]. This is still an ideal situation because every corresponding pixel is conjugate—the geometric distortion is identical in both images.

If the surface is tilted about an axis perpendicular to the air base, as shown in Fig. 10.5(b), the effect is very different because the surface appears as a small rectangle in the right image. This situation is referred to as *foreshortening*. If the surface is tilted further, then a critical angle is reached where it is no longer visible in the right image; we have the case of an occlusion.

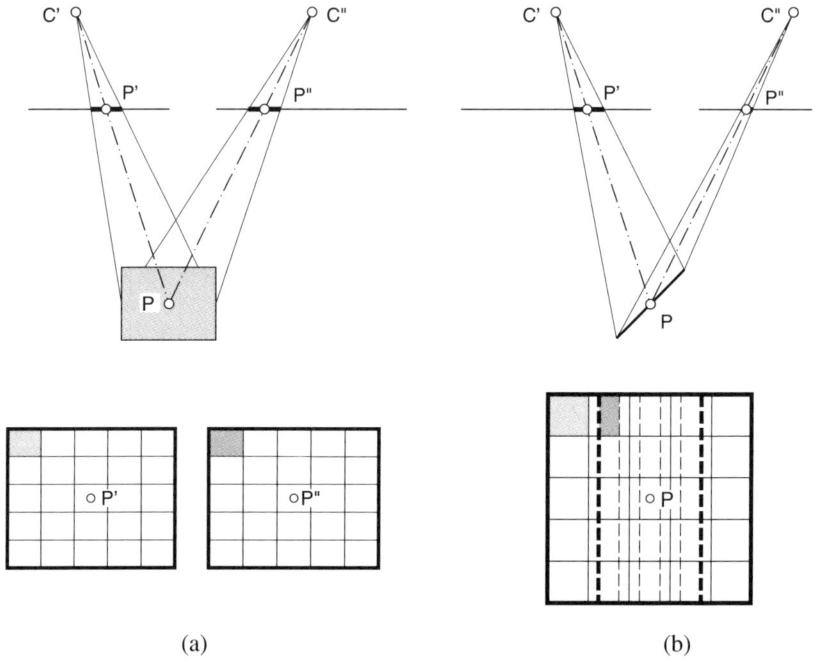

Figure 10.5: Effect of a tilted surface on the similarity measure in area-based matching. In (a) the surface is tilted about the air base (C', C''). There is no ill effect because the pixels still refer to the same ground pixel. This is quite different if the surface is tilted perpendicular to the air base (b), a situation referred to as foreshortening. Here, "conjugate" pixels point to different areas in object space and the similarity measure becomes problematic.

A closer examination of Fig. 10.5 reveals another consequence of a tilted

[2]The air base is defined by the two perspective centers C', C''

surface, namely the decreased resolution in the direction perpendicular to the rotation axis. In the second case, the resolution of the foreshortened image is smaller in x-direction.

Effect of Relief on Geometric Distortions

Unlike the previous cases, we now assume a real surface patch with elevation differences. As before, the image patches at their conjugate locations are projected onto the "mountainous" surface patch. As Fig. 10.6 demonstrates, the individual pixels are not conjugate even in the case of a perfectly vertical stereopair.

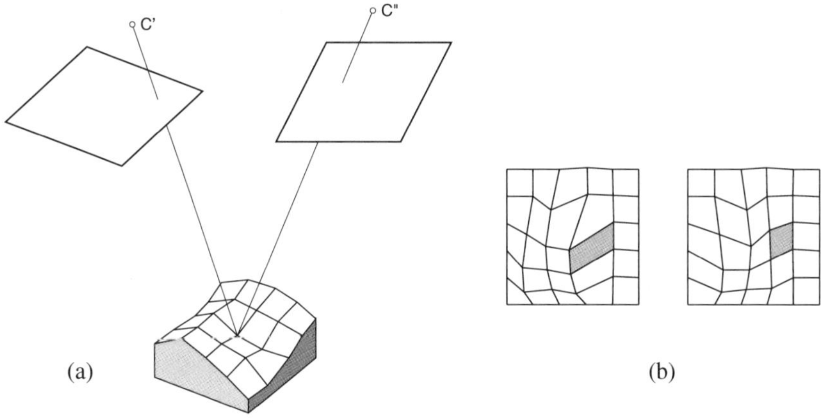

Figure 10.6: Effect of relief on the similarity measure in area-based matching. Pixels that are believed to be conjugate, e.g., the shaded pixels in (b), cover different areas in object space. Consequently, they are not conjugate.

The previous discussions studied the effect of geometric distortions on image patches used as entities in area-based matching. Are feature-based methods unaffected by geometric distortions? In general, no. However, they are much less sensitive to most geometric distortions.

As an example of feature-based matching, let us focus on edge matching. Edges are derived quantities from the original image. Let E be a square-shaped edge operator. Its response at a certain image location depends entirely on the gray levels within the spatial extent of E. Suppose the operator is positioned at a conjugate location in both images. The response will be different, even if the illumination and reflectance in the two images remained constant. The reason for the different response is the geometric distortions. As in area-based matching, the pixels within the operator mask refer to different locations in the object space. As a consequence, the edges will be displaced. Moreover, their shape will be distorted—a fact that must be taken into account when designing the cost function of feature-based matching.

10.3 Solutions to the Fundamental Problems

10.3.1 Search Space and Approximations

In Sec. 10.2.1 we considered the entire image as the search space for finding conjugate pixels. This blind search approach generates an insurmountable computational problem. Moreover, the search space is far too large because the ill-posed nature of image matching requires very good approximations. This is particularly true for area-based methods. Consequently, the search space must be confined to a rather small area which we earlier called the pull-in range. This section focuses on methods to reduce the search space.

As before, the following discussions assume a standard aerial stereopair. Also, a distinction is made between oriented and unoriented stereopairs. We begin with the easier case of an oriented stereopair and evaluate several geometric conditions for the location of conjugate entities.

Epipolar Lines

Epipolar lines are a powerful constraint for conjugate entities. Fig. 10.7 illustrates the concept. The epipolar plane is defined by the two projection centers C', C'', and the object point P. The epipolar lines e', e'' are the intersections of the epipolar plane with the image planes. The epipolar plane contains conjugate points which must lie on the corresponding epipolar lines.

Epipolar lines reduce the search space dramatically. If a matching entity in one image is chosen then the epipolar line in the other image can be computed, provided the stereopair is oriented.

Usually, epipolar lines are not parallel to the x-axis of the photo-coordinate system. Quite often it is desirable to transform the images in such a way that epipolar lines become parallel to the rows of the digital images. Such stereopairs are called *normalized images* in this book. Chapter 12 provides a detailed description of the transformation of original images into their normalized position.

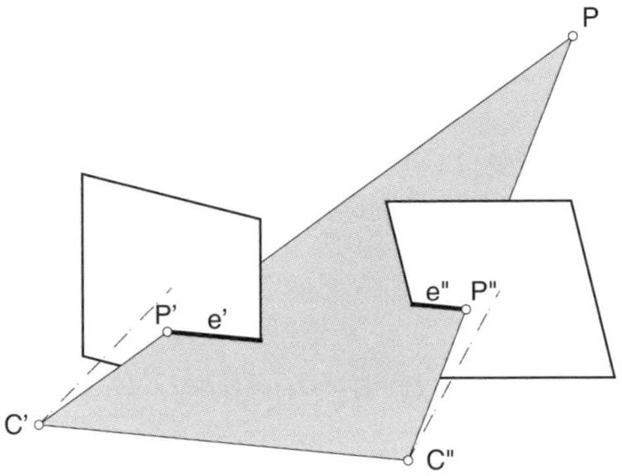

Figure 10.7: Epipolar geometry of a stereopair. The epipolar plane, defined by the base $C'C''$ and a point P in object space intersect the images in the epipolar lines e' and e''.

For the following discussion, normalized images are assumed. Suppose we pick the matching entity P' in one image. Then the search space is restricted to the same row in the other image. Do we need to search along the entire epipolar line, or can we perhaps further reduce the linear search space? We can certainly do better by predicting the conjugate location. Fig. 10.8 shows ray $C'P'$ where P is the feature in object space with an estimated elevation Z_P. This ray intersects the true surface at point S. In an oriented model some elevations exist (from the orientation process). Thus, a reasonable approximation of Z_p is obtained, for example, by taking the average of all known elevations. With Z_P and the orientation parameters, the conjugate feature P'' is computed as the image of the estimated object feature P.

Next, the uncertainty Δz of the predicted elevation must be estimated. This may turn out to be more difficult than predicting Z_p. Usually, some a priori knowledge about the area is available, e.g. mountainous or flat. Also, the maximum and minimum values of the existing orientation points may serve as a basis to estimate Δz.

For near vertical stereopairs, the following equations can be used to determine the approximate location x''_p of the conjugate feature.

$$x''_p = x'_p - px_P \tag{10.1}$$

$$px_P = \frac{B}{H_D - Z_P} f \tag{10.2}$$

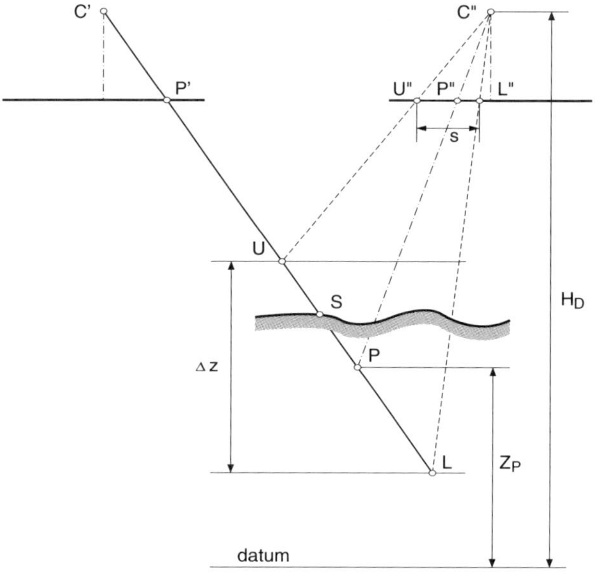

Figure 10.8: Estimating the matching location and the search space on the epipolar line. Point P is an estimated elevation and Δz expresses the uncertainty of this estimation. This defines a range U, L that is projected to the right image where it defines the search space s.

with px_P the x-parallax of point P, B the air base, f the focal length, and H_D the flying height above datum.

By substituting $b_0 H_D / f$ for the air base B in the above equation, where b_0 is the *image base*, the following expression for the location of the conjugate entity is obtained

$$\boxed{x''_P = x'_P - b_0 \frac{H_D}{H_D - Z_P}} \tag{10.3}$$

Now we determine the search interval s and again assume true vertical images. Referring to Fig. 10.8 and Eq. 10.3 we have

$$s = x''_U - x''_L \tag{10.4}$$

$$x''_U = x'_U - b_0 \frac{H_D}{H_D - Z_U} \tag{10.5}$$

$$x''_L = x'_L - b_0 \frac{H_D}{H_D - Z_L} \tag{10.6}$$

Recognizing that $x'_U = x'_L$ we have for s

$$s = b_0 H_D \frac{\Delta z}{(H_D - Z_L)(H_D - Z_U)} \tag{10.7}$$

Approximating the denominator in the above equation by $(H_D - Z_P)^2$ (neglecting $(\Delta z/2)^2$), we have for the search interval s

$$\boxed{s \approx b_0 \frac{H_D \Delta z}{(H_D - Z_P)^2}} \tag{10.8}$$

With Eq. 10.8, the size of the search space, s, is expressed as a function of the variables Z_P (estimated elevation of point P and Δz (uncertainty of the estimate). These two variables change as a function of location. Consequently, for every new match to be performed, the search space changes.

The following procedure describes the essential steps for implementing a matching scheme along epipolar lines:

1. Select matching entity in one image, say P';
2. Estimate the elevation, Z_P, of the matching entity and its uncertainty range Δz;
3. Compute approximate location, P'', using Eq. 10.3;
4. Compute search interval, s, with Eq. 10.8;
5. Perform matching within search interval; selected matching entity P' remains constant;
6. Analyze the similarity measures obtained in step 5 for determining conjugate location P''.

Example 10.1:

Problem: *Given a normalized digital stereopair, resolution* $4K \times 4K$ *(pixel size* $\approx 55\ \mu m$*). The flying height above datum is* $H_D = 700\ m$*, the air base is* $B = 360\ m$ *and the focal length is* $152\ mm$*. Predict the search space for finding the conjugate pixel in the right image when the pixel selected in the left image has the address* $(2876, 3244)$*. The average elevations in the model are about* $100\ m$*.*

Solution: *Not knowing any better we accept the average elevation as the estimate for* $Z_P = 100\ m$ *and assign an uncertainty of* $\pm 50\ m$*, i.e.* $\Delta z = 100m$*. Let us first convert the focal length from metric units to pixels:* $152/0.055 = 2764$ *pixels. With this we obtain for the parallax* px_P *of pixel P the value* $px_P = 360/(700 - 100) \cdot 2764 = 1658$ *pixels. Now, Eq. 10.3 yields* $x''_P = 3244 - 1658 = 1586$*, the center of the search window. To compute the search interval we use either Eq. 10.8 or Eqs. 10.5 and 10.6. In the latter case we obtain for* $x''_U = 1531$ *pixels and for* $x''_L = 1809$*. Thus,* $s = 278$ *pixels. The search space is now defined by (2876,1447—1725).*

Vertical Line Locus

Another useful geometric constraint for the search space is the so-called *vertical line locus*, see, e.g. *Gyer* (1981), *Bethel* (1986). Fig. 10.9 illustrates the concept.

Let P be the estimate of the position of a feature in object space and let Δz be the uncertainty range of the initial guess. This determines the lower (L) and upper (U) bounds of P which are projected back to the images. The distances L', U' and L'', U'' are labeled with v' and v'', respectively. The conjugate entities in the two images are confined to the two lines v' and v'' which are found by intersecting the triangle, defined by the projection center and the lower and upper bounds L and U, with the image planes. Because the triangles are in a plane which also contains the plumb line from the perspective centers, the projected vertical lines in the images are radial with respect to the nadir. Moreover, L' and L'' as well as U' and U'' are on epipolar lines.

Note that v' and v'' have different lengths. In fact, they may collapse to a point if the extension of the vertical line U, L passes through a projection center. Fig. 10.10 shows some examples of vertical line locus pairs in the stereopair.

Several approaches exist to implement the concept of the vertical line locus schema. Fig. 10.11 on page 248 illustrates one possibility. Here, the matching begins at the lower bounds where the template and the matching window are centered at L' and L'', respectively. The matching procedure

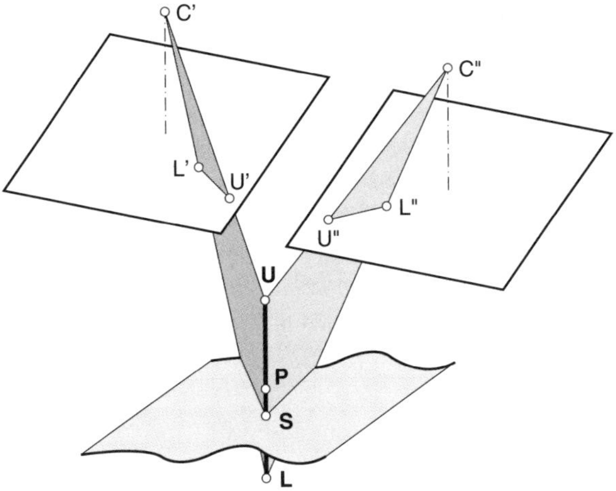

Figure 10.9: Concept of vertical line locus. The search space is constrained to the projection of a vertical line into both images. P is the estimated elevation, and S is the true (but unknown) elevation. The search is conducted within the uncertainty range U, L.

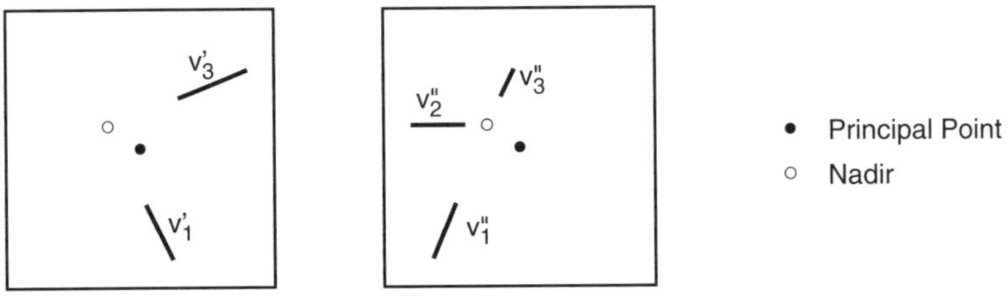

Figure 10.10: Examples of vertical line loci pairs v' and v''. These lines are radial with respect to the nadir. The vertical line through the projection centers degenerate to the principal point.

repeats for points between L and U. Of all the positions matched in this fashion, only the one at S is correct, indicated by the maximum response of the similarity measure.

The vertical line locus method can also be combined with the epipolar line method. *Krupnik and Schenk* (1994) propose a combined approach for

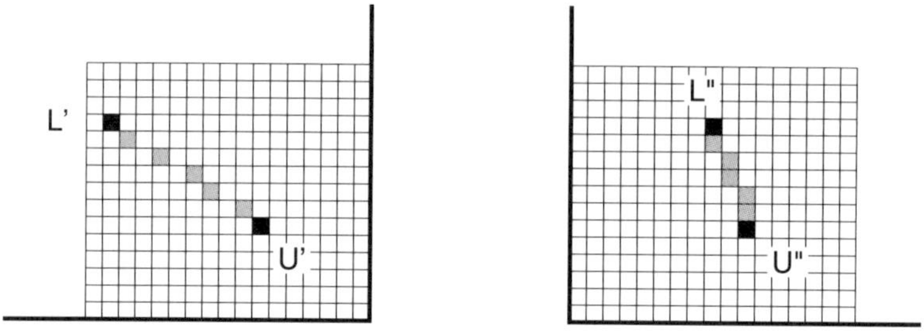

Figure 10.11: Concept of matching along the vertical line loci v' and v''.

increasing the reliability of matched points. Fig. 10.12 shows the concept. Suppose the matching begins at L. Before moving to the next epipolar line several matches are performed at the epipolar line through L' and L''. In Fig. 10.12 L'' is kept constant and matching is performed by moving L'. The true match is found at E'_L. The true conjugate position is always to the right of L' because the true surface is above L in object space. Now we move on to the next epipolar line and repeat the same procedure. When we reach U, the conjugate entity to U'' is in E'_U—always to the left of U' because the true surface is lower (see Fig. 10.12). The line of the true matches E'_i intersects v' at the location where U, L pierces the surface—the point we want to find. The analysis of the similarity measure for all matching positions should reflect this situation by a maximum response.

The following procedure may be followed when implementing the concept of the vertical line locus approach.

1. Select the X-Y-position of a point P in object space and estimate the elevation Z_P and its lower and upper bounds Z_L, Z_R.

2. Begin the matching process by projecting X_P, Y_P, Z_L back to both images. Perform a similarity measure at the image positions x'_L, y'_L and x''_L, x''_L.

3. Move along the vertical line in suitable steps $\Delta z = (Z_U - Z_L)/n$. Project the object points $X_P, Y_P, Z_L + \Delta z$ back to the images and repeat the similarity measure.

4. Repeat step 3 until Z_U is reached.

5. Analyze the similarity measures obtained in every step and determine the maximum/minimum for the conjugate position.

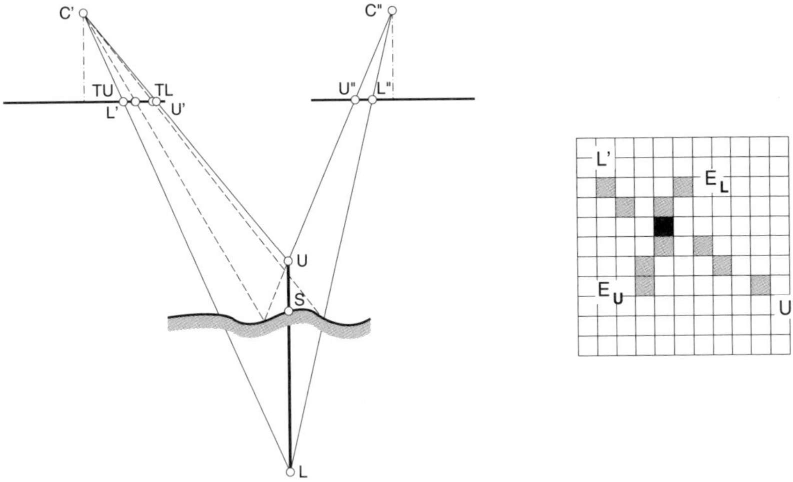

Figure 10.12: Combining vertical line locus and epipolar line method for a more robust matching procedure. On every pixel along the vertical line locii, additional matches are performed along the epipolar line. See text for detailed explanation.

The vertical line locus method is useful for applications where conjugate entities at certain positions in object space are required. One such application is the automatic generation of DEMs. Here, the $X - Y$ positions are given by the grid posts.

In contrast to the epipolar line method, the matching position in both images is changed (along the images of the vertical line). This may be a disadvantage for certain implementations because for every matching position, both image patches are new.

It is worth pointing out that it is advantageous to combine the vertical line locus method with the epipolar line method. In this scheme, additional matches along the epipolar lines complement the matches along the vertical lines. This will render more data for analyzing the cost function, generally resulting in a more reliable match.

Hierarchical Approach

Even though we have made great progress in reducing the search space from the overlapping area to a few millimeters along an epipolar line or vertical locus line we are still far from having sufficiently close approximations for the matching process to start. Another way to reduce the pixel search space is to increase the pixel size. Take example 10.1 with 6 mm search interval along the epipolar line. With the original resolution of $32K \times 32K$ this amounts to a pixel search space of $\approx$ 800 pixels. With a pixel size of 1 mm the search space becomes 6 pixels—a good approximation to begin the matching procedure. What are the options?

Obviously, the solution to drastically reduce the search space and at the same time improve the approximations (in pixels), is to begin with a coarse resolution and then to project the results to ever finer resolution images until we are at the original resolution. This is best accomplished by generating image pyramids for the stereopair. Referring again to our example with images at a resolution of $32K \times 32K$, an image pyramid with intermediate levels of $16K, 8K, 4K, \ldots$ is generated, for example by smoothing the original image with a Gaussian filter of appropriate σ. How many levels should we consider? In other words, on what resolution level can we begin? This depends on the matching method[3] and the uncertainty of the estimated elevations.

Example 10.2:

> *The estimated position of the conjugate entity must be within 6 pixels. The uncertainty of estimating the elevation is* ± 25 *m, photo scale 1 : 6,000, wide-angle camera. With the parallax equation (Eq. 10.8) we obtain a 4.7 mm search interval. Since the tolerance is 6 pixels, a pixel size of* 0.8 *mm is appropriate. Image pyramids are usually generated as multiples of two. Thus the matching procedure should begin at a resolution of* 256×256.

Now we have a clue as to how to begin, but how do we proceed further? Let us assume that the matching on the coarse level renders an accuracy of 1 to 2 pixels. If we project this result from the coarsest level to the finest one, say from 256 to $32K$, we will be way off since the magnification is 2^7. One pixel accuracy on the coarse level amounts to an accuracy of 128 pixels on the finest resolution. We must proceed through all levels or every other level of the image pyramid to assure convergence. On every level, the matching process must be repeated to improve the accuracy. Tracking positions through the image pyramid is essential in the hierarchical approach, also called *coarse-to-fine strategy*. Fig. 10.13 illustrates the concept.

In conclusion, the hierarchical approach serves the purpose of reducing the search space. To be precise, the dimension of the search space in metric units (e.g. expressed by photo-coordinates) is a function of the estimated elevation of point P and its uncertainty Δz. This does not yet

[3] Area-based matching requires much better approximations than feature-based and symbolic matching.

have anything to do with the hierarchical approach. The implementation of the hierarchical approach in the form of image pyramids is related to the pixel space. That is, metric quantities, such as the dimension of the search space, are expressed in units of pixels. Selecting an ever increasing pixel size requires fewer and fewer pixels to cover the same search space.

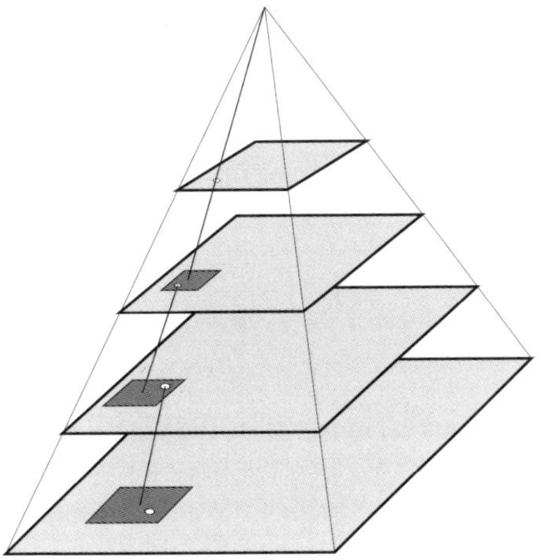

Figure 10.13: Tracking matched entities through the image pyramid. The matching process must be repeated on every level to refine the conjugate position. Matched points are used on the lower level as centers of search windows.

10.3.2 Uniqueness of Matching Entity

In the previous section we discussed some simple but powerful methods to reduce the search space. Considering the entire overlapping area of an image as a potential search space, the epipolar line or vertical line locus constraints result in an impressive reduction, e.g. from $0.6 \times 32K \times 32K$ pixels to a few hundred pixels. Nevertheless, we are still faced with the problem of multiple matches. To prevent this, the matching entities must be unique—at least within the search space.

If gray levels are used as matching entities, then the uniqueness of the image patches increases with their size. Unfortunately, the geometric distortion problems discussed earlier, increase as well. The crux is to find a good compromise between these conflicting factors. Obviously, it is desirable to measure the uniqueness so that we do not increase the image patch size once it is unique enough. Several methods may be used, but we mention just three.

variance of the image function provides a measure of how different the gray levels are. A small variance indicates a rather flat (homogeneous) image while a large variance points to gray level distribution over a larger interval.

autocorrelation of the image function provides a "self-comparison" measure of the image patch. A high autocorrelation factor indicates repetitive patterns within the sub-image—something which reduces the uniqueness. In a totally random image neighboring pixels are not correlated which would be the most unique function, but it does not exist in reality—real images are more deterministic than random.

entropy measures the randomness of the image function. A high entropy, e.g. 8 for images with $2^8 = 256$ gray levels, indicates more randomness than low numbers (e.g. 1 for binary images).

Note that the uniqueness of the gray level function depends on the location in the image. It should be determined for every new location.

Features are derived properties of the original gray level images. Consequently, they are inherently unique. This is one of several reasons why feature-based matching is more robust.

10.4 Area-Based Matching

The entities in area-based matching are gray levels. Here, the idea is to compare the gray level distribution of a small sub-image, called an *image patch*, with its counterpart in the other image. Fig. 10.14 illustrates the concept and introduces a frequently used terminology. The *template* is the image patch, which usually remains in a fixed position in one of the images. The *search window* refers to the search space within which image patches (sometimes called *matching windows*) are compared with the template. The comparison is performed with different similarity measure criteria. The two best known criteria are *cross-correlation* and *least-squares matching*. The position and size of the search window can be determined with the methods described in Sec. 10.3.1.

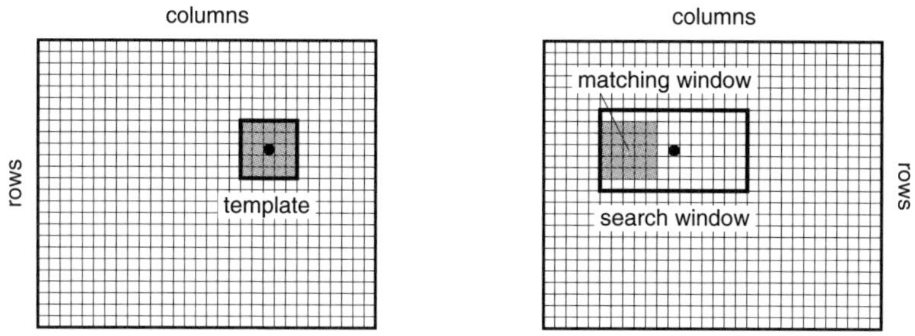

Figure 10.14: Concept of area-based matching.

A number of issues must be addressed independently of the similarity measure method. These issues are discussed briefly.

Location of template The question of where to select the center of the template may appear trivial. Theoretically, the template's center can be placed within an area which is half the template size smaller than the image. On closer examination we probably want to be more selective within the accepted boundary. Certain conditions may cause area-based matching to fail, for example placing the template on areas which are occluded in the other image, selecting an area with low SNR[4], or repetitive pattern, selecting an area containing breaklines—the list of poor conditions can be continued.

Size of template The size of the template and the matching window is an important parameter. With increasing size, the uniqueness of the gray level function usually increases but so do the geometric distortion problems. A compromise must be found, e.g. by computing a uniqueness measure for different template sizes. This also serves the purpose of checking the usefulness of the template location.

Location and size of search window Since area-based matching requires very close approximations, the location of the search window is crucial. Its size, however, does not play an important role because the necessity for close approximations limits the size to a few pixels. A matching strategy based on a hierarchical approach ensures good approximations.

Acceptance criteria The factors obtained for measuring the similarity of the template and matching window must be analyzed. Acceptance/rejection

[4]A low Signal to Noise Ratio of the image function occurs in areas with low contrast.

criteria often change, even within the same image. Threshold values or other criteria should be locally determined.

Quality control The quality control includes an assessment of the accuracy and the reliability of the conjugate locations. Moreover, the consistency of matched points must be analyzed, including the compatibility with expectations or knowledge about the object space.

10.4.1 Correlation

Correlation techniques have a long tradition of finding conjugate points in photogrammetry. In fact, the first experiments with analog correlation methods were conducted in the fifties. The idea is to measure the similarity of the template with the matching window by computing the correlation factor. Kreiling was among the first to employ digital correlation for generating DEMs automatically for the purpose of producing orthophotos, see e.g. *Kreiling* (1976). *Makarovic* (1980) presents an overview of image correlation algorithms.

Cross-Correlation Factor

The correlation coefficient ρ is defined by

$$\boxed{\rho = \frac{\sigma_{LR}}{\sigma_L \sigma_R}} \tag{10.9}$$

If ρ is normalized then $-1 \le \rho \le +1$. We have

σ_{LR}	covariance of image patches L and R
σ_L	standard deviation of image patch L (template)
σ_R	standard deviation of image patch R (matching window)

Introducing the image functions $g_L(x, y), g_R(x, y)$ for the left and right image patches (or the template and matching window for that matter) and their means $\overline{g}_L, \overline{g}_R$, we have the following defining equations

$$\overline{g}_L = \frac{\sum_{i=1}^{n}\sum_{j=1}^{m} g_L(x_i, y_j)}{n \cdot m} \quad (10.10)$$

$$\overline{g}_R = \frac{\sum_{i=1}^{n}\sum_{j=1}^{m} g_R(x_i, y_j)}{n \cdot m} \quad (10.11)$$

$$\sigma_L = \sqrt{\frac{\sum_{i=1}^{n}\sum_{j=1}^{m}\left(g_L(x_i, y_j) - \overline{g}_L\right)^2}{n \cdot m - 1}} \quad (10.12)$$

$$\sigma_R = \sqrt{\frac{\sum_{i=1}^{n}\sum_{j=1}^{m}\left(g_R(x_i, y_j) - \overline{g}_R\right)^2}{n \cdot m - 1}} \quad (10.13)$$

(10.14)

$$\sigma_{LR} = \frac{\sum_{i=1}^{n}\sum_{j=1}^{m}\left(\left(g_L(x_i, y_j) - \overline{g}_L\right)\left(g_R(x_i, y_j) - \overline{g}_R\right)\right)}{n \cdot m - 1} \quad (10.15)$$

The cross-correlation factor is determined for every position r, c of the matching window within the search window. The next problem is to determine the position u, v which yields the maximum correlation factor.

Maximum Correlation Factor

The normalized cross-correlation factor ρ assumes values in the interval ± 1. A value of unity is obtained if the template and the correlation window are identical. If there is no correlation between the two image patches, that is to say no similarity at all, then $\rho = 0$. A value $\rho = -1$ would indicate an inverse correlation. Such is the case with a diapositive and a negative of the same image.

If the search window is constrained to the epipolar line then the correlation factors can be plotted in a graph like Fig. 10.15. The maximum is found by fitting a polynomial, e.g. a parabola, through the correlation values. Note that the maximum only rarely coincides with the integer values of the pixel addresses.

The next figure demonstrates less than ideal cases. Although the correlation factors in the example of Fig. 10.16(a) are fairly high, the maximum is not determined very well at all. It corresponds to two image patches with little contrast. The reliability of the conjugate point is low. Fig. 10.16(b) shows another common problem, namely the difficulty of deciding which of the two peaks is the correct maximum.

A comparison between Figs. 10.15 and 10.16(a) suggests using the "flatness" of the correlation function as a quality measure for the distinctness

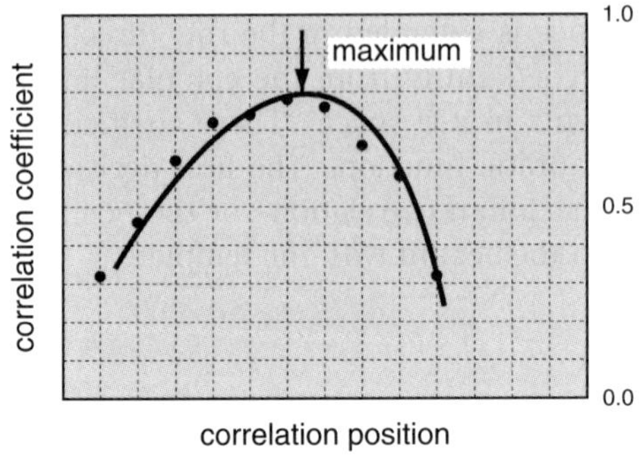

Figure 10.15: The solid circles are the correlation factors obtained at different locations of the matching window. Fitting a parabola through the points and determining its maximum would allow determination of the matched position to subpixel accuracy.

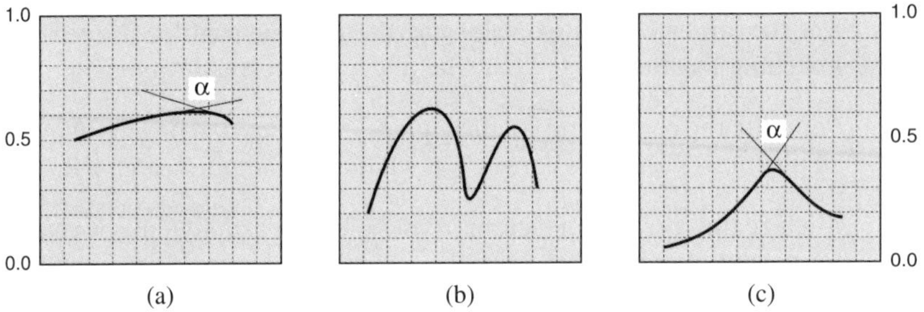

Figure 10.16: Typical problems occurring in the determination of the maximum correlation value. In (a) the SNR of the template and correlation window is low, while (b) may be caused by repetitive patterns in the windows. In (c) the difficulty with global thresholds is manifest: although a distinct peak, the maximum is far less than in (a).

of the maximum. The flatness can be measured, e.g. by the angle between the tangents to the parabola next to the maximum. Using this criterion we would consider the situation in (c) a better match than the one sketched in (a), even though the latter has a higher correlation factor. The correlation function depicted in Fig. 10.16(b) illustrates a problematic case. Here, two relatively high responses are close together which may indicate the occurrence of repetitive patterns in the template and matching window.

If the search space is extended in the row direction (allowing for deviations of the matched entity from the epipolar line), we would obtain correlation coefficients in 2-D space. These coefficients can be approximated by a 2-D quadratic function. The first derivative of this function would render the maximum correlation—or the position of the matching window that yielded the best fit with the template.

Procedure

The following procedure describes the general steps of area-based matching with correlation as a similarity measure.

1. Select the center of the template in one image.

2. Determine approximate locations for the conjugate position in the other image (see Sec. 10.3.1).

3. For both the template and the correlation window, determine the minimum size which passes the uniqueness criterion (see Sec. 10.3.2). Select the larger of the two values as the window size for the current matching location.

4. Compute the correlation coefficients $\rho_{r,c}$ with Eq. 10.9 on page 254 for all positions r, c of the correlation window within the search window.

5. Analyze the correlation factors. A minimum threshold must be reached for a valid match. Apart from the maximum, determine the distinctness as a quality measure.

6. Repeat steps 2—5 for a new template location until all positions to be matched are visited.

7. Analyze the matching results on a global base for consistency and/or compatibility with a priori scene knowledge.

10.4.2 Least-Squares Matching

First experiments with least-squares matching (LSM) were reported by researchers in the early eighties, for example by *Ackermann* (1984); *Grün* (1985); *Grün and Baltsavias* (1987); *Förstner* (1982, 1986); *Rosenholm* (1987); *Thurgood and Mikhail* (1982). In *Förstner* (1984) the author assesses the quality of image correlation techniques. The idea of least-squares matching is to minimize the gray level differences between the template and the matching window whereby the position and the shape of the matching window are parameters to be determined in the adjustment process. That is, the position and the shape of the matching window are changed until the gray level differences between the deformed window and the (constant) template reach a minimum. While we immediately grasp the idea of moving the matching window until the conjugate position is found, the deforming aspect may be less obvious at first sight. After discussing the various geometric distortions caused by either unknown orientation parameters, tilted surface patch, or surface patch with relief, etc., it becomes clear that the shape of the matching window must be changed so that all pixels in the window become conjugate with their counterparts in the template.

Mathematical Model

In this section, the mathematical model is derived and some variations are discussed. This is a helpful enterprise because it not only helps us to appreciate the advantages of this elegant approach but also to recognize the limitations in order to avoid potential problems of LSM.

Let $t(i,j)$ be the image function of the template and $m(i,j)$ the image function of the matching window. The template is defined by $t(i,j), i = R_T-(n-1)/2, \ldots, R_T+(n-1)/2$; $j = C_T-(m-1)/2, \ldots, C_T+(m-1)/2$ with n, m the template size and R_T, C_T the row and column of its center within the digital image (see also Fig. 10.14 on page 253). Likewise, the matching window is defined by $m(i,j), i = R_M - (n-1)/2, \ldots, R_M + (n-1)/2$; $j = C_M - (m-1)/2, \ldots, C_M + (m-1)/2$ with R_M, C_M the row and column of its center as the initial approximation of the conjugate position with respect to the template.

Suppose for a moment that the matching window is centered precisely at the conjugate location. The gray level differences between corresponding pixels of the template and matching window are caused by a number of factors, including

- illumination and reflectance differences between the two images;
- the camera (e.g. vignetting, cosine to the fourth power law);
- the photographic development process and scanner in case of digitized photographs;
- geometric distortion of the regular image tessellation due to orientation, slanted surface patch, surface patch with relief.

To compensate for differences in brightness and contrast, a radiometric transformation T_R of the matching window is introduced. We have

$$T_R\{m(i,j)\} = r_0 + r_1 \cdot m(i,j) \tag{10.16}$$

This transformation would perform a general brightness shift r_0 and a contrast stretching r_1 to adjust the matching window radiometrically to the template. Although it is possible to introduce the radiometric transformation in the least-squares matching process it is customary to adjust the radiometry separately as a pre-process, sometimes for the entire image. This also eliminates the problem of parameter dependency which may occur in the combined approach.

For the following derivations, $m(i,j)$ refers to the original matching window and $m(x,y)$ to the transformed window, with x, y real coordinate values, and i, j the integer values of the pixel location in the original image. The relationship between the two windows is symbolically indicated by $m(x,y) = T_G[m(i,j)]$ where $T_G = [T_x, T_y]^T$ is the geometric transformation operator. To summarize we have

$$m(x,y) = m[T_G(i,j)] \tag{10.17}$$

$$x = T_x(i,j) \tag{10.18}$$

$$y = T_y(i,j) \tag{10.19}$$

Now the question arises as to what kind of geometric transformation T_G is appropriate. Suppose the regular grid tessellation of the template is projected onto the surface patch in object space. Then conjugate pixels are obtained if the surface patch is projected to the matching window. This procedure would require known orientation parameters (exterior orientation) and detailed surface information. Because the latter condition is hardly ever satisfied, certain assumptions must be made.

Since the two image patches are related via the common surface patch by the central projection we should use the central projection model. The matching windows are rather small and so is the corresponding surface patch. It may be approximated by a plane. With this simplification, the image patches and the surface patch are related by a projective transformation (8-parameter transformation). When restricting the matching to aerial cases and "reasonable" surfaces, we can further simplify the problem by approximating the projective transformation by an affine transformation.

In the case of an affine transformation we obtain the transformed matching window $m(x,y)$ as follows

$$x = t_0 + t_1 \cdot i + t_2 \cdot j \tag{10.20}$$
$$y = t_3 + t_4 \cdot i + t_5 \cdot j \tag{10.21}$$

With the geometric transformation in place we can write the following generic observation equation

$$r(i,j) = t(i,j) - m(x,y) \tag{10.22}$$

where the residuals $r(i,j)$ are the gray level differences between the template and the transformed matching window. The linearization of $m(x,y)$ with respect to the transformation parameters follows next

$$m(x,y) \approx m^0(x,y) + \frac{\partial m(x,y)}{\partial T_x}\left[\frac{\partial T_x}{\partial t_0}\Delta t_0 + \frac{\partial T_x}{\partial t_1}\Delta t_1 + \frac{\partial T_x}{\partial t_2}\Delta t_2\right]$$
$$+ \frac{\partial m(x,y)}{\partial T_y}\left[\frac{\partial T_y}{\partial t_3}\Delta t_3 + \frac{\partial T_y}{\partial t_4}\Delta t_4 + \frac{\partial T_y}{\partial t_5}\Delta t_5\right]$$

The partial derivatives are listed below

$$\frac{\partial m(x,y)}{\partial T_x} = g_x \qquad \frac{\partial m(x,y)}{\partial T_y} = g_y$$
$$\frac{\partial T_x}{\partial t_0} = 1 \qquad \frac{\partial T_y}{\partial t_3} = 1$$
$$\frac{\partial T_x}{\partial t_1} = x \qquad \frac{\partial T_y}{\partial t_4} = x$$
$$\frac{\partial T_x}{\partial t_2} = y \qquad \frac{\partial T_y}{\partial t_5} = y$$
$$m^0(x,y) = m(i,j) \qquad c = t(i,j) - m(i,j)$$

g_x	gradient in x-direction
g_y	gradient in y-direction

Table 10.2 contains the design matrix.

For the solution of the transformation parameters **t**, $n \times m$ observation equations are available where n, m is the size of the template and matching window. The observation equations, the solution and the estimated variance of unit weight are expressed by the following three equations

Table 10.2: Design matrix for least-squares matching.

pixel	Δt_0	Δt_1	Δt_2	Δt_3	Δt_4	Δt_5	const
1, 1	g_{x_1}	$g_{x_1} \cdot x_1$	$g_{x_1} \cdot y_1$	g_{y_1}	$g_{y_1} \cdot x_1$	$g_{y_1} \cdot y_1$	$t(1,1) - m(1,1)$
2, 1	g_{x_2}	$g_{x_2} \cdot x_2$	$g_{x_2} \cdot y_1$	g_{y_1}	$g_{y_1} \cdot x_2$	$g_{y_1} \cdot y_1$	$t(2,1) - m(2,1)$
$\vdots$	$\vdots$	$\vdots$	$\vdots$	$\vdots$	$\vdots$	$\vdots$	
n, m	g_{x_n}	$g_{x_n} \cdot x_n$	$g_{x_n} \cdot y_m$	g_{y_m}	$g_{y_m} \cdot x_n$	$g_{y_m} \cdot y_m$	$t(n,m) - m(n,m)$

$$\mathbf{r} = \mathbf{A}\hat{\mathbf{t}} + \mathbf{c} \tag{10.23}$$

$$\hat{\mathbf{t}} = \left(\mathbf{A}^{\mathrm{T}}\mathbf{P}\mathbf{A}\right)^{-1} \mathbf{A}^{\mathrm{T}}\mathbf{P}\mathbf{c} \tag{10.24}$$

$$\hat{\sigma}^2 = \frac{\mathbf{r}^{\mathrm{T}}\mathbf{P}\mathbf{r}}{n \cdot m - 6} \tag{10.25}$$

P is the weight matrix which is usually approximated by the identity matrix.

Note the difference in the spatial variables of the matching window; $m(i, j)$ refers to the regular grid tessellation of the digital image (pixel addresses). While $m(x, y)$ expresses a new tessellation due to the geometric transformation $m\{T_G(i, j)\}$ where x, y are real variables, in contrast to the integer variables i, j. Compare also Fig. 10.17 where the two windows are labeled with m^o, m^1, respectively.

For certain applications it may suffice to determine only the shift parameters. The affine transformation expressed by Eqs. 10.20 and 10.21 reduces to

$$x = t_0 + i \tag{10.26}$$

$$y = t_3 + j \tag{10.27}$$

and the design matrix contains two columns with the gradients in x- and y-direction. Thus the shift parameters $\Delta t_0, \Delta t_3$ can be determined from the following equation system

$$\begin{bmatrix} \Delta t_0 \\ \Delta t_3 \end{bmatrix} = \begin{bmatrix} [g_x]^2 & [g_x \cdot g_y] \\ [g_x \cdot g_y] & [g_y]^2 \end{bmatrix}^{-1} \begin{bmatrix} [g_x\,(t(i,j) - m(i,j))] \\ [g_y\,(t(i,j) - m(i,j))] \end{bmatrix}$$

Adjustment Procedure

Least-squares matching is a non-linear adjustment problem. The solution is found in an iterative fashion. There are some differences from the usual iteration cycle, however. The first iteration begins with the approximate location of the matching window at R_M, C_M. The coefficients of the design matrix are computed with the coefficients of Table 10.2 and the transformation parameters Δt^1 (superscripts refer to the iteration) result from solving Eq. 10.24.

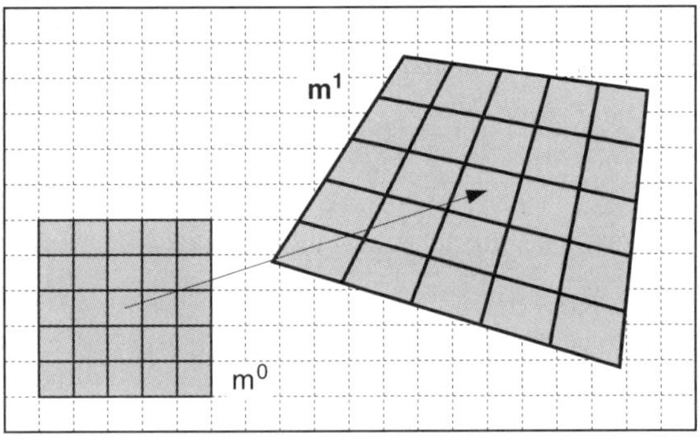

Figure 10.17: The original matching window m_0 is transformed to a new position and new shape after every iteration. The gray values of the transformed window m_1 are found by resampling, such as bilinear interpolation.

Fig. 10.17 shows the matching window $m^1(x^1, y^1)$ after its transformation with $\mathbf{\Delta t}^1$. Before the second iteration begins, the gray values for all positions x^1, y^1 must be determined. This resampling process amounts to interpolating the gray values from the neighboring pixels of the original image. Usually, a bi-linear interpolation is chosen. Resampling is not without problems, however. Fig. 10.17 illustrates one potential problem. After the transformation, the pixels in $m^1(x, y)$ are considerably larger than the original tessellation $m^0(i, j)$. This is caused by foreshortening (see Section 10.2.3). In order for $m^1(x, y)$ to cover the same surface patch as the template, the pixel size must be enlarged in the direction of the base. In this case, the gray values must be interpolated from a larger area, not only from the direct neighbors. But even when the resampling is performed correctly, there remains the problem of comparing two image patches with vastly different resolution. Unfortunately, the process of resampling also correlates the pixels within the transformed matching window. That is, the gray levels in the new tesselation can no longer be considered as independent observations. Fortunately, neglecting the correlation has no major impact on the estimated parameters. However, the variance components will be too favorably estimated. Thus, σ_o of LSM is not a realistic quantity to express the matching success.

After resampling, the next iteration begins and the cycle is repeated until a termination criterion is satisfied. The maximum number of iterations is used as a safeguard in case of slow convergence or—worse—divergence. Under normal circumstances, the criterion about non significant shifts is reached before the maximum iteration criterion. The conjugate point to the template's center is the center of the matching window after it has been transformed into its final position and shape.

The procedure in least-squares matching is summarized by the following steps.

1. Select the center of the template in one image (R_T, C_T).
2. Determine approximate locations for the matching window (R_M, C_M) (see Sec. 10.3.1).
3. Determine the minimum size of template and matching window which passes the uniqueness criteria (see Sec. 10.3.2). Select the larger of the two values as the window size.
4. Start first iteration with matching window at location R_M, C_M.
5. Transform matching window and determine the gray values for the tessellation (resampling).
6. Repeat adjustment and resampling sequence until termination criteria is reached. The conjugate point to the template's center is the sum of the shift parameters obtained in each iteration. This is identical with the center of the matching window in its final position.
7. Assess quality of conjugate point.
8. Repeat steps 1—7 for new position of template.

As mentioned before, radiometric adjustments, such as histogram equalization, should be performed prior to the matching procedure.

Problems

1. Image matching is an ill-posed problem. Name a few measures that make it well-posed.

2. Compare correlation and least-squares matching. List advantages and disadvantages associated with each method.

3. How does foreshortening affect area-based matching? How does it affect feature-based matching?

4. Assume the template and the matching window are both perfectly centered on a conjugate point. You correlate the windows and the cross-correlation coefficient turns out to be 0.8. How do you interpret this result?

5. Describe the geometric relationship between the template, the matching window, and the surface patch in least-squares matching.

6. Let us apply least-squares matching to determine a conjugate point. Suppose we change the overall brightness of one image and match again at the same template location. What differences do you expect? Will you obtain a different conjugate point?

7. Suppose the template in area-based matching is centered on top of a saddle roof. What is the impact if the roof edge is oriented parallel to the base of the stereopair? What is it if the orientation is perpendicular to the base?

8. Given a digital stereopair, scale 1:5000, focal length 152 mm, pixel size 15 μm, average elevation about 600 m and elevation range 120 m. On what level of the image pyramid must area-based matching begin to assure convergence?

9. Given a template and a matching window in area-based matching. How do you select a suitable size? Describe what happens if the window size is increased (until it reaches the image size)?

10. How are geometric distortions considered in least-squares matching? How does correlation cope with geometric distortions?

11. What is the advantage of combining the epipolar line and the vertical line locus constraint?

References

Ackermann, F. (1984). Digital image correlation: performance and potential application in photogrammetry. *The Photogrammetric Record*, **11**(64), 429–439.

Bethel, J. (1986). The DSR11 image correlator. In *Proc. ACSM/ASPRS Ann. Convention*, **4**, 44–49.

Förstner, W. (1982). On the geometric precision of digital correlation. In *International Archives for Photogrammetry and Remote Sensing*, **24**(3), 176–189.

Förstner, W. (1984). Quality assessment of object location and point transfer using digital image correlation techniques. In *International Archives for Photogrammetry and Remote Sensing*, **25**(3), 197–220.

Förstner, W. (1986). Digital image matching techniques for standard photogrammetric applications. In *Proc. ACSM/ASPRS Ann. Convention*, **1**, 210–219.

Grimson, W. E. L. (1981). *From Images To Surfaces: A Computational Study of the Human Early Vision System*. M.I.T. Press, Cambridge, MA.

Grün, A. (1985). Adaptive least-squares correlation: a powerful image matching technique. *South Africa Journal of Photogrammetry, Remote Sensing and Cartography*, **14**(3), 175–187.

Grün, A. and E. Baltsavias (1987). High-precision image matching for digital terrain model generation. *Photogrammetria*, **42**, 97–112.

Gyer, M. (1981). *Automated stereo photogrammetric terrain elevation extraction*. Tech. Report, Gyer and Saliba, Inc.

Helava, U. (1978). Digital correlation in photogrammetric instruments. *Photogrammetria*, **34**, 19–41.

Hobrough, G. (1959). Automatic stereoplotting. *Photogrammetric Engineering and Remote Sensing*, **25**(5), 763–769.

Hobrough, G. (1978). Digital on-line correlation. *Bildmessung und Luftbildwesen*, **46**(3), 79–86.

Horn, B. K. P. (1983). Noncorrelation methods for stereo matching. *Photogrammetric Engineering and Remote Sensing*, **49**(4), 535–536.

Kreiling, W. (1976). *Automatische Herstellung von Höhenmodellen und Orthophotos aus Stereobildern durch digitale Korrelation*. Diss. Fakultät für Bauingenieur- und Vermessungswesen, Universität Karlsruhe.

Krupnik, A. and T. Schenk (1994). Prediciting the reliability of matched points. In *Proc. ACSM/ASPRS Ann. Convention*, **1**, 337–343.

Makarovic, B. (1980). Image correlation algorithms. In *International Archives for Photogrammetry and Remote Sensing*, **23**(B2), 139–158.

Rosenholm, D. (1987). Least-squares matching method: some experimental results. *The Photogrammetric Record*, **12**(70), 493–512.

Thurgood, J. and E. Mikhail (1982). Photogrammetric analysis of digital images. In *International Archives for Photogrammetry and Remote Sensing*, **24**(3), 576–590.

Chapter 11

Advanced Image Matching Methods

The previous chapter introduced image matching, the most fundamental procedure in digital photogrammetry. The fact that we perform matching—finding conjugate points—without conscious effort should not lure us into assuming that it is simple. In fact, matching is quite a complex problem and utterly ill-posed. To gain a broader view we found it useful to consider the relationship between object space and the images. This helped us to greatly reduce the search space and to alleviate the ambiguity problem.

The matching methods discussed so far, correlation and least-squares matching, operate directly on the gray levels. As alluded to, gray values contain little explicit information about the object space. As a consequence, area-based matching methods are not reliable enough. In this chapter we discuss matching methods that are more robust and also more general than area-based methods.

The next sections are devoted to feature-based matching. Here, the matching entities are extracted features, such as interest points, edges, and regions. With the attempt of matching edges as entities, we encounter the problem of dealing with shape. Two approaches to this general problem are discussed: the $\psi - s$ method and the generalized Hough transform.

To increase the general scope of image matching, we introduce the notion of relational matching and demonstrate how relationships among matching entities help to reduce the ambiguity problem. The chapter ends with template matching. The emphasis is on an application—finding targets in calibration test fields.

There is much more to image matching than presented in this and the previous chapter. We will indeed address other aspects in subsequent chapters. For example, multiple image matching will be discussed in the aerial triangulation chapter (*Agouris*, 1992; *Krupnik*, 1994; *Li*, 1989). The chapter about DEM generation will provide a good opportunity to extend the methods to object space matching, see, e.g. *Doorn* (1991), *Heipke* (1990), *Nevatia* (1996), and *Wrobel* (1987). We will also introduce matching by trial and error in Chapter 14.

11.1 Feature-Based Matching

As the name suggests, feature-based matching (FBM) employs as conjugate entities features extracted from the original gray level image. Such features include points, edges, and regions. Edges are by far the most widely used features, although in photogrammetry, feature points (interest points) are more popular.

FBM gained popularity in computer vision in the late seventies when it was realized that the remarkable stereovision ability of humans is based on finding conjugate edges rather than finding similar gray level distributions in a stereopair (*Grimson*, 1981; *Horn*, 1983).

As in the previous sections we restrict the discussion to aerial stereopairs. The next section provides a concise summary of feature extraction. Subsequent sections deal with matching the features, that is, matching interest points and various ways of matching edges.

11.1.1 Feature Extraction

Feature extraction is a monocular process, performed independently on every image. The subject is treated in more detail in other chapters. Here, a short summary is provided.

Interest Points

The basic idea of extracting distinct points in an image or image patch is to identify areas with high variance—areas which are potentially useful for matching. An interest operator performs the extraction process; points with distinct features are called interest points.

Moravec (1976) developed an interest operator which measures the distinctness of an image patch compared to its surroundings. This is accomplished by computing the variances of small areas and by thresholding them locally and globally.

The Förstner interest operator is rotation invariant and renders subpixel accuracy, see *Förstner* (1986b); *Förstner and Gülch* (1987). Different point features can be detected, such as corner points and circular features (blobs).

These two interest operators are most commonly used in digital photogrammetry. Fig. 11.1 shows an example of extracted interest points superimposed on the original image.

Edge Detection

Chapter 5 deals with the subject of edge detection. Numerous edge operators are in use, each with its own distinct characteristics. Often, the choice of one operator over another is driven by personal preference or special experience. Edge detection involves the identification of edge pixels, sometimes also called edgels, and the grouping of edge pixels into entire edges, also called edge contours. There is an important difference between the two steps. To appreciate the benefit of grouping, consider convolving the image with an edge operator followed by thresholding the response, for example, by assigning black to an edgel and white to all other pixels. If this binary image is displayed, you clearly see all edges, but the connection between neighboring edgels is not yet explicitly defined. That is, the edge contours are only implicitly represented, which is not suitable for most edge matching methods.

Edges correspond to brightness differences in the images. Such differences may be abrupt ("sharp" edge) or may occur over an extended area ("smooth" edge). Ideally, an edge operator should be capable of detecting sharp and smooth edges. Also, edges usually occur at all orientations which requires a direction independent operator. Detecting brightness differences amounts to determining derivatives or, in the case of discrete functions such as digital images, n-order differences. Taking differences amplifies noise, hence edge operators are noise sensitive. Another prerequisite then, is to reduce the impact of noise by smoothing the image.

Fig. 11.2 on page 272 shows a stereopair, resolution 512×512 which was convolved with an LoG operator, $w = 3$ pixels. Note how the edges, here zero-crossings, delineate object boundaries in many cases.

Extracting Regions

Regions can be extracted in different ways. As discussed in Sec. 3.5, images can be segmented, for example by histogram thresholding or texture segmentation. The ideal situation for matching regions is to find areas in an image which are rather small, distinct, and invariant to illumination differences and perspective views.

Such an ideal situation is hardly ever present in black and white images. However, with the increasing use of color and multispectral imagery, the chances of extracting regions that are very distinct in their spectral characteristics, are much higher.

11.1.2 Matching Interest Points

The easiest way to match interest points is by area-based matching. The original motivation for determining interest points was to find image patches of sufficient information so that correlation has a higher chance to succeed. As discussed in Sec. 10.4.1, image patches with uniform intensity lead to a flat correlation function without a distinct maximum. Therefore, such areas should be avoided.

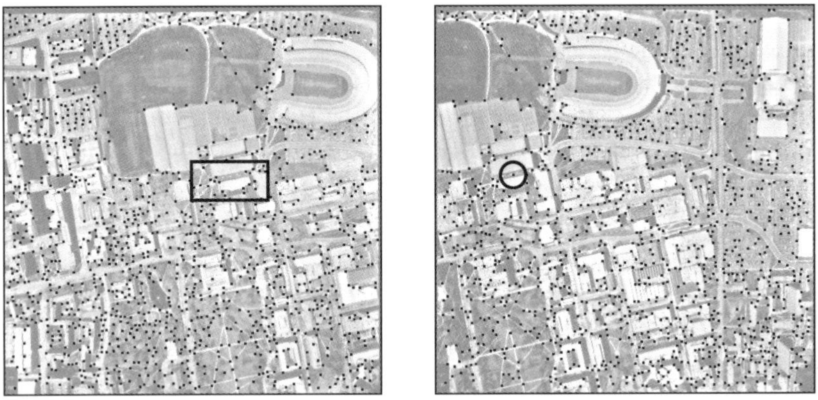

Figure 11.1: Stereopair with interest points superimposed. The template is centered on an interest point in one image, indicated by a circle in the right image. A search window is determined which contains potential conjugate interest points (rectangle in the left image). The matching window is centered on all these candidates. The most likely conjugate interest point is the one with the highest correlation factor. Note that the interest point selected in the right image has no conjugate interest point in the left image.

Fig. 11.1 demonstrates the concept of matching interest points. The stereopair shows interest points extracted by the Förstner operator. The template is centered on one of the interest points, say in the right image. Now, with the method described in Sec. 10.3.1, the search window in the other image is computed. As Fig. 11.1 demonstrates, several interest points fall within the search window. The matching window is centered on all these points and correlation coefficients are computed. The point yielding the highest correlation factor is usually considered the conjugate point to the template's center. The entire procedure repeats until the template has been centered on all interest points in the right image[1].

It is clear that matching interest points in this fashion is a more logical approach to area-based matching in that the template is not arbitrarily placed somewhere in the image but at locations with a distinct image function. However, the problems associated with foreshortening, breaklines, and occlusions persist. Moreover, not every interest point has a corresponding interest point in the other image as demonstrated in Fig. 11.1. Yet, the correlation coefficient may be above the threshold for accepting a match. This would result in a wrong match. Therefore, a subsequent check, such as relative orientation, must be performed to identify wrong matches.

As an alternative to cross-correlation, LSM can be applied. An interesting question arises, however. During the iteration cycle, the matching window is translated and deformed. Is the conjugate point the original interest point or the center of the translated matching window?

Other approaches to matching interest points are based on comparing the structural descriptions of the two sets of interest points in both images. *Tsingas* (1994) describes a method with a graph theoretical underpinning.

11.1.3 Matching Edge Pixels

After the theory of stereovision was established (see Sec. 4.5), FBM was first implemented by an approach which became known as the Marr-Poggio-Grimson algorithm (*Grimson* , 1985). In this method, zero-crossing contours are first matched pixel by pixel; then multiple matches are sorted considering regional criteria. It is assumed that the stereopair is normalized. That is, conjugate entities are found in the same rows of the digital stereopair. Since normalized images require a known orientation (relative or absolute), the application of this approach is limited to the automatic determination of DEMs.

Fig. 11.3 shows a small portion of the stereopair depicted in Fig. 11.2. Suppose we want to match the edge pixel highlighted in the left image patch of Fig. 11.3. The first task is to determine the search window in the right image patch. Its center and dimensions are computed with Eq. 10.3

[1]If the template is chosen in the right image, as in our example, then the left image is matched against the right one. There is no rule about whether the right or left image is selected for the template. Often, the entire procedure is repeated with the other image. That is, both images are matched against each other.

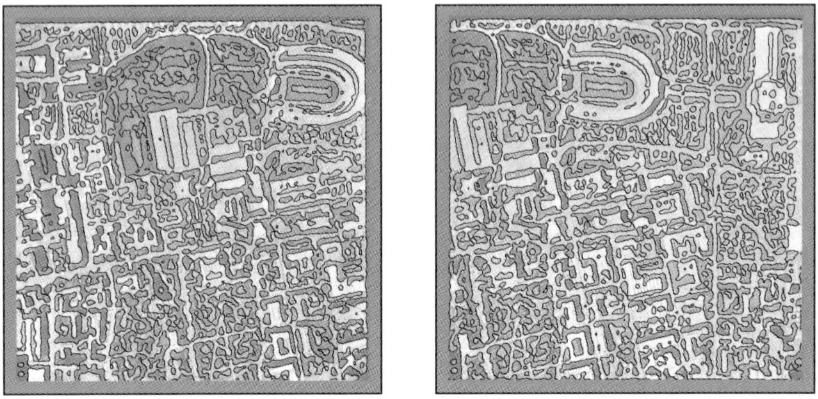

Figure 11.2: Example of a stereopair with zero-crossing contours superimposed. The contours were obtained with an LoG operator, $w = 3$. The resolution of the images is 512×512 pixels.

and Eq. 10.8 on page 245. Although normalized images are assumed, the search is not strictly confined to the epipolar line, but a few rows above and below. Therefore, the height of the search window, indicated by the white rectangle in Fig. 11.3, is seven. It contains nine edge pixels on the epipolar line and approximately the same number on the six neighboring rows. Some of these potential matches can be discarded immediately, based on incompatible attributes. In the case of zero-crossings, the sign is a very useful attribute. It implies that if the image appears brighter to the left of the edge pixel (as in Fig. 11.3) then you would expect the same situation in the other image[2].

Other edge pixel attributes, for example orientation and strength, can also be used to further reduce the number of candidate matches. Observing the sign reduces the number of matches on the epipolar line; in our example from nine to four. Including the edge orientation further reduces the pool of candidate matches.

The more potential matches we find, the more difficult will it be to determine the correct one. Hence, the cardinal question is how many candidates will we have on an average and—more importantly—how can this number be influenced and kept to a maximum of, say three to five per row. The number of zero-crossings depends on the size of the LoG operator. Chap-

[2]This general assumption is rarely violated. An example might be a surface with strong spectral reflectance properties (e.g. glass, certain metals, water). The brightness of such surfaces may change enough between two adjacent images to reverse the sign.

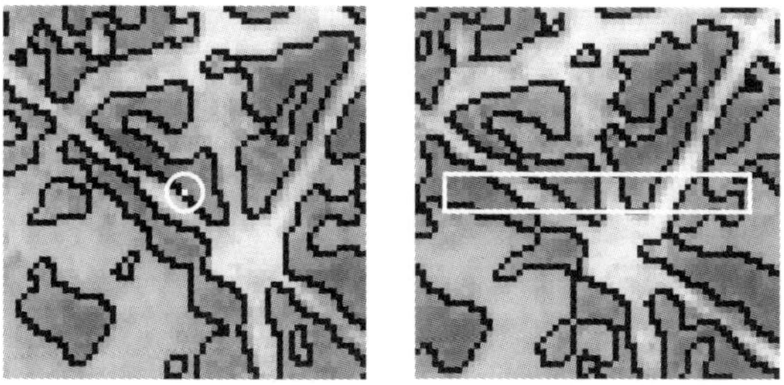

Figure 11.3: Small portion of stereopair shown in Fig. 11.2. The edge pixel in the center of the circle in the left image is selected and a search window is predicted (rectangle in the right image). The height of the search window is 3 pixels below and above the epipolar line to allow for vertical shifts. The length depends on the uncertainty of the elevation (depth) of the conjugate edge pixel. The number of candidate matches depends on the edge pixel's attributes, such as sign, orientation, and strength.

ter 14 provides a more detailed account of how to select suitable sizes and how to proceed with the matching through the image pyramid.

Selecting the correct conjugate pixel from a pool of candidate matches is governed by the principle of continuity along discontinuities. An edge signals a discontinuity in the gray levels across the edge. Edges may correspond to boundaries of real objects and such boundaries are usually continuous, at least piecewise. As a consequence, the parallax along matched edges changes only gradually, hardly ever abruptly. A possible scenario is to determine all possible parallaxes, for example between the pixel to be matched and the candidate matches. The dominant parallax can easily be found by analyzing the histogram of all possible parallaxes. It is conceivable to consider more than one peak thus allowing a discontinuity along the edge. Further refinements of this method would include a trend analysis of the histogram reflecting gradual changes of the parallax.

Finally, additional edge attributes, not yet considered in the original match, may come to the rescue when the histogram analysis cannot completely resolve ambiguous associations. For example, if for several adjacent epipolar lines two conjugate matches are equally likely, then it may help to check their strength attributes. A sharp edge in one image is unlikely to be fuzzy in the other image. By the same token, it is quite unlikely that an edge is conjugate to several edges. Therefore, the association of pixels of the same edge would cast more votes in the disambiguation process.

11.1.4 Matching Entire Edges

A more general solution of feature-based matching is obtained when the matching entity is an entire edge rather than individual edge pixels (*Medioni and Nevatia* , 1984; *Greenfeld and Schenk* , 1989). An obvious advantage is that unoriented stereopairs can be matched. Hence, this approach is suitable for determining the relative orientation.

Matching entire edges strongly depends on a suitable representation of the shape characteristics of edges. We discuss two approaches to elucidate the notion of extracting, representing, and comparing shapes. The two approaches, $\psi - s$ curve and generalized Hough Transform, are good examples of finding more suitable representations for solving shape problems.

Before delving into details let us examine Fig. 11.4 for the purpose of defining the problem we want to solve. The edge contour labeled A in Fig.11.4 occurs in both images with a very similar shape. Its location differs roughly by the image base. This is obviously the easiest case of matching edges. A slightly more complex situation arises when one contour separates into two or more in the other image (case B). Another challenge looms much larger; look closely at the edges within the square C. Several edge segments look quite different or belong to different edge contours.

With these observations we can formulate the edge matching problem as follows. Let r_i be an edge in one image and let $q_j, j = 1, \ldots, n$ be a set of n candidate edges in the other image, selected within an estimated search window. The problem is to find one edge, a segment of one edge, or segments of several edges which match r_i in terms of similar shape.

$\psi - s$ Approach

The $\psi - s$ curve is a functional representation of a line where the arc length s is the parameter of the tangent ψ, see e.g. *Ballard and Brown* (1982). Fig. 11.5 shows some examples of lines in the traditional x, y representation as well as in the $\psi - s$ representation. Straight lines in the spatial domain correspond to horizontal straight lines in the $\psi - s$ domain (derivative is constant). The ψ value of horizontal lines is proportional to the orientation of the straight line, see Fig. 11.5(a). Circles are represented as straight lines with a slope proportional to the curvature as illustrated in Fig. 11.5(c). Note that a rotation in the spatial domain amounts to a vertical shift in the $\psi - s$ representation. Fig. 11.5(b) is a rotated version of Fig. 11.5(a). Likewise, Fig. 11.5(d) is a rotated version of Fig. 11.5(c).

The examples presented in Fig. 11.5 illuminate the advantages of the $\psi - s$ representation for matching edge contours. For one, the representation

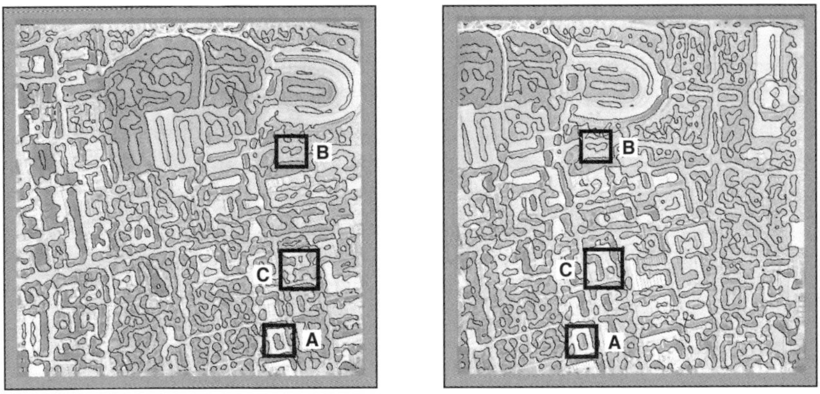

Figure 11.4: Illustration of the edge matching problem. The same stereopair as in Fig. 11.2 is used. The three areas highlighted demonstrate some of the fundamental problems of edge matching. Area A illustrates an ideal case. The edge contour is very similar in both images. Area B represents a slightly more complex situation because one contour appears as two separate contours in the other image. The most challenging situation is depicted in C where segments of edges match segments of several different edges in the right image.

is invariant with respect to the edge's position in the image. A rotation between the two images of the stereopair amounts to a simple shift of the zero position. Also, the degree of the original line is reduced by one.

Edges are represented as a sequence of edge pixels, for example by chain codes. Here, the beginning of an edge is given by the image coordinates, but all subsequent pixels are linked to the previous pixel by the direction. The chain code is the first difference of the discrete line representation (corresponding to the first derivative of the continuous representation). Thus, the discrete $\psi - s$ representation is in essence a chain code representation.

The motivation for using the $\psi - s$ representation for shape matching is rooted in its ability to extract distinct shape features. One example of a distinct shape feature is change of curvature. In terms of shape, there is nothing more boring than straight lines. In contrast, lines with frequent changes of curvature are rich in shape information. How do we make this information explicit so that it can be exploited for matching edges?

Suppose we segment the $\psi - s$ curve into a sequence of straight lines. Since horizontal straight segments correspond to straight lines in the x, y - domain with an orientation equal to ψ, and sloped straight lines are circular arcs, the segmentation of the $\psi - s$ curve amounts to a curvilinear

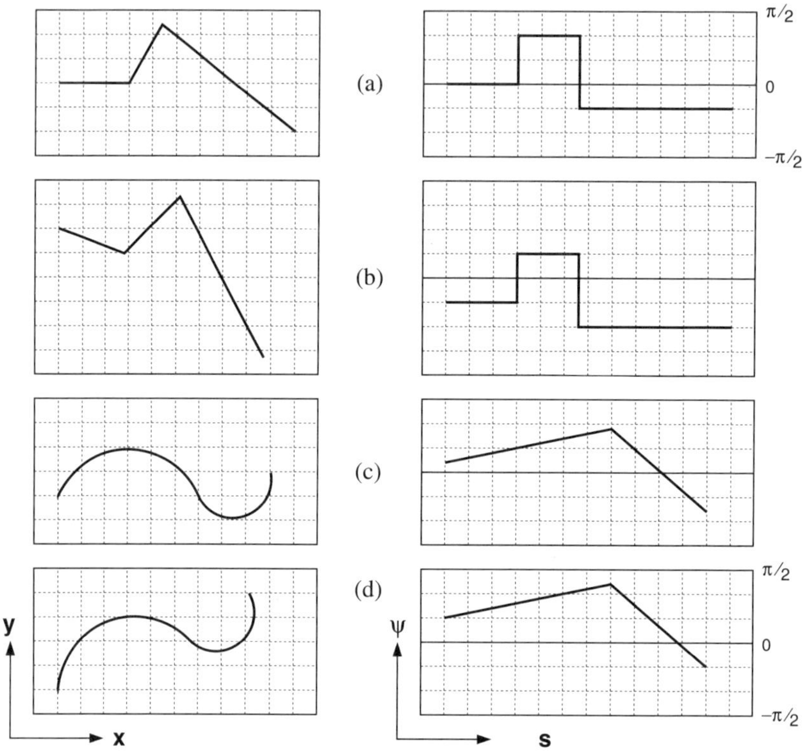

Figure 11.5: Examples of lines in the x, y domain and their representation in the $\psi - s$ domain. The top two figures show straight lines that are represented as horizontal straight lines. The ψ value is proportional to the line orientation, e.g. the angle between the x -axis and the line. Circles are represented as straight lines. Their slope is proportional to the curvature. Note that a rotation in the x, y domain amounts to a vertical shift in the $\psi - s$ domain with an amount proportional to the rotation angle.

segmentation in the spatial domain. Now, we are in a position to extract salient shape features by analyzing the segmented $\psi - s$ curve. For example, vertices indicate changes of curvature and edges of similar shape are characterized by similar vertices.

One strategy is to find similar vertices followed by local and global consistency checks as suggested by *Schenk et al.* (1991). Let P_{ij} be vertex j of edge L_i in one image. Now, find all vertices in the search window of the other image with similar angles and orientations, resulting in a set of candidate matches M_{jk}.

Since for opaque surfaces, such as the Earth's surface, only one match can exist, the candidate matches generated in the previous step must be analyzed to identify the most likely match. This process is governed by applying local and global consistency checks. A powerful global consistency check is the average model parallax ($x-$ and $y-$parallaxes). They can be easily determined by generating a 2-D histogram. True matches are clustered within an elongated ellipse, the major axis of which is parallel to the flight direction (base). Its size is a function of the topography. The size of the minor axis is determined by differences in the exterior orientation.

The remaining multiple matches are now analyzed on a more local basis. Preference is given to matches belonging to the same edge. This ensures figural continuity as discussed in the previous section. *Li and Schenk* (1990) describe a matching system based on the $\psi - s$ approach.

Generalized Hough Transform

The Generalized Hough Transform (GHT) is an extension of the Hough technique discussed in Chapter 5. The traditional Hough method is based on parametric shape description. The GHT works for arbitrary shapes and is therefore suitable for matching edges.

The main application of the GHT, developed by *Ballard* (1981), is to find objects in an image. It is assumed that the object boundaries can be extracted and that they have a known, but arbitrary, shape. The shape is described by a template.

The task of finding instances of the template in the image by GHT is best explained by way of an example. Fig. 11.6 depicts a template. The idea is to represent the pixels of the template as polar coordinates with respect to a reference point R. Let us concentrate on pixel P for a moment. The vector $\boldsymbol{r}$ is expressed by the polar coordinates r, α, with $r = \|\boldsymbol{r}\|$ and $\alpha = \arctan(r_x / r_y)$. The template's shape is sufficiently described by the triplet (φ, r, α) for every pixel. These values are stored in a table, called an $R-$table. Depending on the implementation, the $R-$table can be organized sequentially, from the starting pixel to the end (Table 11.1.4 exemplifies a sequentially organized $R-$table) or it may be advantageous to sort the $R-$table in ascending order of the tangents, φ. This particular ordering facilitates the search process explained next.

Take an image from which three line features L_1, L_2, L_3 have been extracted (Fig.11.7). The task is to determine if any of these three shapes is similar to the template. For this purpose we check every entry in the $R-$table for a matching tangent in the image. Fig.11.7 shows all pixels that

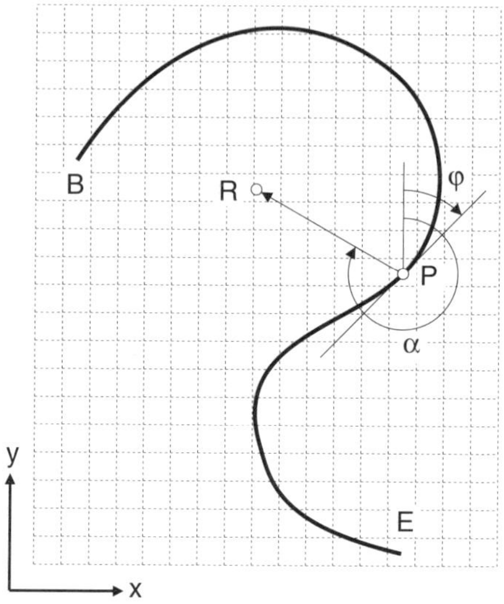

Figure 11.6: Shape description of a template in the Generalized Hough Transform. The reference point R is arbitrarily chosen. For every pixel P of the template, the triplet φ, r, α is computed and entered into the R–table. φ is the azimuth of the tangent; r and α are the polar coordinates of the reference point with respect to the pixel.

have a tangent $\varphi = 0^0$ and $\varphi = 55^0$. Next, the reference point for every pixel with $\varphi = 0^0$ is constructed by adding the vector $\boldsymbol{r}_0 = [r, \alpha]T$ to the pixel location. The same procedure is repeated for pixels with $\varphi = 55^0$. Fig.11.7 illustrates the construction of the reference points: the vector $\boldsymbol{r}_0$ is added to the pixel locations with tangents at zero degrees. Likewise, $\boldsymbol{r}_{55}$ is added to pixels with a tangent at 55^0. The vectors are stored in the R–table; the subindex refers to φ.

The sequence of operations just described is repeated for every entry in the R–table. As a result, we obtain many reconstructed reference points in the image. An accumulator array keeps track of the reference points. Imagine the accumulator array as being an image, pixel size equal to the precision of the reconstructed reference points; size of the image equal to the image containing the extracted features. Initially, all gray levels of this accumulator image are set to zero. Whenever a reference point is constructed, its gray level is increased by one. How does the image look after all entries in the R–table have been processed?

Table 11.1: Sequentially organized R-table of the template shown in Fig. 11.6.

Pixel	*Tangent*	*Polar Coordinates*	
	φ [^{0}]	r [mm]	α [^{0}]
B	20	46	100
1	21	45	103
.	.	.	.
P	43	39	301
.	.	.	.
E	281	85	338

After a moment's thought we realize that the reference points are clustered. For an exact copy of the template, e.g. L_1, we find the reconstructed reference point n times at the same location (n number of pixels of L_1). The other two lines have some parts which are similar to the template's shape. Thus, we find smaller clusters of reference points for those parts as well. The number of entries (gray value) of the accumulator array tells how many pixels of the same shape have been found. It does not tell, however, if these pixels are consecutive or scattered along the line. To ensure connectivity we must also keep track of the pixel locations.

To apply the GHT to matching edges we must further generalize the technique as suggested by *Zahran* (1997). Most importantly, the orientation of the pixels should be independent of the image coordinate system: it must be rotation invariant. This can be accomplished by changing from a tangent to a curvature representation. Another aspect concerns the shape difference of conjugate edges, resulting in a cluster that is spatially spread. The dimension of the cluster is directly related to the shape differences. Topographic surfaces cause shape differences of conjugate features.

11.2 Relational Matching

The matching methods discussed so far share the common characteristic that individual matching entities are compared with entities in the other image on a one to one basis. For example, an edge contour is selected in one image and compared with edges within the search area of the other image, based on shape similarity. This process is repeated more or less independently for all edges. After all entities are matched in this fashion, a separate process selects the most plausible match from candidate matches

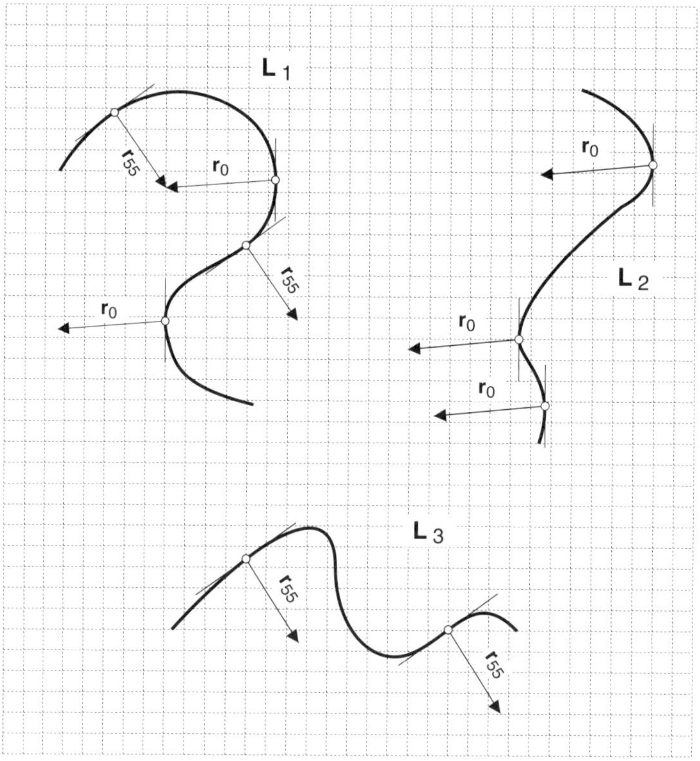

Figure 11.7: Three lines extracted in an image. Pixels with an azimuth of $\varphi = 0^0$ are matched and the reference point is reconstructed with the polar coordinates from the R–table (vector $\boldsymbol{r}_0$). A second example indicates reconstructed reference points for pixels whose tangent has an azimuth of 55^0. The location of all reconstructed reference points is stored in the accumulator array. The number of reference points per accumulator cell is proportional to the number of pixels that have the same shape as the template.

and ensures local and global consistency with a priori expectations of the object space surface.

A close examination of Fig. 11.4 on page 275 reveals that the matching process could be improved by incorporating relationships among entities. For example, if you try to identify corresponding edge fragments in area C

of Fig. 11.4 then you strongly rely on the relationship between neighboring edges.

Relational matching provides a mechanism to consider relationships among the matching entities. Relational matching can be characterized as finding the best mapping between two relational descriptions (*Vosselman* , 1992). This more general matching scheme has been developed by computer vision researchers, e.g. *Shapiro and Haralick* (1987) and *Boyer and Kak* (1988). It comes in the guise of consistent labeling, relaxation, and structural matching. More recently, relational matching is successfully used in photogrammetric applications by, for example *Vosselman* (1994), *Zilberstein* (1992), *Cho* (1996), and *Wang* (1996).

There are three equally important aspects to consider in relational matching: the description, the evaluation function, and the search tree. The following sections briefly describe these aspects, emphasizing the concept by way of an example. A formal treatment of the theory is beyond the scope of this book; interested readers are referred to the literature cited above.

11.2.1 Description of Primitives and Relations

The basic elements of the relational description are called *primitives*, characterized by *attributes*. Primitives can be almost anything. In our application of matching images, the primitives are the matching entities. Let us take the example of edges. Fig. 11.8 depicts two edges. Suppose we segment the edges into curvilinear parts, for example with the $\psi - s$ method described earlier in this chapter. In this case, the length of a segment and its curvature are suitable attributes to describe the primitive. We have

$$p_i = \{\text{centroid}(x_i, y_i), \text{length}(l_i), \text{curvature}(c_i), \text{arc}(\alpha_i)\}$$

In addition to the length, l, the location of the centroid (x_i, y_i) and the angle α_i, corresponding to the arc length, can be used to describe a primitive. Depending on the edge detection scheme used, additional information might be available, for example edge strength, edge sign, perhaps even spectral information, lending itself to a richer set of attributes. Imagine an edge characterized as a strong, closed circle with known radius, red, and positive[3]. Thousands of edges are extracted and must be matched. Of all those edges, some are closed, a few have a small radius, and fewer are red. The richer the description, the easier the matching.

It should be pointed out that matching can be performed with the description of primitives only. The attributes would be stored in a vector,

[3]A positive sign can be assigned if the convolution value inside the circle is positive, for example.

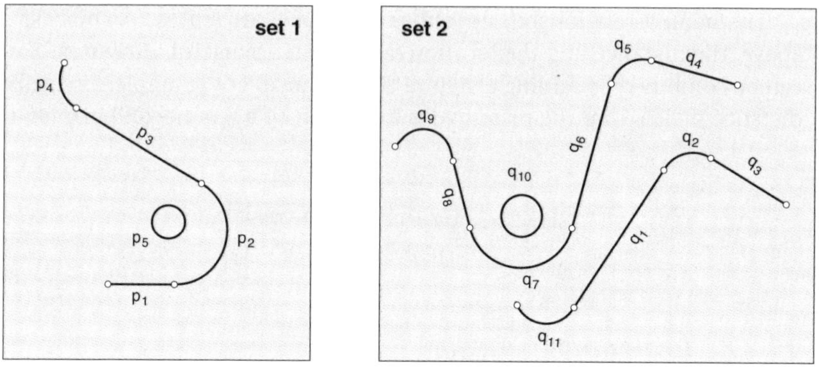

Figure 11.8: Example for relational matching. The left figure contains two segmented edge contours (set 1). The right figure shows three contours (set 2). The best mapping between the two sets is found by a tree search, explained in the text. The corresponding tree is shown in Fig. 11.9. The relations used are `connectivity` and `neighbor`.

called a *feature vector.* Of course you realize that this is precisely what we discussed in the previous section under feature-based matching.

Let us now turn to the problem of introducing relations among the primitives. Considering our simple example depicted in Fig. 11.8 `connection` is a useful relationship. We simply specify which line segments are connected. For example, p_2 is connected to p_1; p_5 is not connected, and so on. An example of an interrelationship between edges would be to say something about the location of p_5 with respect to the primitives making up the other edge. Many possibilities exist. As we trace an edge there is a sense of what is "left" or "right" of the edge. Beginning with p_1 and following the edge until p_4 (based on the connectivity), the primitive p_5 would always be left. Moreover, we could introduce a distance, for example centroid to centroid. To specify even further, it is reasonable to restrict this relationship to immediate neighbors only. Thus, we end up with the powerful relationship `neighbor`.

Relations are described by the name of the relation and a set of relation tuples.

$$\text{(connection)}\ \{p_2p_1\},\ \{p_3p_2\},\ \{p_4p_3\})$$
$$\text{(neighbor)}\ \{p_5p_1\},\ \{p_5p_2\},\ \{p_5p_3\})$$

The named relationship `neighbor` may have attributes. As discussed above, the distance and the location (left, right) are useful attributes. Both can be combined by adding a sign to the distance. For example, a negative distance means that the primitive is to the left of a (connected) primitive.

$$\text{(neighbor)} \quad \{p_5 p_1\}\,(\text{distance} - 18), \quad \{p_5 p_2\}(\text{distance} - 12),$$

11.2.2 Evaluation Function

Image matching means comparing two images and establishing a list of corresponding items, called entities here. In the case of relational matching, the entities are descriptions of primitives and relations among primitives. Comparing such relations is accomplished by the evaluation function. It is usually assumed that the attributes and relational tuples are independent of each other. Then, the measure of similarity between two relational descriptions is determined by adding a number (cost) expressing the degree of dissimilarity between descriptions.

The cost function is zero if two relational descriptions are identical. Such is rarely the case; we must deal with measuring the dissimilarity. The cost function is estimated either by a distance measure or by a probability approach. An example of a distance measure is to take the absolute difference of attributes, both for the primitives and relational tuples. This simple approach is not without problems, however. First, some attributes are symbolic, such as color. How do we determine the cost between red and purple? Second, the attributes have different units, for example angles, pixels, or square millimeters. The latter problem can be solved by normalizing the attribute values so that they are all in the same range of $[0, 1]$.

As suggested by *Boyer and Kak* (1988), the probability approach is better suited to deal with attributes of a symbolic nature, such as color, shape (straight, curved), and topology (above, left). Here, the conditional probability function expresses how well the primitives and relations of one relational description correspond to another relational description. The closer the two descriptions are, the closer the probability function will be to unity. What actually enters as a cost measure is the negative logarithm of the conditional probability function. Vosselman expands this approach further, proposing a merit function that estimates the support which attributes of corresponding primitives and relational tuples provide to the mapping (*Vosselman* , 1992).

11.2.3 Tree Search

After having established the relational description of the matching entities and having decided on the similarity measure, we finally need to execute the matching scheme. The most common way to find a solution is by way of a tree search. Tree search methods are successfully used in computer vision and artificial intelligence (*Bender*, 1996).

Trees consist of nodes and arcs, starting from a root node, descending through parent nodes and on to leaf nodes. Pairs of nodes are connected by directed arcs. The primitives $\{p_1, p_2, \ldots, p_n\}$ of one relational description are called units and the primitives of the description to be matched, say $\{q_1, q_2, \ldots, q_m\}$, are termed labels. The depth of the tree is defined by the number of units, n. The best mapping is found by determining the maximal path with the lowest cost.

Fig. 11.9 illustrates the principle. The units are the primitives of the two edge contours of Fig. 11.8 on page 282 and the labels are the primitives of the second set. The tree has a depth of five. We begin expanding the tree on the first level by associating compatible labels with units. That is, p_1 is compared with $\{q_1, q_2, \ldots, q_{11}\}$. It turns out that q_3, q_4, q_8 have similar attributes, hence these three labels become the nodes on the first level. All three nodes are further expanded on the second level by matching unit p_2. Continuing in this fashion through all units we obtain a tree with 120 leaf nodes that represents the search space of our matching problem, $P \rightarrow Q$.

Each path from the root to a leaf constitutes a solution. Obviously, most of the 120 solutions are geometrically wrong. How do we determine the most plausible solution? Suppose for a moment that costs are determined for every match unit/label based on the differences between the attributes. Adding up the costs through one path renders its total cost. The optimal solution is represented by the path with minimal cost.

So far we have only used the description of the primitives and the matching just described is simply feature-based matching, implemented as a tree search (as opposed to, say, $\psi - s$ approach). The next step considers the relations among the primitives. Let us go back to the first level where we should consider any binary relationship $(p_1 p_2)$. The `connection` relationship of p_2 with p_1 exists and we should only allow labels to be added to the tree having the same relationship. In other words, of the five labels originally matched with p_2, only one satisfies the `connection` relationship. As a result, only one node is further expanded.

Considering the `connection` relationship on all nodes, we obtain three solutions as Fig. 11.9 demonstrates. The cost of solution $q_4, q_5, q_6, q_7, q_{10}$ will be highest because the attributes differ more between units and labels as compared to the other solutions. This is also evident by comparing the

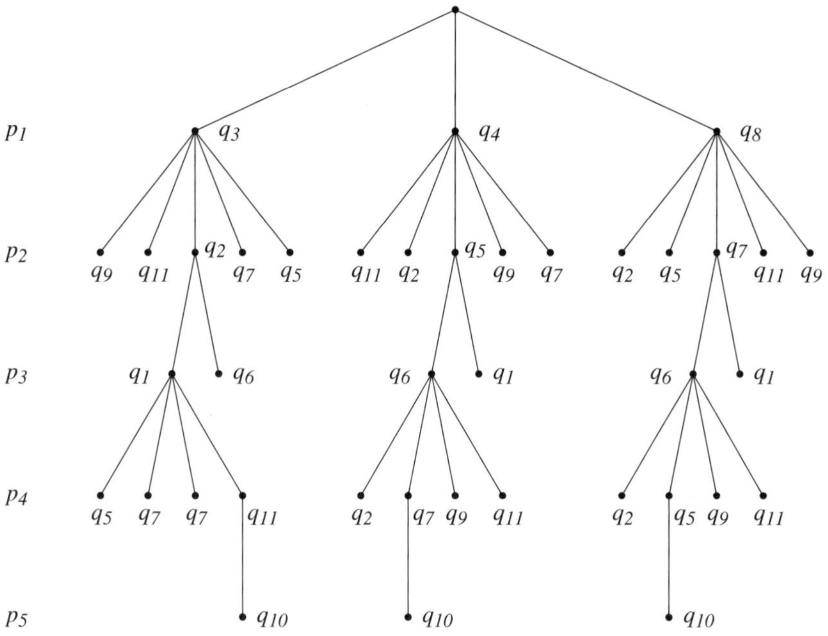

Figure 11.9: The tree shown corresponds to the example depicted in Fig. 11.8 on page 282. The set of primitives $p_1, \ldots, p_5$ (units) is compared with the set $q_1, \ldots, q_{11}$ (labels). Matches are accepted based on similar attributes. Only those nodes are further expanded which satisfy the relationships among the units. For example, p_1 and p_2 have a `connection` relationship. Of all the possible matches to p_2, only q_2 satisfies this relationship with q_3 in the leftmost branch. Likewise, only q_5 can be expanded further in the central path.

contours in Fig. 11.8 on page 282: the shape difference between the two contours is obvious.

Which of the two remaining solutions is more plausible? Since the costs are similar, a conclusive answer can only be reached by considering additional information—here in the form of the relationship `neighbor` that has not yet been used. The label q_{10}, paired with unit p_5, can only be associated with the path q_8, q_7, q_6, q_5 because of the `neighbor` relationship of p_5 with p_1, p_2, p_3. This simple example demonstrates vividly that a unique mapping $P \to Q$ can be reached by introducing proper relations.

11.3 Template Matching

Template matching is not a matching method in the sense discussed in the previous sections. Rather, template matching is an application, usually associated with the problem of finding a particular feature in an image. Suppose the feature can be suitably described, then template matching means finding instances of this description in the image. The description of the feature is called a template. 'Finding' the template in the image entails detection, precise localization, and verification.

Typical template matching applications in photogrammetry include finding fiducial marks for the purpose of automatic interior orientation (see Chapter 13), finding targets that have been placed on tools in industrial close range applications, and finding control points in test fields for calibration purposes (see Fig. 11.10). The latter two applications are examples of automatic target location, involving all aspects of matching discussed so far. Therefore, you may consider this section as an extended summary of matching.

Figure 11.10: Example of a test site for calibrating digital cameras. Targets are placed on a concrete wall of a building. Photo courtesy of TransMap.

As stated earlier, matching encompasses the analysis of the matching environment, the design of a matching strategy, the execution of the matching process, and the assessment of the matching results. Let us briefly review these processes with target location in mind.

11.3.1 Matching Environment

A careful analysis of the environment within which matching takes place is important for several reasons. For one, there is no general purpose matching scheme that can be employed to solve all matching applications in photogrammetry. The more specific a system, the higher the success rate. We usually know quite a bit about the application of locating targets—so why not take advantage of it and transform the utterly ill-posed matching problem into a solvable one. Yet another advantage of this approach is that in the end you have a much clearer understanding of under what conditions the devised matching scheme is supposed to work.

The matching environment consists of the object space, illumination, image formation and image space. If we focus for a moment on calibrating a digital camera using an outdoor test field, then the knowledge about the matching environment may include:

1. *Illumination:* range of acceptable outdoor illumination for operational calibration procedures is known. If these conditions are not met then one does not calibrate.

2. *Object Space:* known is the precise location of targets and their description, the depth range, and the scene contrast.

3. *Image Space:* interior and exterior orientation approximately known.

With this specific domain knowledge we can already make very useful assumptions; imagery of good quality, approximate location of the targets in the image, whether or not the targets are partially occluded, and how much the targets may be distorted due to relief and perspective transformation.

Other factors to be considered in the matching environment are project related, such as time requirements (real-time, or more relaxed response), precision, accuracy, and reliability.

11.3.2 Matching Strategy

Simply put, the goal of the matching strategy is to devise a scheme that performs reliably within the conditions specified in the matching environment. Specifically, the matching strategy must answer three basic questions: how will the targets be detected, how will they be precisely located, and how will the results be verified.

The distinction between detection and localization may come as a surprise because often the two processes are combined. The goal of the two processes is quite different, however, each requiring its own solution. The goal of the detection process is to find all instances of the template in the image. The emphasis is on completeness, not on precision. Consider the detection as putting forward a hypothesis that the feature detected at a certain location may be the template. The hypothesis is then checked during the precise localization and either confirmed or rejected.

Do not take precise localization of the target in too narrow a sense. In addition to the location, other parameters may be of interest, such as rotation, shape deformations, and perhaps radiometric deformations. If shape deformations are to be expected, for example due to relief and perspective transformation, then we need a matching method which is either insensitive to shape differences between target and feature, or includes shape parameters in its mathematical model. In the latter case, the shape parameters can be used for local consistency checks.

Another important decision involves the description of the target. Should it be pictorial or symbolic? Should it be the same for the detection and localization process? Many factors influence the decision. A pictorial description, that is, an image of the object, either generated synthetically or by using a camera, may be suitable if the images are well behaved. Such is the case in our example of target location with a fairly controlled environment. If targets are partially occluded (e.g., by shadows, extreme foreshortening, or other objects), a more explicit description of the target in terms of its parts may be more advantageous.

In summary, the matching strategy tells us how the template is described, how we find instances of the template in the image, and how we determine its precise localization, possibly including its shape deformation.

11.3.3 Target Detection

Target detection aims at finding occurrences of the template in the image. To find all possible instances, the conditions imposed on target detection are not very stringent. If we find a candidate that will be rejected later, for example during precise localization or verification, then nothing happened, except that we wasted some computer resources. On the other hand, if we miss a target, then later processes will not recover it—it is lost. We discuss now several approaches that may serve the purpose of identifying an image patch containing the features which match the template.

Computational Approach

In the case of a calibration test field it is reasonable to assume that the target locations are known. They may have been precisely surveyed or their location is known from previous calibration procedures. Also, the interior and exterior orientation of the camera is approximately known (from in-flight calibration by GPS/INS and from terrestrial calibration by the test field setup). The potential location in the image is easily obtained by the collinearity equations. It is important to consider the uncertainty of the location, however. Since the interior and exterior orientation parameters are only approximately known, we must estimate their variances and perform an error propagation through the collinearity equations to obtain variances of the estimated image locations. If the values exceed the tolerances, the computational approach may not work or should be followed up by another method to refine the location.

What are reasonable tolerances? It depends on the matching method used in the precise localization. Area-based methods only work reliably if the approximate location is as good as three pixels or so. Feature-based matching methods are more forgiving. Here, the tolerance is determined more by practical considerations. An example is the size of the search window. Its size should be as small as possible but still contain the entire target, including its uncertain location.

Histogram Thresholding

Given the relatively controlled environment in which target location takes place, it may well be possible to find potential target sites in the image by thresholding the histogram. As discussed in Sec. 3.5.1, histogram thresholding is successful if the radiometric properties of the target are distinct. If the gray level distribution of all targets is different from the rest of the image then we can simply determine an appropriate threshold value. A quick quality control measure of such an approach is to count the number of thresholded pixels—it should more or less agree with the number of pixels per target times the number of expected targets.

In many cases such a bold approach will fail on the global level of the entire image. That is not yet the end, however, because we may be able to execute the same scheme locally since the gray level distribution of local regions may permit thresholding. Now, the question of defining local regions arises. Often, the approach of negative selection is quite successful—eliminate those regions which definitely do not contain targets. We may know this in advance (domain specific knowledge), or we may exclude homogeneous regions as they are unlikely to contain targets. Fig. 11.11 shows the result of this approach with the test field shown in Fig. 11.10 on page 286.

An interesting approach is to cascade the computational method with local histogram segmentation. We begin by estimating the location of tar-

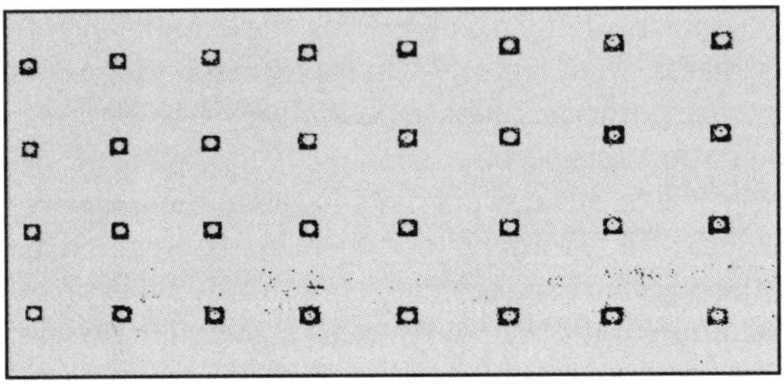

Figure 11.11: Result of thresholding the test field of Fig. 11.10. First, the wall with the targets has been extracted based on approximate location of the camera. Thresholding was performed locally.

gets as discussed in the previous section. The uncertainty of the location gives a good indication of the local region to be used for the subsequent segmentation.

Cross-Correlation

As described by *Ballard and Brown* (1982), the traditional approach in template matching is to combine target detection with localization by cross-correlation. The template is an image patch of an ideal target. Normalized correlation coefficients are determined for every location of the template in the image (or sub-image) (see Sec. 10.4.1). Whenever the 2-D correlation function reaches a local maximum above a threshold value, the occurrence of a target is hypothesized.

This approach is computationally expensive. Therefore it may be advisable to run it locally, for example only in regions where we suspect targets. We will examine cross-correlation as a precise localization method in the next section.

Feature-Based Matching

All the detection methods discussed so far are performed on the pixel level. The template description may be raster instead of vector. This is simple and works well in many instances, because of the rather controlled environment of this particular application. There may be situations where the gray level distribution does not allow for reliable detection, however. In this case we must resort to matching higher level descriptions. Now we can get very sophisticated with describing the target—the limiting factor is the image, that is, what features can we extract and group to match the template, or what properties can we possibly deduce for a region using an image segmentation approach.

One approach is to extract edges and to group them together to find the target's boundary. The grouping process benefits greatly from the knowledge of how the boundary should look. Only those edge pixels which satisfy the shape condition of the boundary within some tolerance are linked. The detection problem can be considered solved if a grouping of edge pixels with the desired shape is possible. It is not necessary to find the complete boundary, however. Depending on the application, a partial boundary reconstruction may suffice.

11.3.4 Precise Localization

After the preliminary step of target detection we can assume that the target location is approximately known, say to a few pixels. Thus, the precise localization becomes a local process, confined to a subimage with the target more or less in the center. The chief advantage of a local process is that parameters and thresholds can be determined individually for every subimage. With a clever control strategy it is even possible to select different algorithms for locating the target as precisely and reliably as possible.

Most of the matching methods discussed in this chapter may be used for precise localization. In the following we briefly discuss their suitability and limitations.

Area-Based Methods

Probably the first method that comes to mind for finding the precise location of a target in an image is correlation. Here, we simply move an image of the template within the search window, determine the cross correlation factor, and accept the position with the highest coefficient as the location of the target. As discussed in Sec. 10.4.1, the location can be determined to sub-pixel accuracy by fitting a polynomial through the correlation coefficients and taking its maximum.

Assuming that the targets are geometrically and radiometrically well defined we can generate a synthetic image and use it as the template. For more complicated targets it may be easier to generate the template by taking an image with a digital camera at a large scale and then resampling it to the scale of the images in which we want to detect the targets.

Frequently, least-squares matching (LSM, see Sec. 10.4.2) is used for precise template matching. Its advantage over cross-correlation is the increased flexibility to cope with more complex situations. If we take our application of detecting targets in a calibration test field then we realize that their size and shape may be quite different[4]. Moreover, some targets may not be rotation invariant. The affine transformation as the mathematical model of LSM takes this into account. Excellent precision of target location determined by LSM has been reported (*Beyer* , 1992; *Mikhail and Mitchell* , 1984).

We now turn to a third method that can be considered to be area-based, namely target detection by determining centroids. The centroid, or center of gravity, is the first central moment of a target. As shown in *Schalkoff* (1989), central moments contain distinct information about regions. In fact, there are seven central moments invariant to scale, rotation and translation between template and image.

The first central moments are defined by

$$m_r = \frac{1}{m_0} \sum_{i=1}^{N} \sum_{j=1}^{M} f(i,j)j \tag{11.1}$$

$$m_c = \frac{1}{m_0} \sum_{i=1}^{N} \sum_{j=1}^{M} f(i,j)i \tag{11.2}$$

(11.3)

with $m_0 = N \cdot M$ (area) of the region and m_r, m_c the row and column coordinates of the centroid.

Let us concentrate on $f(i,j)$, the image function in the search window, for a moment. Suppose a black circular target, radius R, is centered on a white square, size $S > 2R$. Fig. 11.12 depicts an image of this target. Perhaps contrary to our expectation, we do not obtain a binary image, although the target (object space) is binary (white and black). Clearly, all pixels fully covered by the circle are black and all pixels entirely outside are white, but pixels which are partially covered are gray, proportional to the area covered by the circle.

In order to distinguish between foreground (target) and background, we threshold. Depending on the threshold value, the circle is now represented by a cross, a square, or a combination as depicted in Fig. 11.12(b)-(d). Since the resulting shapes are symmetrical, we still obtain the correct location by computing their centroids.

Let us now shift the target by one fourth of a pixel towards the left upper corner and repeat the procedure. Fig. 11.13 depicts the situation. First, we

[4]To alleviate the problem of parameter dependency between interior and exterior orientation, oblique exposures are often taken and targets appear quite deformed.

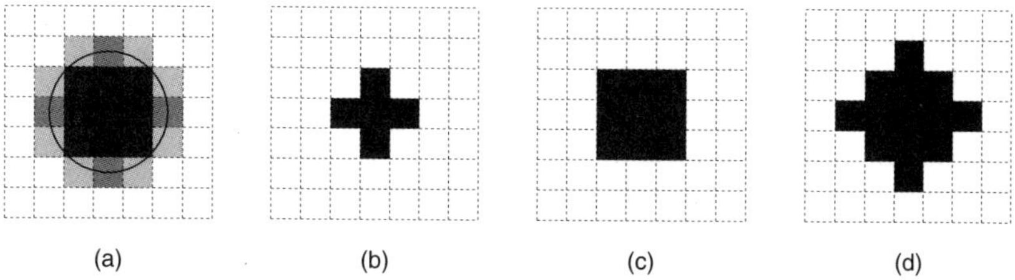

Figure 11.12: A circular black target on a white background is imaged as shown in (a). Although the object space is binary, the image contains several different gray values. If the image is thresholded, different shapes of the target occur, shown in (b)-(d). The shape depends on the threshold value.

realize that more mixed pixels are obtained, but more annoyingly, some of the thresholded images are not symmetric any more. As a result, the centroid of the thresholded template is no longer identical with the center of the target. The correct location is only obtained if every pixel enters the computation with its correct gray value. Consequently, thresholding is not appropriate. Using the gray values, however, is difficult because of noise. The pixels in the white background region are not perfectly white, nor are the target pixels perfectly black. If we do not threshold, then noise will influence the result.

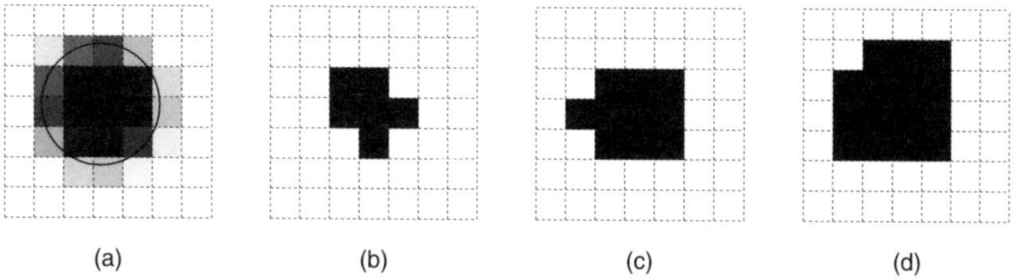

Figure 11.13: Same situation as in Fig. 11.12, except that the target is shifted by one fourth of a pixel toward the left-upper corner. Note that more gray levels occur. When thresholding is performed, the target becomes asymmetrical and the centroid changes. This, in turn, causes a positional error. The three examples in (b), (c) and (d) all give a different centroid.

Whatever approach is chosen to threshold the image, ultimately some gray value ranges are determined and pixels are classified accordingly.

Considering noise, it is quite likely that some of the mixed pixels will be wrongly classified, resulting in a wrong target location.

Obviously, the dislocalization of the target's centroid due to wrong thresholding depends on the ratio of mixed pixels to pure pixels. Mixed pixels are present along the boundary of the target. The ratio decreases with increasing target size. For our example with a circular target we have $2R\pi/(R^2\pi) = 2/R$ as the ratio with R the radius of the circle. Therefore the precision of the target localization is functionally related to the target size by $2/R$. As a consequence, the more pixels the target contains the higher the precision of its location. Moreover, the precision increases with higher quantization level. Experimental studies confirm these problems. Trinder concludes that at least five bits per pixel (32 gray values) should be used (*Trinder*, 1989; *Trinder et al.*, 1995).

Are LSM and correlation immune to these problems? Both methods are affected because of the generation of the template which is matched with the target in the image. To understand the effect imagine that Fig. 11.12(a) is the synthetically generated template and Fig. 11.13(a) is the image containing the target. Even under ideal conditions (noise free), the two images differ and in the case of asymmetry, for example caused by thresholding, the result is slightly wrong, just as demonstrated above.

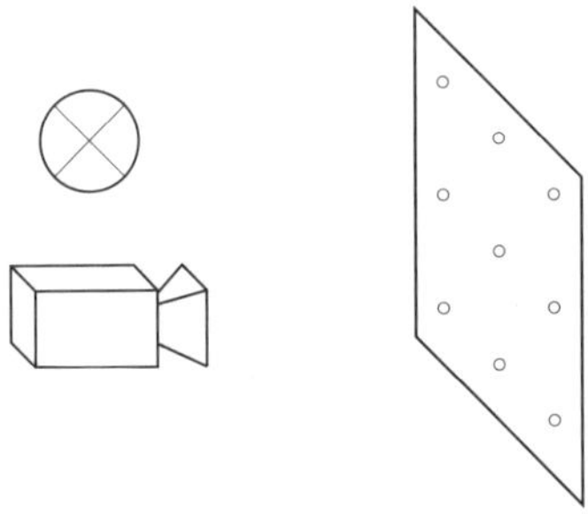

(a)

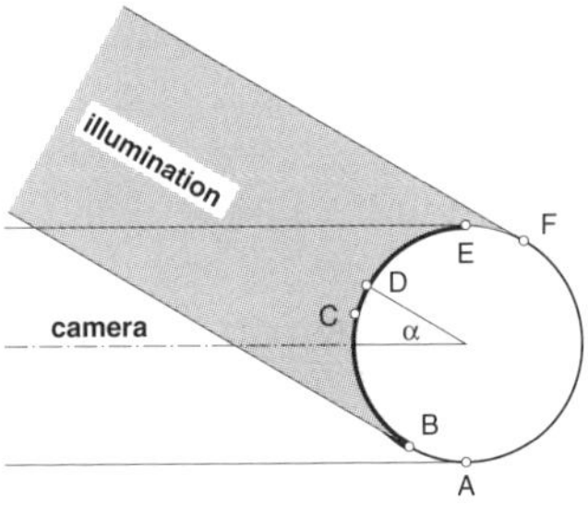

(b)

Figure 11.14: Camera, illumination and test field with ball-shaped targets is depicted in (a). One target is shown in (b) with camera and illumination at an angle of 30^0. With this arrangement, the hemisphere from B to F is illuminated. From the camera, the hemisphere from A to E is visible, but the arc A, B is not illuminated. The center of brightness is at D as seen from the illumination source. From the camera's viewpoint, point C appears to be the center when the target is measured manually or automatically. Hence, an error in the position of the target occurs.

Another challenge looms for area-based methods; the shift between the geometric and the radiometric center of the target. Fig. 11.14(a) depicts a target of spherical shape, a camera, and an illumination source. Spherical targets are often chosen because their images appear as circles over a wide range of viewing directions. Unfortunately, the images of ball-shaped targets are sensitive to illumination as illustrated in Fig. 11.14(b). The center of the image of the spherical target is displaced by $e = r\tan(\frac{\alpha}{2})$. To reduce the displacement, the illumination source and the camera should be close together, or a diffuse (Lambertian) illumination of the object space should be chosen.

Example 11.1:

> *For a target size of 10 pixels and an illumination source that is at a* 10^0 *angle with the camera axis, the displacement e is* 0.87 *pixels.*

Feature-Based Methods

As discussed in Sec. 11.1, feature-based methods are generally more robust than area-based methods. Advocates of area-based methods are quick to add that feature-based methods are less precise and therefore less suited for template matching.

Trinder carried out extensive experiments with comparing target location with LSM and FBM (*Trinder et al.* , 1995). He developed a method for extracting edges with sub-pixel accuracy. Edge pixels are detected based on a modified interest operator, then linked together to form an edge. This process benefits greatly from the known target shape; false edge pixels are readily eliminated. Once all correct edge pixels have been identified, a circle (or ellipse) is fitted by a least-squares adjustment.

The experiments reported by *Trinder et al.* (1995) reveal that the highest precision of target location was achieved by edge matching. The authors report precisions of 0.01 pixel for edge based matching and 0.02 to 0.04 pixel for LSM, provided that at least six pixels are used for quantization.

We will return to the feature-based approach in Chapter 13 for the detection and precise localization of fiducial marks.

Problems

1. Compare area-based and feature-based matching. List advantages and disadvantages associated with each method.

2. How does foreshortening affect feature-based matching?

3. What is the major difference between matching interest points, edge pixels, and edge contours?

4. Explore the possibilities of finding structures in a set of interest points and ways of using these structures to guide the matching of interest points.

5. Suppose a straight edge segment that is parallel to the image base is matched. How would you verify the match?

6. What knowledge about the object space can be used to select a plausible match among several candidate matches?

7. Suppose you have transformed a line (sequence of pixels) from the x, y domain to the $\psi - s$ representation. The $\psi - s$ curve is approximated by a polyline. What is the geometric interpretation of the vertices and what properties of the segmented curve are rotation invariant?

8. Prove the statement made on page 279 that by switching from tangent to curvature representation, the template in the GHT becomes rotation invariant.

9. Show the functional relationship between the number of pixels covering a target and the precision of the target location, assuming that correlation is used.

10. Suppose we have a perfect image and object space. The object space contains a black square on a white background. The size of the square is 2×2 pixels. Explain why we obtain more than two gray values in the image, even assuming an ideal imaging system.

11. Consider the fiducial mark of a camera as a template. Sketch a procedure of detecting fiducial marks by way of template matching.

References

Agouris, P. (1992). *Multiple Image Multipoint Matching for Automatic Aerotriangulation.* PhD dissertation, Dept. of Geodetic Science and Surveying, The Ohio State University, Columbus, OH.

Ballard, D.H. (1981). Generalizing the Hough transform to detect arbitrary shapes. *Pattern Recognition,* **13**(2), 111–222.

Ballard, D.H. and C.M. Brown (1982). *Computer Vision.* Prentice-Hall, Inc.

Bender, E. A. (1996). *Mathematical Methods in Artificial Intelligence.* IEEE Computer Society Press, Los Alamitos, California.

Beyer, H. (1992). *Geometric and radiometric analysis of a CCD-Camera based on a photogrammetric close-range system.* Tech. Report, No. 51, ETH Zürich.

Boyer, K.L. and A.C. Kak (1988). Structural Stereopsis for 3-D Vision. *IEEE Transactions on Pattern Analysis and Machine Intelligence*, **10**(2), 144–166.

Cho, W. (1996). Relational matching for automatic orientation. In *International Archives of Photogrammetry and Remote Sensing*, **31**(B3), 111–119.

Doorn, B. (1991). *Multi-Scale surface reconstruction in the object space.* PhD dissertation, Report No. 413, Dept. of Geodetic Science and Surveying, The Ohio State University, Columbus, OH.

Dhond, U.R. and J.K. Aggarwal (1989). Structure from stereo—a review. *IEEE Transactions on Systems, Man, and Cybernetics*, **19**(6), 1489–1510.

Förstner, W. (1986). Digital image matching techniques for standard photogrammetric applications. In *Proc. ACSM/ASPRS Ann. Convention*, **1**, 210–219.

Förstner, W. (1986). A feature based correspondence algorithm for image matching. In *International Archives of Photogrammetry and Remote Sensing*, **26**(B3/3), 13–19.

Förstner, W. and E. Gülch (1987). A fast operator for detection and precise location of distinct points, corners and centers of circular features. In *Proc. ISPRS Intercommission Workshop on "Fast Processing of Photogrammetric Data"*, 281–305, Interlaken.

Greenfeld, J. and T. Schenk (1989). Experiments with edge-based stereo matching. *Photogrammetric Engineering and Remote Sensing*, **55**(12), 1771–1777.

Grimson, W.E.L. (1981). *From Images To Surfaces: A Computational Study of the Human Early Vision System.* M.I.T. Press, Cambridge, MA.

Grimson, W.E.L. (1985). Computational experiments with a feature based stereo algorithm. *IEEE Transactions on Pattern Analysis and Machine Intelligence*, **7**(1), 17–43.

Grün, A. (1985). Adaptive least-squares correlation: a powerful image matching technique. *South Africa Journal of Photogrammetry, Remote Sensing and Cartography*, **14**(3), 175–187.

Grün, A. and E.P. Baltsavias (1988). Geometrically constrained multiphoto matching. *Photogrammetric Engineering and Remote Sensing*, **54**(5), 633-641.

Heipke, C. (1990). *Integration von Bildzuordnung, Punktbestimmung, Oberflächenrekonstruktion und Orthoprojektion innerhalb der digitalen Photogrammetrie.* Dissertation. DGK-C.

Horn, B.K.P. (1983). Noncorrelation methods for stereo matching. *Photogrammetric Engineering and Remote Sensing*, **49**(4), 535-536.

Krupnik, A. (1994). *Multiple-patch matching in the object space for aerotriangulation.* PhD dissertation, Report No. 428, Dept. of Geodetic Science and Surveying, The Ohio State University, Columbus, OH.

Li, M. (1989). *Hierarchical Multipoint Matching and Simultaneous Detection of Breaklines.* PhD dissertation, Dept. of Photogrammetry, The Royal Institute of Technology, Stockholm.

Li, J.C. and T. Schenk (1990). *Aerial Image Matching Based on Zero-Crossings and Least-Squares Correlation.* Technical Notes in Photogrammetry, No. 9, Dept. of Geodetic Science and Surveying, The Ohio State University, Columbus, OH.

Medioni, G. and R. Nevatia (1984). Matching images using linear features. *IEEE Transactions on Pattern Analysis and Machine Intelligence*, **6**(6), 675-685.

Mikhail, E.M., M.L. Akey and O.R. Mitchell (1984). Detection and sub-pixel location of photogrammetric targets in digital images. *Photogrammetria*, **39**, 63-83.

Mohan, R., G. Medioni, and R. Nevatia (1989). Stereo error detection, correction, and evaluation. *IEEE Transactions on Pattern Analysis and Machine Intelligence*, **11**(2), 113-120.

Moravec, H.P. (1976). Towards automatic visual obstacle avoidance. *Proc. 5th Int. Joint Conference on Artificial Intelligence*, page 584.

Nevatia, R. (1996). Matching in 2-D and 3-D. In *International Archives of Photogrammetry and Remote Sensing*, **31**(B3), 567-574.

Schalkoff, R. (1989). *Digital Image Processing and Computer Vision.* John Wiley & Sons, New York.

Schenk, T., J.C. Li and C. Toth (1991). Towards an autonomous system for orienting digital stereopairs. *Photogrammetric Engineering and Remote Sensing*, **57**(8), 1057-1064.

Shapiro, L.H. and R.M. Haralick (1987). Relational Matching. *Applied Optics*, **26**, 1845–1851.

Stefanidis, A. (1993). *Using scale-space techniques to eliminate scale differences across images.* PhD dissertation, Dept. of Geodetic Science and Surveying, The Ohio State University, Columbus, OH.

Trinder, J.C. (1989). Precision of digital target location. *Photogrammetric Engineering and Remote Sensing*, **55**(6), 883–886.

Trinder, J.C., J. Jansa and Y. Huang (1995). An assessment of the precision and accuracy of methods of digital target detection. *ISPRS Journal of Photogrammetry and Remote Sensing*, **50**(2), 12–20.

Tseng, H.Y. (1992). *Digital photogrammetric approach to ice-flow determination in Antarctica.* PhD dissertation, Report No. 422, Dept. of Geodetic Science and Surveying, The Ohio State University, Columbus, OH.

Tsingas, V. (1994). A graph-theoretical approach for multiple feature matching and its application on digital point transfer. In *International Archives of Photogrammetry and Remote Sensing*, **30**(3/2), 865–871.

Vosselman, G. (1992). *Relational Matching*, Springer-Verlag Berlin Heidelberg.

Vosselman, G. (1994). On the use of tree search methods in digital photogrammetry. In *International Archives of Photogrammetry and Remote Sensing*, **30**(3/2), 886–893.

Wang, Y. (1996). Structural matching and its applications for photogrammetric automation. In *International Archives of Photogrammetry and Remote Sensing*, **31**(B3), 918–923.

Wrobel, B. (1987). Digitale Bildzuordnung durch Facetten mit Hilfe von Objektraummodellen. *Bildwesen und Luftbildmessung*, **55**, 93–101.

Zahran, M. (1997). *Shape matching using a modified generalized Hough transform.* PhD dissertation, Dept. of Geodetic Science and Surveying, The Ohio State University, Columbus, OH.

Zilberstein, O. (1992). Relational matching for stereopsis. In *International Archives of Photogrammetry and Remote Sensing*, **29**(B3), 711–719.

Chapter 12

Computing Normalized Images

Many stereo algorithms in computer vision and digital photogrammetry assume that stereopairs are registered such that conjugate features appear in the same row. This configuration is often referred to as "registered in epipolar geometry". It expresses the fact that epipolar lines are identical with the rows of the images. Since epipolar geometry is independent of how images are oriented, I prefer the term *normalized images* for stereopairs that satisfy the stated conditions.

The computation of normalized images requires that the interior and exterior orientation of a stereopair is known. Thus, the interior orientation discussed in Chapter 13, and the relative or direct orientation (Chapter 14) must be performed first, unless a digital camera with known interior orientation and a platform orientation system, such as GPS/INS, is used.

We shall explore the relationship between original and normalized images in more detail in the following sections. First we begin with the definition of epipolar geometry, followed by a problem statement for computing normalized images. Next, we will derive the rotation matrices that transform the original images into their normalized position. Different scenarios with relative and exterior orientation are considered. We will discover that the projective equations are a (better) alternative to the collinearity equations for computing normalized pixel locations. Consequently, we derive the coefficients of the projective equations. Finally, we shall summarize the procedure for obtaining normalized aerial stereopairs and include an example.

As always, emphasis is on aerial imagery obtained with a central projection camera. The concept presented here is also applicable to other sensors. However, the interior orientation and most likely the exterior orientation must be modeled accordingly.

12.1 Introduction

Many stereo algorithms in computer vision and digital photogrammetry are based on the assumption that digital stereopairs are registered in epipolar geometry. That is, the scan lines (rows) are epipolar lines. This condition is satisfied when the two camera axes of a stereo vision system are parallel to each other and perpendicular to the camera base. Mobile mapping systems and robot vision systems are examples of stereo data acquisition systems that satisfy this condition. In both cases, the epipolar condition is imposed mechanically by mounting the two cameras at the ends of a horizontal bar, with the camera axes perpendicular to the bar. Vibration and other forces from moving platforms may change the attitude of the cameras. Hence, frequent calibration of the vision system is necessary.

In conventional aerial photogrammetry, it is very unlikely that stereopairs are recorded "in epipolar geometry" since the attitude of the camera at the instant of exposure is different at every exposure station. Of course you realize that this is true of conventional aerial cameras and of digital cameras.

When comparing Fig. 12.2 on page 305, we recognize that the frequently used term "images in epipolar geometry" doesn't quite express what is meant because the concept of epipolar geometry is independent of the particular orientation of a stereopair. To express the fact that epipolar lines should be parallel to rows (confining conjugate entities to the same row), I use the term *normalized images* in this book.

Kreiling (1976) described a method for computing normalized images from the parameters of an independent relative orientation. The method is restricted to relatively oriented models, using only angles as parameters. In many instances it is desirable to obtain normalized images with respect to object space. The method presented here is based on the exterior orientation parameters (*Cho et al.* , 1992).

It is perhaps useful at this point to summarize the relationship between the two images. Fig. 12.1 illustrates the different stages, beginning with aerial photographs that are scanned. The result is raw images. Sometimes, the raw images are transformed after the interior orientation such that rows and columns become parallel to the fiducial coordinate system. This process is justified if camera distortions, film shrinkage, and scanner errors are considered at the same time. Note that these images are not yet in their normalized position. Where do images acquired by a digital camera

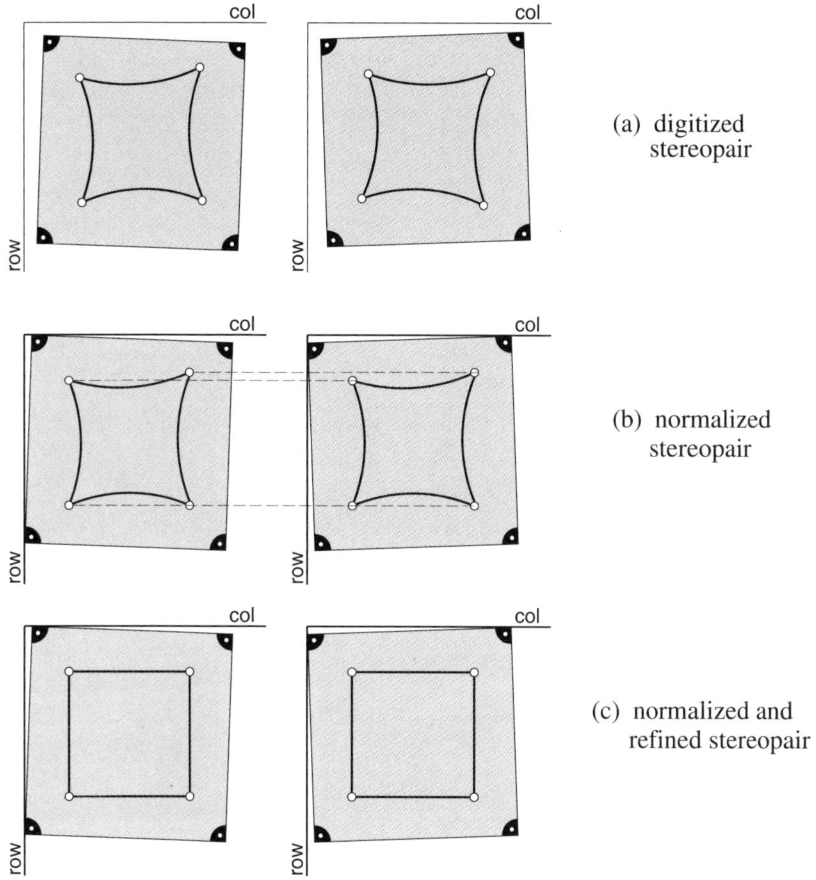

Figure 12.1: Relationship of images obtained at different stages. In (a), a digital stereopair is shown, obtained by scanning aerial photographs. With the exterior orientation parameters, normalized images are shown in (b). As an optional step, the images may be corrected at the same time for camera errors (e.g. radial distortion), atmosphere (e.g. refraction), and scanner errors (c).

fit in this scheme? Such images are aligned such that the image and photo-coordinate systems are parallel. Again, this does not constitute normalized images, however.

Normalized images simplify the process of matching stereopairs because conjugate entities are confined to the same row. This was a distinct computational advantage in the past, but considering the ever increasing processing power of computers, the advantage becomes less significant. Moreover, normalized images are related to stereopairs; every new stereopair requires new normalized images. Now suppose we have an entire strip. Every image within a strip belongs to two different stereopairs (except the first and last one). It follows that every image has two normalized positions. It does not make much sense to store normalized images for later computations as this would double the required precious storage capacity. A more realistic scenario is to compute normalized images on the fly. *Graham et al.* (1997) describe this approach.

A distinct advantage of normalized images is related to the display of stereopairs on digital photogrammetric workstations (Chapter 9). It is easier for an operator to work with normalized images, particularly in nonconventional applications that may involve convergent imagery.

12.2 Epipolar Geometry

Fig. 12.2 shows a stereopair in both the original (a) and normalized position (b). Assuming two images with central projection, the following definitions hold:

air base joins the two projection centers C', C'';

epipolar plane all planes which contain the air base;

epipolar lines e', e'' intersections of epipolar plane with photographs;

epipoles intersections of the line through the projection centers (extended airbase) with image planes. Except for convergent imagery, the epipoles are not within the physical boundary of an image.

The epipoles are the centers of bundles of epipolar lines that result from intersecting the photographs with all possible epipolar planes. The projections P', P'' of a point P on the images are on two corresponding (conjugate) epipolar lines. The conjugate epipolar lines in Fig. 12.2(b) are parallel and identical to the rows of both images. To obtain normalized images from their original position, a transformation must be devised such that the camera axes become perpendicular to the airbase. Furthermore, the axes must be contained in a plane that contains the base. This leaves one degree of freedom. You can easily imagine the infinite number of normalized positions by rotating the plane with the camera axes about the air base. A reasonable choice is a position that minimizes the size of the normalized images (see Sec. 12.4.1).

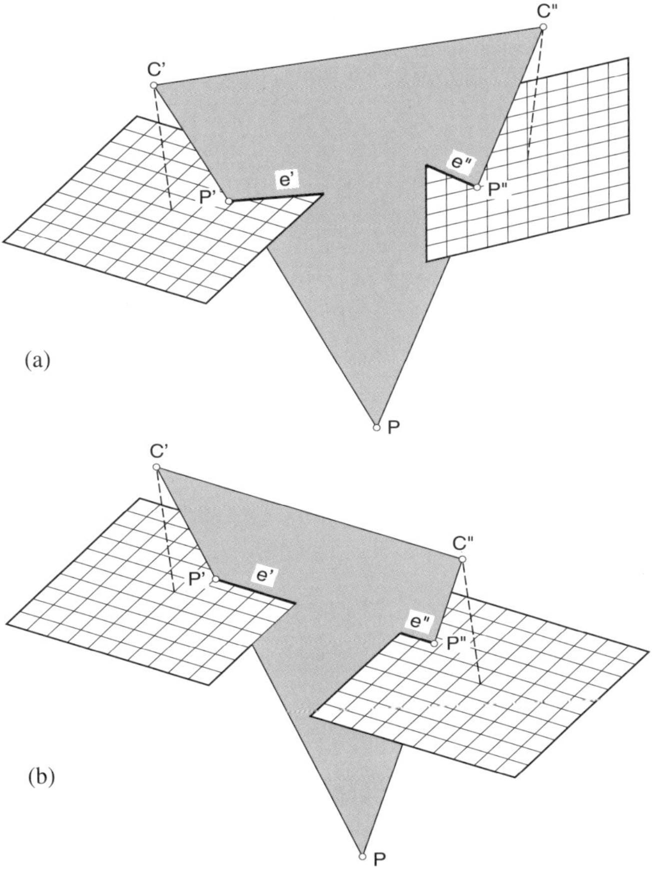

Figure 12.2: Aerial stereopair shown in its original position (a), as captured during the instant of exposure. In (b), the normalized position is shown. This position is characterized by the fact that the rows of both images are identical to epipolar lines (see text for detailed explanations).

12.3 *Transformation from Original to Normalized Image*

In this section, we establish the relationship between the original stereopair and its normalized position. It is more convenient to work with photo coordinates. With the interior orientation parameters we can always switch easily between image and photo coordinates. O' and O'' are the left and right image of the original stereopair; similarly, N' and N'' denote the normalized stereopair. Quantities related to these images are subscripted accordingly.

The exterior orientation of the original images consists of three rotation angles and the positions of the perspective centers for both images. The transformation from original images (O', O'') to normalized images (N', N'') can be thought of as proceeding in two steps. First, the images are transformed to their true vertical position and from there to the normalized position. Fig. 12.3 shows the relationship between the two positions.

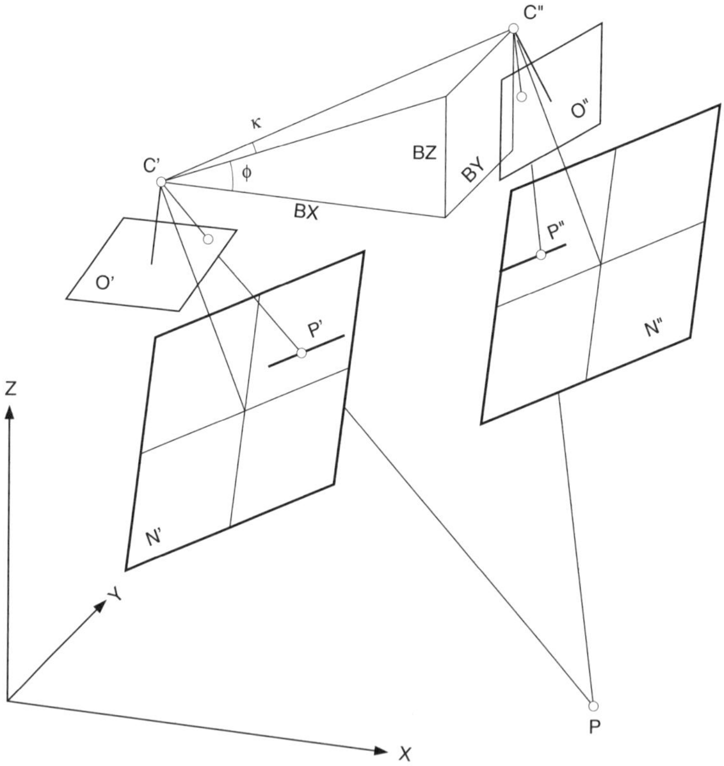

Figure 12.3: Relationship between original and normalized stereopair.

The first transformation from the original position to true vertical simply involves a rotation with transposed rotation matrix $\mathbf{R}^T$ of the exterior orientation.

Next, the image is transformed from true vertical to its normalized position. This involves rotating the true vertical images about the airbase. Thus, we need the rotation matrix of the base, denoted by $\mathbf{R_b}$. Two angles are required to determine the spatial direction of the base in object space. As illustrated in Fig. 12.3, the two angles are ϕ and κ. They can be determined from the base components $[BX, BY, BZ]^T$. First, we rotate the true vertical image about the Y – axis by ϕ, followed by the rotation about the Z – axis (κ). We have

$$
\begin{aligned}
BX &= X_C'' - X_C' && (12.1)\\
BY &= Y_C'' - Y_C' && (12.2)\\
BZ &= Z_C'' - Z_C' && (12.3)\\
\phi &= \arctan\left(\frac{BZ}{BX}\right) && (12.4)\\
\kappa &= \arctan\left(\frac{BY}{(BX^2 + BZ^2)^{1/2}}\right) && (12.5)
\end{aligned}
$$

At this point we still can rotate the normalized image about the base. To provide a unique solution, this rotation must be fixed. A reasonable choice is to define the rotation angle as follows

$$\Omega = \frac{\omega' + \omega''}{2} \qquad (12.6)$$

with ω', ω'' the rotation angles about the X – axis from the exterior orientation. Whenever we rotate images, the net area becomes larger. If we maintain the resolution (number of pixels) between original and rotated image, then the pixel size will be larger. If we are more interested in preserving the pixel size, then a higher resolution is required. The choice of Ω in Eq. 12.6 minimizes the increase in size of the normalized image.

Let us combine the three elementary rotations into the base rotation $\mathbf{R_b}$.

$$\boxed{\mathbf{R_b} = \mathbf{R}_\kappa \mathbf{R}_\phi \mathbf{R}_\Omega} \qquad (12.7)$$

where

$$
\begin{aligned}
\mathbf{R}_\Omega &= \begin{vmatrix} 1 & 0 & 0 \\ 0 & \cos(\Omega) & \sin(\Omega) \\ 0 & -\sin(\Omega) & \cos(\Omega) \end{vmatrix} \\
\mathbf{R}_\phi &= \begin{vmatrix} \cos(\phi) & 0 & \sin\phi \\ 0 & 1 & 0 \\ -\sin(\phi) & 0 & \cos(\phi) \end{vmatrix} \qquad (12.8) \\
\mathbf{R}_\kappa &= \begin{vmatrix} \cos(\kappa) & -\sin(\kappa) & 0 \\ \sin(\kappa) & \cos(\kappa) & 0 \\ 0 & 0 & 1 \end{vmatrix}
\end{aligned}
$$

The normalized rotation matrix $\mathbf{R_n}$ is the product of two matrices: the rotation matrix from image to true vertical, and the base rotation matrix. We combine the two rotations and obtain

$$
\boxed{\mathbf{R_n} = \mathbf{R_b}\mathbf{R}^T} \qquad (12.9)
$$

Since $\mathbf{R}^T$ is the transposed rotation matrix of the exterior orientation, it is different for each image and $\mathbf{R_n}$ must be determined for both images of the stereopair. We may use one of two transformations from original image to normalized image; transformation using collinearity condition or projective transformation.

12.3.1 Transformation Using Collinearity Equations

The relationship between the images can be conveniently expressed by the collinearity equations. The transformation is represented by the following equation.

$$
\begin{aligned}
x_n &= -f_n \frac{r_{11}x_o + r_{12}y_o - r_{13}f_o}{r_{31}x_o + r_{32}y_o - r_{33}f_o} \\
y_n &= -f_n \frac{r_{21}x_o + r_{22}y_o - r_{23}f_o}{r_{31}x_o + r_{32}y_o - r_{33}f_o} \qquad (12.10)
\end{aligned}
$$

where x_o, y_o are the photo coordinates of the original images, x_n, y_n those of the normalized images, and $r_{11} \ldots r_{33}$ are the elements of $\mathbf{R_n}$.

12.3.2 Projective Transformation

The projective transformation can be applied since both original image and normalized image are planar.

$$
\begin{aligned}
x_n &= \frac{c_{11}x_o + c_{12}y_o + c_{13}}{c_{31}x_o + c_{32}y_o + 1} \\
y_n &= \frac{c_{21}x_o + c_{22}y_o + c_{23}}{c_{31}x_o + c_{32}y_o + 1}
\end{aligned}
\tag{12.11}
$$

By comparing the coefficients in the projective transformation with those in the collinearity equations, we find the following identities:

$$
\begin{aligned}
c_{11} &= \frac{f_n r_{11}}{f_o r_{33}} & c_{21} &= \frac{f_n r_{21}}{f_o r_{33}} \\
c_{12} &= \frac{f_n r_{12}}{f_o r_{33}} & c_{22} &= \frac{f_n r_{22}}{f_o r_{33}} \\
c_{13} &= -\frac{f_n r_{13}}{r_{33}} & c_{23} &= -\frac{f_n r_{23}}{r_{33}} \\
c_{31} &= -\frac{r_{31}}{f_o r_{33}} & c_{32} &= -\frac{r_{32}}{f_o r_{33}}
\end{aligned}
\tag{12.12}
$$

When performing the transformation from original to normalized image, the square tesselation of the original image will become distorted. It is usually easier to project the tesselation of the normalized image back to the original image to determine the gray levels.

If the focal lengths of the original and normalized images are the same, that is $f_o = f_n$, then the coefficients $c'_{11} \cdots c'_{32}$ for the backward projection are obtained in the same fashion by $\mathbf{R_n^T}$.

$$
\begin{aligned}
c'_{11} &= c_{11} & c'_{21} &= c_{12} \\
c'_{12} &= c_{21} & c'_{22} &= c_{22} \\
c'_{13} &= c_{31} f_o^2 & c'_{23} &= c_{32} f_o^2 \\
c'_{31} &= c_{13} \frac{1}{f_o^2} & c'_{32} &= c_{23} \frac{1}{f_o^2}
\end{aligned}
\tag{12.13}
$$

For the more general case of different focal lengths ($f_o \neq f_n$), the backward projection is obtained by inverting $\mathbf{R_n}$ because $\mathbf{R_n^{-1}} \neq \mathbf{R_n^T}$.

12.4 Normalized Digital Images

The previous section described the procedure to transform images into their normalized positions. We have assumed that the images were represented in the photo-coordinate[1] system. This section describes how to perform the same operation with image coordinates[2].

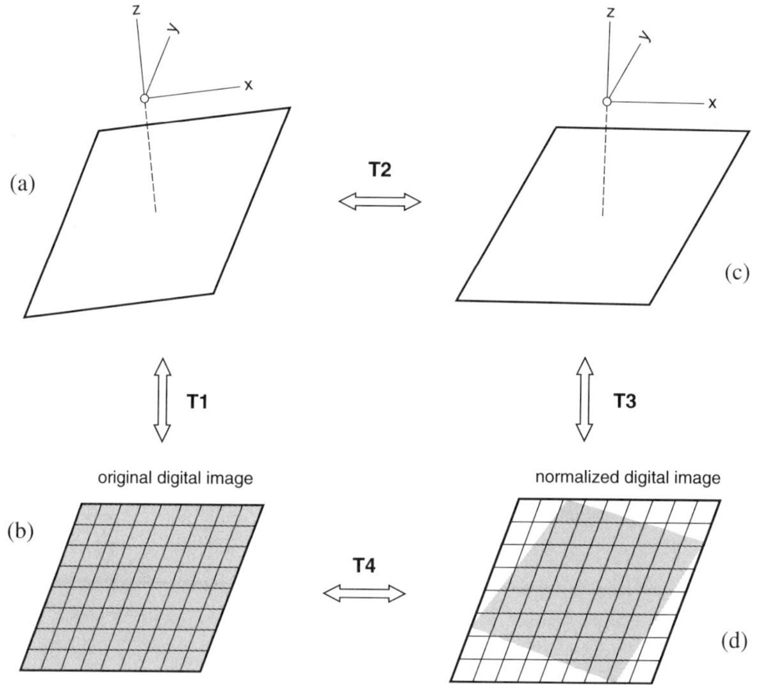

Figure 12.4: Relationship between original and normalized stereopair.

Fig. 12.4 shows the different images involved. The original aerial photograph is shown in (a), including the photo-coordinate system. Its digitized version is depicted in (b). Fig. (c) depicts the normalized aerial photograph while (d) is the normalized digital image. The relationship between these different images is symbolically shown by the transformations $\mathbf{T}_1, \ldots, \mathbf{T}_4$.

Let us now further examine the relationship between the different images.

$\mathbf{T}_1$ Transformation between original photograph (diapositive) and digital image. The transformation parameters are determined during the

[1]The origin of the photo-coordinate system is in the perspective center and the positive x – axis points toward the flight direction.

[2]As used throughout this book, the image coordinate system has its origin in the left upper corner and points are expressed as pixel addresses.

process of interior orientation. $\mathbf{T}_1$ enables the computation of pixel addresses from photo-coordinates and vice versa.

$\mathbf{T}_2$ is the projective transformation between photographs in original and normalized position. Sec. 12.3 explains the procedure in detail.

$\mathbf{T}_3$ establishes the origin and size of the normalized digital image. The procedure is explained below.

$\mathbf{T}_4$ transformation between original and normalized digital image. It corresponds to $\mathbf{T}_2$. This relationship is necessary for performing the resampling of the normalized digital image from the original one.

12.4.1 Definition of Normalized Digital Image

We now turn to the problem of defining the size of the normalized digital images. Starting point is the normalized stereopair one exposure of which is depicted in Fig. 12.4(b). It includes the four corner points $P_1, \ldots, P_4$ of the original digital image. For example, P_2 is the origin $(0, 0)$, and P_4 has the pixel address $(N-1, M-1)$. Fig. 12.5 shows the normalized stereopair with the corner points.

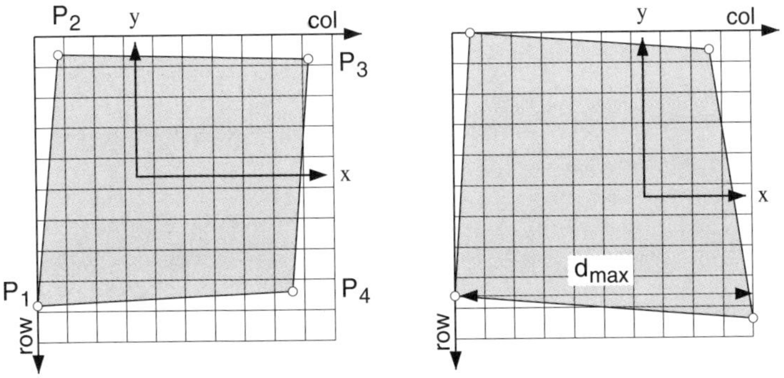

Figure 12.5: Definition of epipolar digital image.

As illustrated in Fig. 12.5, the origin of the normalized digital image is found by determining the maximum y – coordinate that determines row $= 0$ for both images. The minimum x–coordinate in both images define col $= 0$. For the size of the normalized digital image we compute the maximum coordinate difference, d_{max}, in either x– or y–direction. Now we have a choice between the following two cases

1. Keep original pixel size, p and compute the number of pixels (resolution) r from $r = d_{\max}/p$.

2. Keep original resolution and compute a new pixel size to $p = d_{\max}/r$.

With this definition of the normalized pixel images, we determine the gray levels of the pixels by resampling. This requires the transformation $\mathbf{T}_4$ that allows the computation of the pixel addresses of identical points between the original and normalized digital image. An easy way to determine the transformation parameters is by computing a perspective transformation with the corner points $P_1, \ldots, P_4$.

We summarize the procedure as follows.

1. The interior orientation establishes $\mathbf{T}_1$ with which points between the pixel system and the photo-coordinate system can be transformed.

2. With $\mathbf{T}_1$ compute the corner points $P_1, \ldots, P_4$ in the photo-coordinate system.

3. Compute normalized images in photo-coordinates (Eq. 12.9 on page 308 and Eq. 12.12 on page 309).

4. Determine maximum y – coordinate and minimum x – coordinate of the corner points. This defines the origin of the normalized digital images.

5. Determine maximum coordinate difference d_{max} between corner points to determine the size (number of pixels) or pixel size of the normalized digital image.

6. Resample normalized digital image from original digital image with $\mathbf{T}_4$ that can be found by a perspective transformation between the two images (using the corner points).

Problems

1. Explain why true vertical photographs are not normalized.

2. The rotation about the base is the one degree of freedom that is inherent to the computation of normalized images. What are reasonable choices for this rotation?

3. Suppose the air base of a stereopair is off by 5° from a perfect east-west direction. Can the stereopair be normalized such that epipolar lines are exactly in an east-west direction?

4. Can more than two images be normalized with respect to each other?

5. What are the advantages of normalized images?

6. At what stage in the normalization process should image refinement (e.g. correction for distortion, atmosphere) take place?

7. Taking images with a digital camera assures that the digital system is parallel to the photo-coordinate system. Explain why these images are still not yet normalized.

References

Cho, W., T. Schenk and M. Madani (1992). Resampling digital imagery to epipolar geometry. In *International Archives of Photogrammetry and Remote Sensing*, **29**(B3), 404–408.

Graham, L., K. Ellison and S. Riddell (1997). The architecture of a softcopy photogrammetry system. *Photogrammetric Engineering and Remote Sensing*, **63**(8), 1013–1020.

Kreiling, W. 1976. *Automatische Herstellung von Höhenmodellen und Orthophotos aus Stereobildern durch digitale Korrelation.* Diss. Fakultät für Bauingenieur- und Vermessungswesen, Universität Karlsruhe.

Part III

Automatic Orientation Procedures

The main task of photogrammetry—analog, analytical and digital alike—is to reconstruct the object space from images. You may consider the reconstruction as the inverse process of image formation. The latter proceeds from the scene to the image while reconstruction begins with images and ends with a suitable description and representation of the scene.

One task of reconstruction deals with determining positions of features in object space from known quantities in image space. Before computing positions in object space, two major problems must be solved—the camera's interior and exterior orientation.

The interior, relative and absolute orientation are the fundamental orientation procedures in analog and analytical photogrammetry. Are these procedures still necessary in digital photogrammetry—and if so—are there perhaps substantial differences in how to perform them in a digital environment?

Pundits of platform orientation systems have long predicted the end of traditional orientation, including aerial triangulation. Recent advances expedite the process but do not supersede it. For one, the relative orientation and aerial triangulation also render information about the surface. Moreover, the orientation parameters compensate model deficiencies to a certain degree, that is, small differences between the mathematical model and the physical reality.

Can we simply duplicate the classical orientation procedures or does digital photogrammetry require new approaches? If you work in an interactive setting, for example on digital photogrammetric workstations, then the orientation tasks are essentially performed in the same fashion as on an analytical plotter. We focus, in this part, on automatic orientation procedures, point out important differences, and modify existing methods accordingly, or derive new solutions.

In traditional photogrammetry, a human operator performs the orientation. The crucial task for the operator is to identify and to measure suitable "points". Economic reasons put a limit on the number of measured points, however. Simple but effective adjustment procedures have been developed to determine the orientation parameters and to assess their quality. The algorithms are point-based.

In contrast to humans, computers are far inferior in terms of selecting (extracting) suitable points and measuring (matching) them. It is easier to extract features instead. Strictly speaking, a point in a mathematical sense does not exist in an image. What the human operator measures as a "point" is in fact the result of an elaborate image interpretation process—an ability that is far beyond the reach of a computer operator. On the other hand, there is virtually no limit to the number of features a computer operator can extract and process. Consider linear features, such as edges, for a moment. Edges are sequences of edge pixels. They are not nearly as accurate as points carefully selected and measured by a human operator, but there are many more, increasing the redundancy manifold. The poorer quality is compensated by the greater number—quantity for quality!

The common theme in the following three chapters is how to use features effectively in the orientation procedure. Research in interior, relative and absolute orientation has primarily focused on extracting and matching features but the computation of the orientation parameters is performed with existing point-based algorithms developed in analytical photogrammetry. We expand this concept and use linear and areal features as entities in the adjustment procedure.

It is amazingly difficult to develop robust automatic orientation systems that work autonomously for fairly general scenarios. It takes a human operator a few minutes at most to perform the interior orientation, for example. Why bother with developing an autonomous system?

The motivation for developing automatic—or autonomous if you prefer—orientation systems is the desire to build an automation chain that would link several photogrammetric processes. The interior orientation is the starting point of this automation chain, followed by relative and absolute orientation or aerial triangulation. Even intermediate stages are of interest. For example, after relative orientation, the model can be displayed and viewed stereoscopically. If we take this automation chain to the extreme, then at the end we have a module that would recognize objects—the ultimate goal of digital photogrammetry; to produce maps automatically. Here is another reason for emphasizing features at as early a stage as possible. Linear and areal features carry more explicit information about the object space than points. For example, edges may correspond to object boundaries, say, a building. Having included them in the computation of orientation parameters increases their "value" in view of object recognition. Again, consider an edge that has been used successfully in the relative orientation. It transforms to a 3-D edge in object space—clearly, a more interesting entity.

Chapter 13

Automatic Interior Orientation

The main task of photogrammetry—analog and digital alike—is to reconstruct the object space from images. You may consider the reconstruction as the inverse process of image formation. The latter proceeds from the scene to the image while reconstruction begins with images and ends with a suitable description and representation of the scene.

One task of reconstruction deals with determining positions of features in object space from known quantities in image space. Before computing positions in object space, two major problems must be solved, however. For one, we need to determine the *exterior orientation* of the camera—its position and attitude referenced in the object space. The other prerequisite is the *interior orientation*, the subject of this chapter.

13.1 Purpose of Interior Orientation

The purpose of interior orientation is to establish a transformation from measured or computed features in the image to a 3-D Cartesian coordinate system that has its origin in the camera's perspective center. We call this system the *image coordinate system* while the reference system for the digital image is the *pixel coordinate system*, or simply the *pixel system*. Fig. 13.1 depicts the two systems. The digital image shown is a digitized diapositive.

The image coordinate system serves as a suitable reference for expressing spatial positions in the image space. Its origin is in the perspective center. The fiducial marks of the diapositive shown in Fig. 13.1 define the fiducial center **FC**. The offset between the fiducial center and the principal point of autocollimation, **PP**, is known from the camera calibration procedure. The point of best symmetry, **PS**, is the origin of the radial distortion.

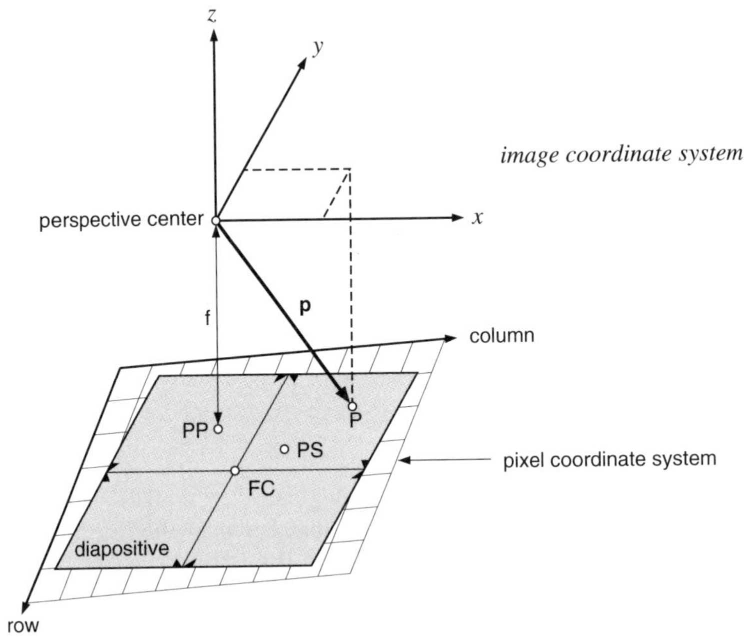

Figure 13.1: The 3-D Cartesian image coordinate system has its origin in the perspective center. The intersection of lines between opposite fiducial marks defines the fiducial center **FC**. The principal point **PP** is the vertical projection of the perspective center onto the pixel system. The point of best symmetry, **PS**, is the origin of the radial distortion.

The x, y coordinate plane is parallel to the pixel image and the positive x−axis points in the flight direction.

Positions in the image space are expressed by point vectors. For example, point vector **p** defines the position of point P in the pixel system. We have for point P

$$\mathbf{p} = \begin{bmatrix} x_p \\ y_p \\ -f \end{bmatrix} \tag{13.1}$$

Note that the third component is negative. This changes to a positive value in the rare case when the film negative instead of the diapositive is digitized.

The geometric reconstruction is a transformation between the image coordinate system and the object space coordinate system. To keep this transformation simple, it is customary in analytical photogrammetry to employ the collinearity model; perspective center, image point, and object point are on a straight line.

Now, the collinearity model only approximately describes the physical image formation process. Deviations between mathematical model and physical reality must be considered. The standard way to accomplish this is by adding "corrections" to the collinearity model— a process known as image refinement. In analogy to approximating a function by expanding it into a Taylor series about initial values, you may consider the collinearity model as the initial approximation of the image formation process and the corrections applied to it as the partial derivatives. To continue with this analogy, we will be discussing only "first order corrections" in this chapter. As a consequence, we ought to extend the definition of interior orientation to include at least the first order corrections, including refraction, radial distortion, film shrinkage, and scanner errors.

13.2 Determining Interior Orientation

13.2.1 Transformation from Pixel to Image Coordinate System

The transformation of a point P in the pixel system (r_p, c_p) to image coordinates is rather simple as the following equations demonstrate (see also Fig. 1.3 on page 6.

$$\boxed{\begin{aligned} x_p &= (c_p - c_o)p_c \\ y_p &= -(r_p - r_o)p_r \\ z_p &= -f \end{aligned}} \qquad (13.2)$$

where r_o, c_o are the pixel coordinates of the principal point and $p_r \times p_c$ is the pixel size in metric units, e.g., microns.

At this point you may wonder where the pixel coordinates of the principal point come from. If the image is obtained by a digital camera, then r_o, c_o are available from the camera calibration.

Perhaps the more typical case is that of digitized diapositives. Here, the principal point is not directly available. First, the fiducial center must be determined by measuring the fiducial marks, followed by applying the shift from **FC** to **PP**.

The fiducial center is determined by way of a 2-D transformation. The set of measured fiducial marks in the pixel system is transformed to the corresponding set in the fiducial system that is available from the camera calibration.

The general form for transforming pixel coordinates to the fiducial system x^f, y^f is

$$\begin{aligned} x_p^f &= a_{11} r_p + a_{12} c_p + a_{13} \\ y_p^f &= a_{21} r_p + a_{22} c_p + a_{23} \end{aligned} \tag{13.3}$$

For a similarity transformation the coefficients a_{ij} are

$$a_{11} = s \cdot \cos\alpha \qquad a_{12} = -s \cdot \sin\alpha$$
$$a_{21} = -a_{12} \qquad a_{22} = a_{11}$$

where the scale factor s is related to the pixel size. Note that in the case of rectangularly shaped pixels ($p_r \neq p_c$), two scale factors must be introduced. Hence, an affine transformation should be chosen, leading to the following coefficients for Eq. 13.3

$$a_{11} = s_x\,(\cos(\alpha - \epsilon \sin\alpha)) \qquad a_{12} = -s_y \sin\alpha$$
$$a_{21} = s_x\,(\sin(\alpha + \epsilon \cos\alpha)) \qquad a_{22} = s_y \cos\alpha$$

Here, s_x, s_y express the scale factors in the $x-$ and $y-$directions. The skew angle ϵ denotes the non-orthogonality of the coordinate axes.

The last step translates the fiducial system into the image coordinate system. The shift parameters x_o, y_o are known from the camera calibration procedure.

$$\boxed{\begin{aligned} x_p &= x_p^f - x_o \\ y_p &= y_p^f - y_o \\ z_p &= -f \end{aligned}} \tag{13.4}$$

13.2.2 *Image Refinement*

To refine the collinearity model, let us consider some of the factors that cause it to deviate from the physical reality of the image formation process. Fig. 13.2 shows the major steps involved in obtaining digital imagery. The left side of the figure depicts the traditional scenario of data acquisition with an aerial camera, followed by scanning the film. The right side illustrates data acquisition by digital cameras. Proceeding from the scene to the digital image, the following factors ought to be considered in first order image refinement.

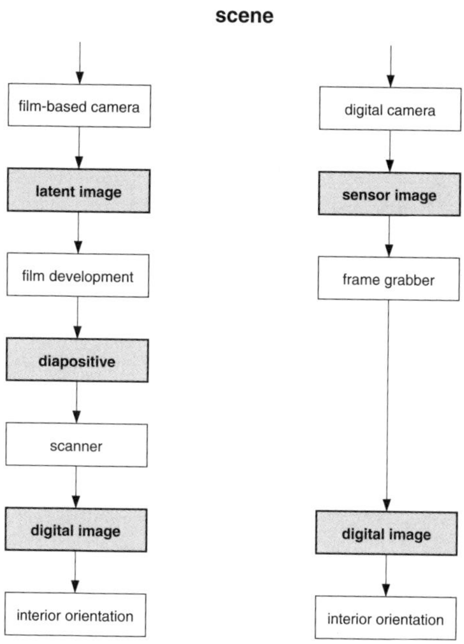

Figure 13.2: Major steps involved in image formation. The left side sketches data acquisition using a film-based camera, followed by digitizing the diapositives with a scanner. The right side of the figure shows image formation with digital cameras.

Refraction

Fig. 13.3 shows how an oblique light ray is refracted by the atmosphere. According to Snell's law, a light ray is refracted at the interface of two different media. The density differences in the atmosphere are in fact different media. The refraction causes the image to be displaced outward.

The radial displacement caused by refraction can be computed based on a model atmosphere.

$$
\begin{aligned}
dr &= K\left(r + \frac{r^3}{f^2}\right) \qquad (13.5)\\
K &= \left(\frac{2410\,H}{H^2 - 6H + 250} - \frac{2410\,h^2}{(h^2 - 6h + 250)H}\right) 10^{-6}
\end{aligned}
$$

with r the distance from the principal point PP to the imaged point, f the

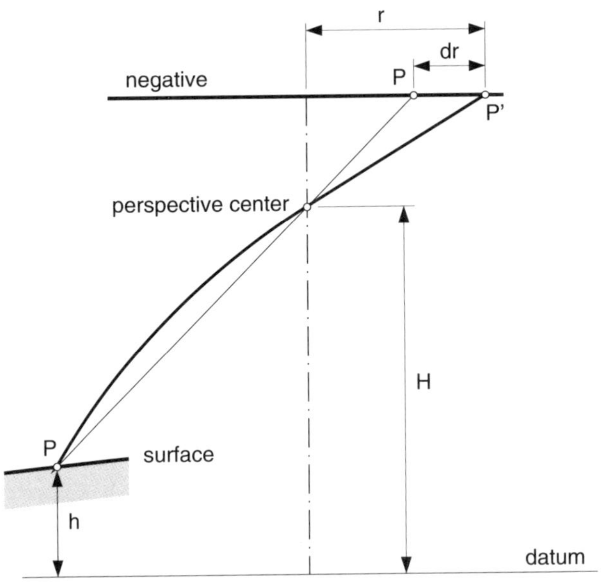

Figure 13.3: Oblique light rays are refracted by the atmosphere. The effect is quite similar to that of radial distortion.

focal length, H the flying height, and h the ground elevation.

With dr computed from Eq. 13.5 we have for the displacement of an image point x, y

$$
\begin{aligned}
dr_x &= \frac{x}{r} dr \\
dr_y &= \frac{y}{r} dr
\end{aligned}
\tag{13.6}
$$

Camera Imperfections

The latent image formed during the time of exposure is not exactly where you would expect it with the collinearity model. First, the light rays are slightly deflected from the path through the lens by *radial distortion*. It displaces off-axial points in a radial fashion. A positive distortion increases the lateral magnification while a negative distortion reduces it.

Distortion values are determined during the process of camera calibration. They are usually listed in tabular form, either as a function of the radius or of the angle at the perspective center. Let dd be the radial distortion value for points on a circle (centered on the point of best symmetry, PS), radius $r = (x^2 + y^2)^{1/2}$. Then the displacement of the image coordinates are

$$dd_x = \frac{x}{r} dd \tag{13.7}$$

$$dd_y = \frac{y}{r} dd \tag{13.8}$$

For aerial cameras, the distortion values are very small. Hence, it suffices to linearly interpolate the values from the table provided in the calibration protocol. For cameras with significant radial distortion, it may be necessary to interpolate dd from an odd-power polynomial of the form

$$dd = p_0 r + p_1 r^3 + p_2 r^5 + \cdots \tag{13.9}$$

The coefficients p_i are found by fitting the polynomial curve to the distortion values.

Other camera imperfections include tangential distortion, non-flatness of image plane, non-perpendicularity of image plane and optical axis. These are examples of second order image refinement.

Film Shrinkage

During the development process, the film undergoes considerable mechanical stress. This, together with bathing and drying, causes deformations. The developed film, say a diapositive, is not an exact geometric copy of the latent image. It is difficult to model these errors. Luckily enough, the base material of aerial films is extremely stable, and under normal circumstances, the film shrinkage is neglected.

There is more mechanical stress along the film roll than across. Moreover, the stress may not be uniform. Therefore, one may model the difference between latent image and diapositive by an affine transformation. Such a transformation would only take global shrinkage into account. To consider local film shrinkage (second order image refinement), a reseau camera[1] is needed.

Scanner Errors

To obtain a digital image from aerial photography, the film is digitized by a scanner. The scanner is not perfect and introduces new errors. That is, the

[1] On the focal plane plate of a reseau camera, an accurately ruled grid is etched. This grid is imaged at the instance of exposure. It allows a more rigorous elimination of systematic errors.

digitized image is not geometrically identical with the diapositive, nor with the latent image. In Chapter 8 we discussed some of the potential errors, and the question is, now, how to model them.

Photogrammetric scanners are accurate and reliable. Systems that use modified carrier stages of analytical plotters for the photo carrier have a typical geometric accuracy of less than 5 μm. A hallmark of these systems is that they closely follow Abbe's principle[2]. This important principle is violated by desktop scanners. Here, the plane defined by the measuring system (cross-slide system carrying the optics and sensor elements) and the diapositive (glass plate on top of measuring system) are not perfectly parallel. Consequently, a projective transformation should be employed to describe the relationship between the diapositive and the digital image. It turns out that an affine transformation approximates a projective transformation quite well if the two planes are nearly parallel.

Unless data is available from a scanner calibration, an affine transformation may sufficiently describe the relationship between diapositive and digital image.

Image Refinement Using Digital Cameras

The previous sections described some of the differences between the mathematical model of central projection (collinearity model) and the physical model in the case of obtaining imagery by film-based cameras. Let us now briefly analyze the situation for the case of digital cameras. Assuming the aerial case, refraction affects digital cameras as much as aerial cameras and we ought to consider it.

Digital cameras suffer the same shortcomings as aerial cameras as far as geometric distortions are concerned. In fact, many digital cameras have considerably larger radial and tangential distortions. Moreover, the rigidity of the camera body is inferior to aerial cameras. As a consequence, the metric properties are more likely to change between calibrations. It appears that digital cameras should be calibrated more frequently than aerial cameras.

Fig. 13.2 on page 323 indicates a difference between the sensor image and the digital image. The sensor image is analogous to the latent image of a film-based camera. It is the image captured by the detector elements. In the case of a CCD array, this image consists of accumulated charge and is not directly accessible as such. The processes of transferring the charge, amplifying and measuring it, and converting it to a digital signal cause differences between the sensor image and the digital image. Chapter 7

[2] Abbe put forward the principle that the object to be measured should be in the same plane as the measuring device, represented by coordinate axes, for example.

describes some potential error sources. The calibration of digital cameras should attempt to capture the nature and the magnitudes of the systematic errors, so that they can be modeled appropriately.

Summary of Image Refinement

We have discussed two aspects of the interior orientation. The first aspect concerns the transformation of pixel coordinates to image coordinates. Eq. 13.2 on page 321 describes the simple translation when digital cameras are used. In the case of digitized aerial photography, two steps are necessary. First, the pixel coordinates are transformed to the fiducial system (Eq. 13.3), followed by a shift to the image coordinate system (Eq. 13.4).

The second aspect of interior orientation relates to the deviations between the real (physical) image formation process and the simple mathematical model of central projection that is used in the reconstruction process. We have taken the approach of adding "corrections" to the position of features in the real image. Fig. 13.4 illustrates this point. The influence of radial distortion and refraction is explicitly expressed in Eqs. 13.7 and 13.5, respectively. The influence of film shrinkage and scanner errors is considered by applying an affine transformation from pixel to fiducial coordinates. It is worth pointing out that a single affine transformation lumps together all sorts of systematic errors related to camera, film, and scanner. It will not be possible to separate them into the individual components, however.

In conclusion, we should clearly distinguish between two images: the real digital image and the virtual image. Conceptually, the latter is a rubber-sheet version of the real image. The real image, of course, is used for human interpretation and image related computations, such as filtering, edge detection, and segmentation. The virtual image is used for the geometric reconstruction—the transformation of features from image to object space.

In fact, it is possible to generate a new digital image that is geometrically identical to the virtual image simply by performing a rubber-sheet transformation. This may be advantageous, particularly when combined with the process of computing normalized images (see Chapter 12).

13.3 Interactive Interior Orientation

Interactive interior orientation is performed on Digital Photogrammetric Workstations (DPWs). These stations offer various degrees of support for measuring the fiducial marks. As a minimum you would expect the same support that analytical plotters offer. The drive to feature drives analytical

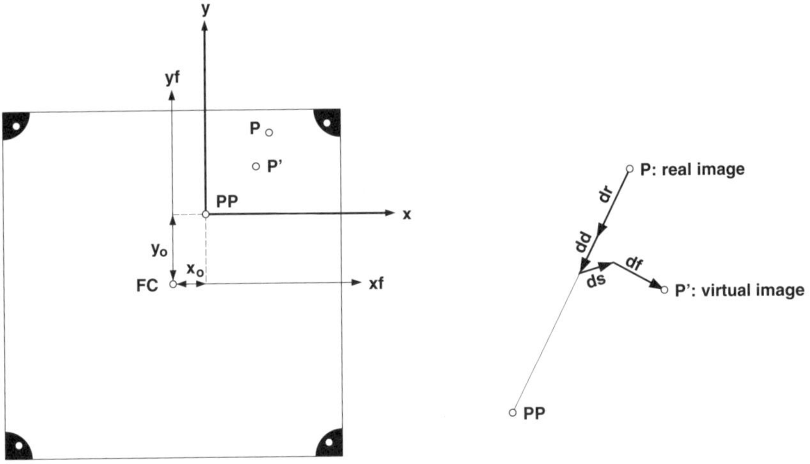

Figure 13.4: Image refinement. The real position of a point is "corrected" and shifted into a new location. If applied to every point in the image, a new, virtual image is obtained. This virtual image is used for transforming features from image to object space. See text for additional explanations.

plotters near the first fiducial mark. After the first measurement, newly computed translation parameters drive the plotter closer to the second fiducial mark. After the second point is measured, a rotation can be computed which drives the analytical plotter even closer to the remaining fiducial marks.

The same procedure is performed on a DPWs, except that the drive to feature is replaced by positioning the cursor accordingly. Since no mechanical parts move, the operation of positioning the cursor to predicted fiducial mark locations is considerably quicker. Yet another advantage is the shape of the cursor that can easily be changed for optimal positioning. Fig. 13.5 shows a typical display of Intergraph's ImageStation Z during the interior orientation. Note the position of the cursor in the overview image and the large display with the cursor centered on the fiducial mark.

Most DPWs offer an option to measure the fiducial marks automatically. The preferred method is correlation through the image pyramid to improve the initial approximations. This is followed by gray level correlation, see, e.g. *Schickler* (1995); Kersten and Haering (1997), or by least-squares matching together with template matching as suggested by Lue (1997). We will examine some of these methods in the next section.

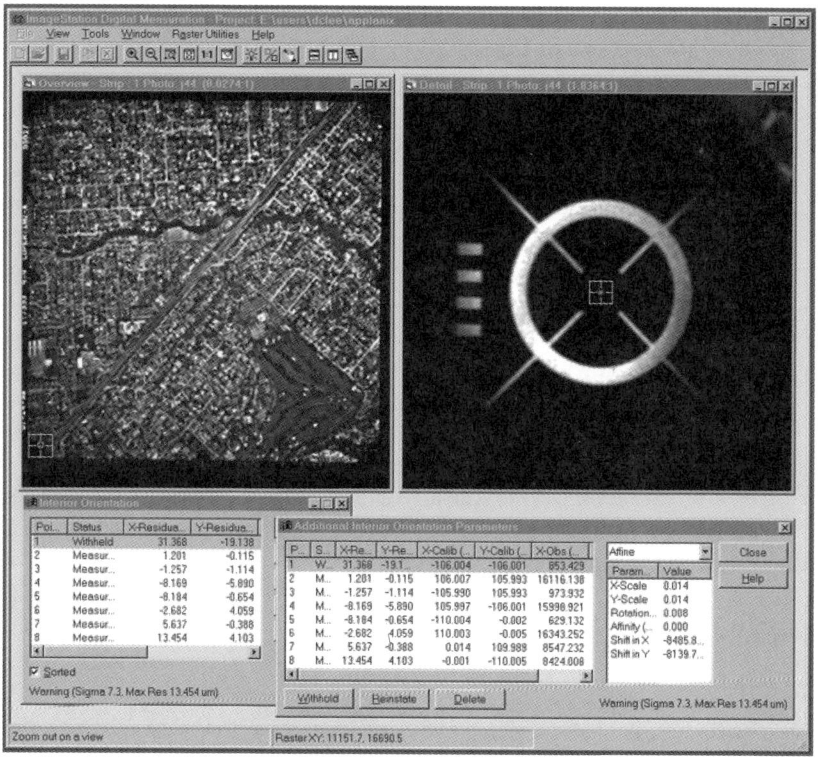

Figure 13.5: Interior orientation performed on Intergraph's ImageStation Z. The system offers an option to measure the fiducials automatically. The operator can accept or reject the result. The screen is divided into three parts. The left display contains an overview image. The highlighted fiducial mark (left lower corner) is displayed in the right field of the display where a square shaped box shows the result of finding the fiducial center automatically. The lower third of the screen shows two windows with measurement and computation results.

13.4 Autonomous Interior Orientation

In this section we contrast interactive interior orientation with an autonomous approach and motivate such a development. Moreover, we formulate the problems to be solved and analyze assumptions that can be made. Further sections elaborate on possible solutions.

13.4.1 *Background and Motivation*

With autonomous interior orientation, subsequently called AIO, I refer to a system that performs the entire orientation process without human interaction—sort of a "black box" process, if you wish. The term "automatic interior orientation" is often used to express various degrees of computer support in measuring fiducial marks on digital photogrammetric workstations. Even if the human operator's involvement is simply to accept or reject the result, there is a substantial difference between the "100% automatic system" and the autonomous system; the latter cannot rely on the safety net of a human operator who may come to the rescue should something go wrong. Hence, autonomous systems need to be foolproof. As a result, the development effort is much higher compared to an "automatic" system that runs in a (protected) interactive environment. You are about to ask the right question: the interactive interior orientation requires only a few minutes—so why bother with an autonomous system?

The motivation for developing an AIO system is the desire to build an automation chain consisting of various photogrammetric processes. Fig. 13.6 illustrates this scenario.

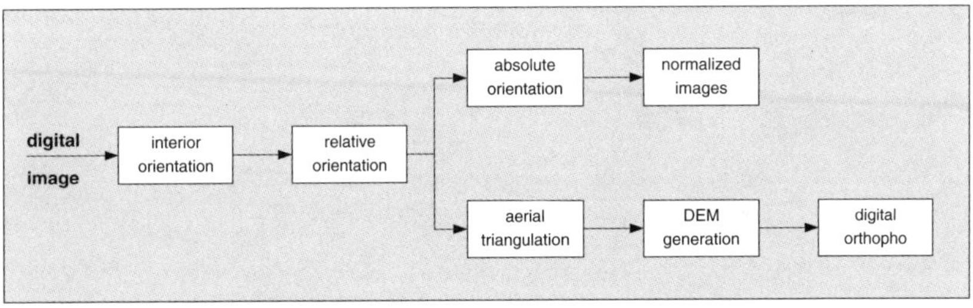

Figure 13.6: Automation chain. After scanning, the interior orientation executes autonomously, followed by the relative and absolute orientation or aerial triangulation. With known exterior orientation data, DEMs and orthophotos can be generated. These processes do not require an interactive environment, such as digital photogrammetric workstations.

The interior orientation is the starting point of the automation chain. Ideally, it executes parallel to the scanning process. Should the system discover a problem, then the image can be digitized again. In a traditional photogrammetry environment, all subsequent processes assume that the interior orientation parameters are known. Such is the case with relative orientation. After relative orientation, normalized images can be determined (see Chapter 12). Now you can display the images for the first time and view the model stereoscopically. Alternatively, you may be interested in determining the exterior orientation of all images involved in a project (aerial triangulation), which is a prerequisite for generating DEMs from more than one model.

13.4.2 Objective of Autonomous Interior Orientation

The following list contains the major objectives of an autonomous interior orientation system.

identification and subpixel localization of fiducial marks is the main objective. Identification includes the task of determining which of the fiducial mark is detected. Subpixel localization is required because the pixel size is most likely larger than the expected precision of the fiducial centers.

autonomous process requires a general and robust solution, accommodating different types of fiducial marks. A system is robust if it can cope with different problems as they may occur in a production environment. For example, the film may be placed upside down in the scanner, some fiducials may be only partly digitized. In addition to operator mistakes, there are film imperfections to consider, such as noise, blemishes, low contrast due to the cosine to the fourth power drop-off law, vignetting, and over exposure when the fiducial mark is projected on the film. Fig. 13.7 on the following page shows some fiducial marks with different degrees of complexity for an AIO system to identify and locate precisely and reliably.

diapositives or negatives may be present and must be handled.

color or black and white film no assumptions can be made, both types may occur.

coarse resolution imagery the system should cope with image resolutions where the fiducial center may have been lost. As long as the fiducial mark is still uniquely identifiable, the location may be determined from the features that describe a fiducial.

13.4.3 Assumptions

Here we list some of the assumptions that one can reasonably make without restricting the generality of the solution. In fact, we describe the domain knowledge that should be brought to bear on the solution. If one or several

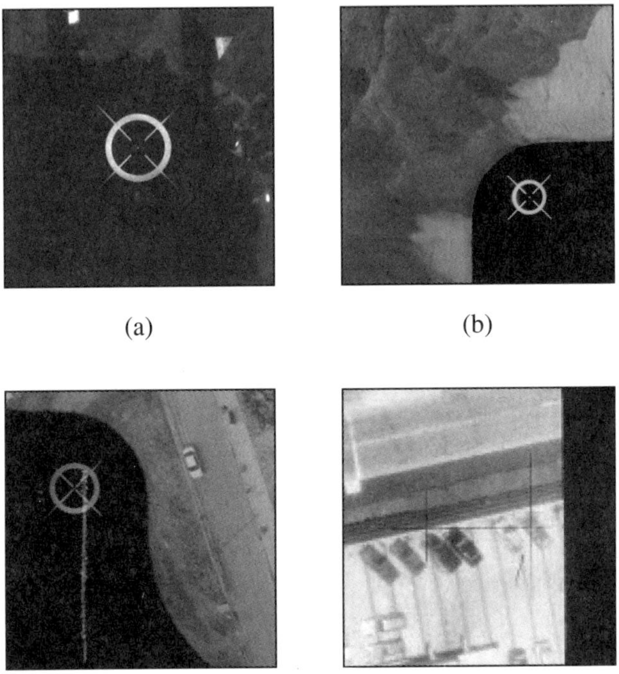

(a) (b)

(c) (d)

Figure 13.7: Different types of fiducial marks under various conditions are shown. Examples (a)–(c) show the typical structure consisting of a small disc marking the center, surrounded by an annulus and cross for easier recognition. The example in (c) shows a typical real world problem with a film scratch running across the fiducial mark. The unusual fiducial mark in (d) is from a reseau camera.

of these assumptions are violated, then the system should recognize this and comment accordingly. If it must terminate the process it should do so gracefully—that is, with proper explanations.

- It is assumed that the scanning system is known together with relevant parameters, such as maximum range and scanner calibration data.

- It is assumed that the camera type is known, including calibration data and suitable fiducial mark descriptions.

From the information about the scanner we can deduce a maximum rotation angle and the range within which the film may be placed, as illustrated in Fig. 13.8. Both help in predicting the search space where fiducial marks are to be expected. If no scanner calibration data is available, then there is still a chance to consider some scanner errors by using an affine transformation (see Sec. 13.2.2).

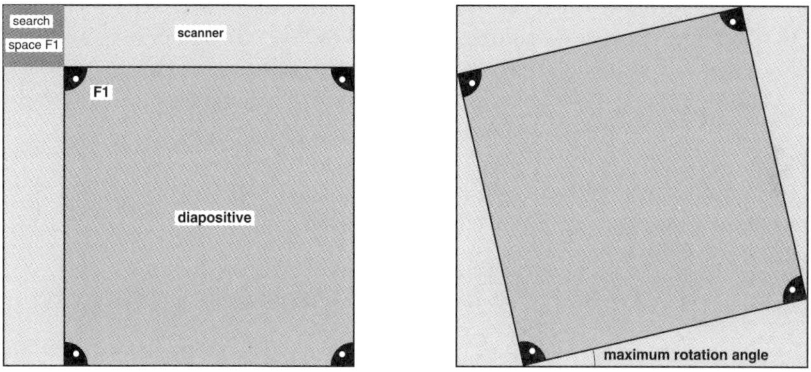

Figure 13.8: Search space for the left upper fiducial mark. The film size is smaller than the net scanning area. Since it is not known exactly where the operator places the film, extreme positions and a possible rotation angle are assumed for predicting the search space.

A traditional camera calibration report contains information about the position of the fiducial center, the offset to the principal point, the radial distortion, and the calibrated focal length. However, additional information is required for AIO. For example, each fiducial mark should have a unique identifier, such as a number or a signature that can be extracted from the imagery. Moreover, an ideal image (template) of the fiducial mark must be available. If relational matching methods are used, then we also need a structural description. It is clear that a great deal of additional information is required. The standard calibration report does not suffice.

13.4.4 Approaches to AIO

The chief problem to be solved is the identification of the fiducial marks and their precise and robust localization. It is reasonable to separate these two tasks. Approaches may pursue one of the following two strategies.

1. **Area-based approach:** The subimage that contains a fiducial mark is binarized. This approach resembles the Gestalt law of figure/ground separation (see 4.4.1). The precise localization is performed by cross-correlating an exact copy of the fiducial mark with the foreground image.

2. **Feature-based approach:** Here, extracted features are matched with the features of the ideal fiducial mark. This approach casts AIO as an object recognition problem. Features to be extracted consists of elements of a fiducial mark, for example straight lines, circles, crosses, and squares. Different matching methods may be employed. Examples include shape matching and relational matching.

The following sections elaborate on the details of the two approaches and highlight advantages as well as drawbacks.

13.4.5 The Anatomy of a Fiducial Mark

The fiducial marks are located on the upper surface of the inner cone of an aerial camera. They are located either at the four corners of the format opening and/or in the center of each of the four sides. Fiducial marks are registered on the film by projecting an image of the mark through a small lens housed inside the cone. Since each fiducial mark is built and projected separately, small variations may occur.

Fiducial marks have typical patterns, made up of geometric structures including straight line segments, crosses, solid squares and circles, and annuli. Most of the structures serve the identification process. The actual center of a fiducial mark, whose coordinates are known from camera calibration, is a small disc, just slightly bigger than the measuring mark of an analytical plotter[3]. The size of the center ranges from 50 to 100 microns.

It is quite straightforward to model fiducial marks geometrically. A simple description of the mark shown in Fig. 13.9 is a list of parts:

```
{(circle1; center=(0,0); radius=r1)}
{(circle2; center=(0,0); radius=r2)}
{(disc; center=(0,0); radius=rd)}
{(line1; length=l1);...(line4; length=l4)}
```

A more elaborate description would continue with relations. For example, the four line segments are located symmetrically about the origin, and they intersect the annulus halfway.

[3]The highest measuring accuracy is obtained when the white measuring mark is positioned within a slightly larger, black disc.

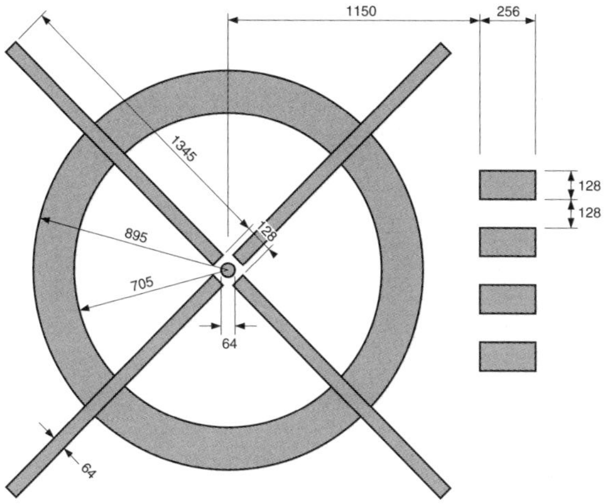

Figure 13.9: Design drawing of a fiducial mark. The design data depict the corner fiducials of the RC 30 camera by LH Systems, Heerbrugg, Switzerland. Dimensions are in units of microns. The data is useful for generating a geometric model that will be used for matching extracted features from the image. Elementary structures are line segments, annulus, and disc. There are relations among these structures that are useful in a relational matching scheme. The four rectangles on the right side indicate the position of the fiducial mark with respect to the image coordinate system. The example shown indicates the right upper corner fiducial.

The radiometric model of a fiducial mark is trivial. The elements making up the mark are evenly bright against a dark background. The first strategy for locating the fiducial marks is based on the radiometric model and proceeds as foreground/background separation, for example. The second strategy builds on the geometric model of the fiducial mark.

13.5 Area-Based Approach

Suppose we have extracted subimages such that each contains a fiducial mark. Now the simple radiometric model of a fiducial mark—bright mark against dark background—suggests binarizing the subimages. A straightforward approach is to define ranges of brightness values, for example by choosing suitable threshold values from the gray level histogram. We have introduced this method in the image processing chapter, Sec. 3.5.1. In fact, the example shown in Fig. 3.14 on page 68 is a fiducial mark.

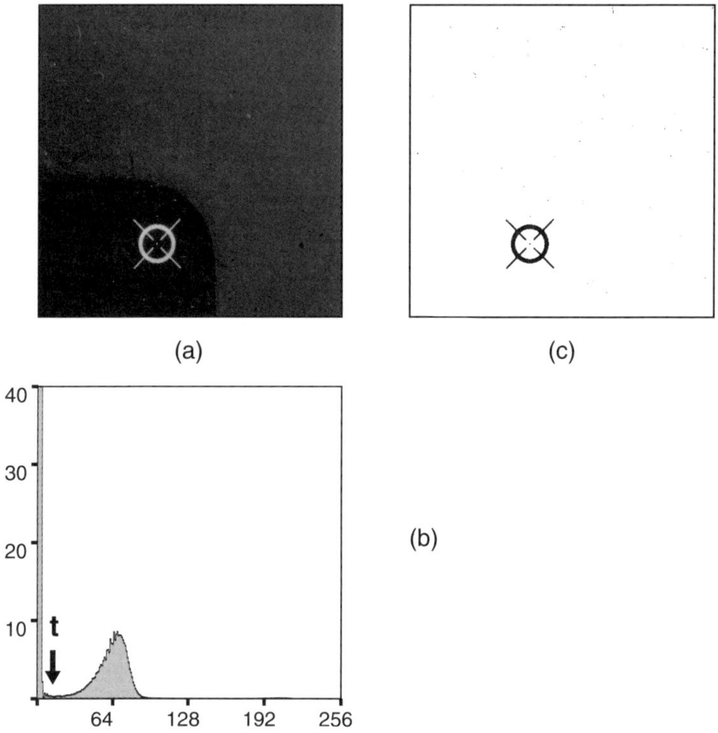

Figure 13.10: The histogram of the fiducial mark in (a) has two widely separated clusters (b). Binarizing the subimage with the threshold value t renders the expected result (c).

What is involved? The histogram of the subimage with the fiducial mark is most likely bi-modal. As Fig. 13.10(b) demonstrates, the dark cluster of gray values is neatly separated from the bright cluster. Choosing the threshold value t in the valley between the two clusters, and assigning all pixels with a gray value smaller than t to binary 0 and gray values greater than t to binary 1, delivers the expected result.

The same procedure, applied to another subimage, is far less successful, as Fig. 13.11 reveals. Although the two clusters are reasonably well separated, choosing the threshold t at the lowest point does not deliver what we were hoping for. Selecting different thresholds leads to quite different binary images as Figs. 13.11(c)-(f) convince you. The problem is rooted in the fact that the gray level distribution of the background and foreground overlap. In the transition zone between the two peaks we have mixed pixels that may belong to either of the two groups.

One approach to determining a suitable threshold value, suggested by *Otsu* (1979), is to minimize the weighted sum of group variances. Let t be the threshold value, dividing the histogram into two groups. Let $\sigma_1^2(t)$ be the variance of the first group and $\sigma_2^2(t)$ the variance of the second group. Then σ_w^2, the weighted sum of the group variances, is defined as

$$\sigma_w^2(t) = s_1(t)\sigma_1^2(t) + s_2(t)\sigma_2^2(t) \tag{13.10}$$

where $s_1(t)$, $s_2(t)$ are the probabilities for group one and group two, respectively.

Now we find the best threshold value t where $\sigma_w^2(t)$ reaches a minimum. To determine the group variances $\sigma_1^2(t)$ and $\sigma_2^2(t)$, we first need the group means $\mu_1(t)$ and $\mu_2(t)$. The following procedure, together with Eq. 13.10, must be repeated for different threshold values

$$s_1(t) = \sum_{i=1}^{t} p(i) \tag{13.11}$$

$$s_2(t) = \sum_{j=t+1}^{N} p(j) \tag{13.12}$$

$$\mu_1(t) = \frac{\sum_{i=1}^{t} p(i) \cdot i}{s_1(t)} \tag{13.13}$$

$$\mu_2(t) = \frac{\sum_{j=t+1}^{N} p(j) \cdot j}{s_2(t)} \tag{13.14}$$

$$\sigma_1^2(t) = \frac{\sum_{i=1}^{t} (i - \mu_1(t))^2}{s_1(t)} \tag{13.15}$$

$$\sigma_2^2(t) = \frac{\sum_{j=t+1}^{N} (j - \mu_2(t))^2}{s_2(t)} \tag{13.16}$$

Fig. 13.11(f) shows the binarized image of the fiducial mark, using the principle of minimizing the group variances. *Haralick and Shapiro* (1992) describe more sophisticated thresholding methods. *Sahoo* (1988) provides a comprehensive review of binarization methods.

Whatever approach we may choose, it is very unrealistic to assume that we can neatly separate the fiducial mark from the background. Some of the pixels labeled with a binary 1 do not belong to the fiducial mark. Moreover, several pixels of the fiducial mark are classified as background (binary 0).

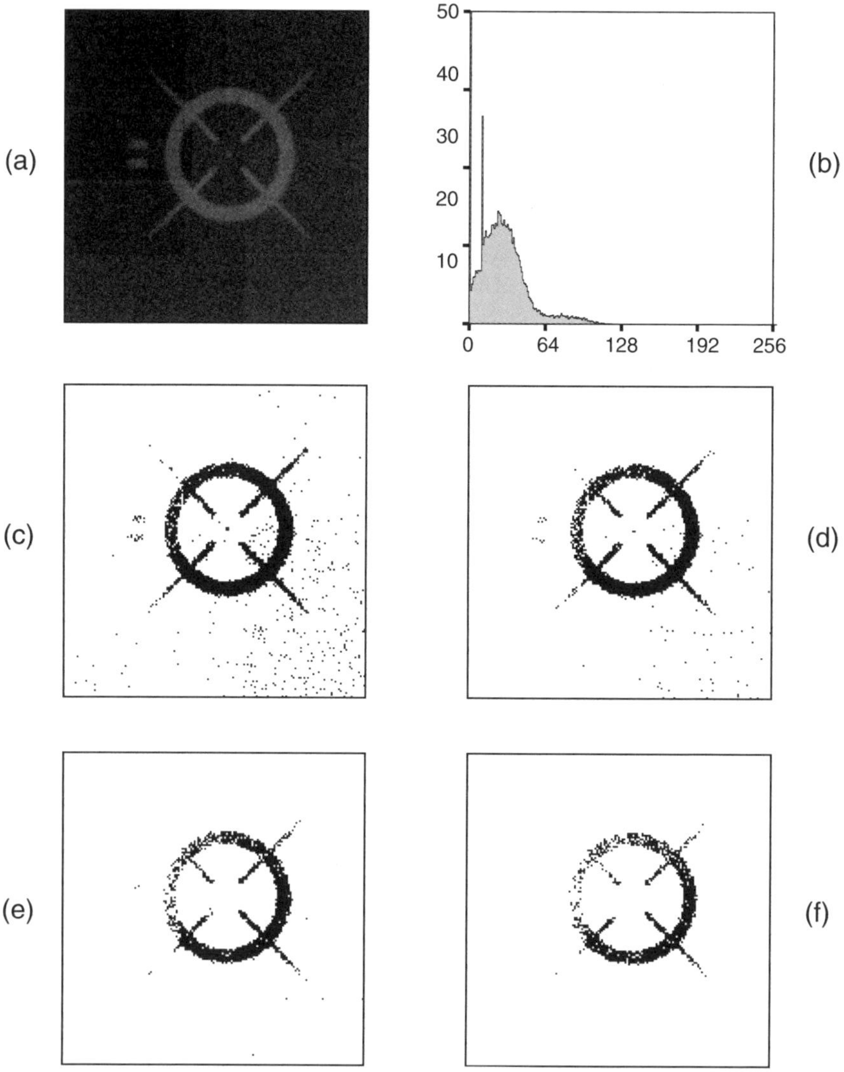

Figure 13.11: The histogram of the fiducial mark depicted in (a) has a transition zone containing pixels that may belong to either of the two clusters. Threshold values at 55, 61, 69, and 74 have been selected. The resulting binary images, shown in (c) through (f) are quite different and may lead to different fiducial center locations.

The only way to sort things out is to follow the binarization with a grouping process, guided by the knowledge about the shape and size of the fiducial mark. Here, the first step is to form binary 0 and binary 1 regions by a connected component algorithm, for example. Once regions are made explicit, their area, shape, moments, and boundaries can be determined. This is necessary for identifying the regions that belong to the fiducial mark. Single, isolated pixels, or small isolated regions that are labeled as foreground probably do not belong to the fiducial mark. Hence, they can be eliminated.

13.5.1 Precise Localization

We now turn to the problem of determining the fiducial center as precisely as possible. To achieve results comparable to analytical photogrammetry, subpixel accuracy is necessary because the pixel size most likely exceeds the expected accuracy by an order of magnitude. Assume a typical pixel size of 15 μm to 30 μm. Then, the fiducial center should be determined to one tenth of a pixel. The larger the pixel size, the higher the subpixel accuracy requirement. But with a larger pixel size, fewer pixels make up the fiducial center, and its position is determined less accurately!

Let us analyze the problem by way of an example. Fig. 13.12 on the following page illustrates the process of sampling the fiducial center. Suppose a square-shaped fiducial mark, size 60 μm, is ideally imaged on the film[4]. Now we scan the film with 20 μm resolution. As indicated in Fig. 13.12(a), the fiducial mark does not align exactly with the pixel tessellation. As a consequence, pixels along the boundary also receive energy and are turned on. The numbers in Fig. 13.12(b) express the area that the fiducial mark covers for in each pixel. The four pixels in the center are fully covered and assigned a value of ten. The pixel at the lower left corner receives only 8% of the energy of the central pixels. Let us turn these numbers into gray values. Then, Fig. 13.12(b) reflects the gray level image of the fiducial mark. After thresholding, Fig. 13.12(c) is obtained. Note that regardless of how the threshold value is chosen, the resulting binary image of the fiducial mark differs from the original in various ways, for example in size and shape. Worse, it also differs in location. We conclude that sampling and thresholding cause dislocalization of the fiducial mark.

Now, the question is how to determine the center of the fiducial mark. Probably, the first method that comes to mind is to consider the centroid as the center. Although simple, this approach is affected by sampling and

[4]An ideal image means a perfect copy of the object. Theoretically we should use the line spread function to obtain a more realistic image.

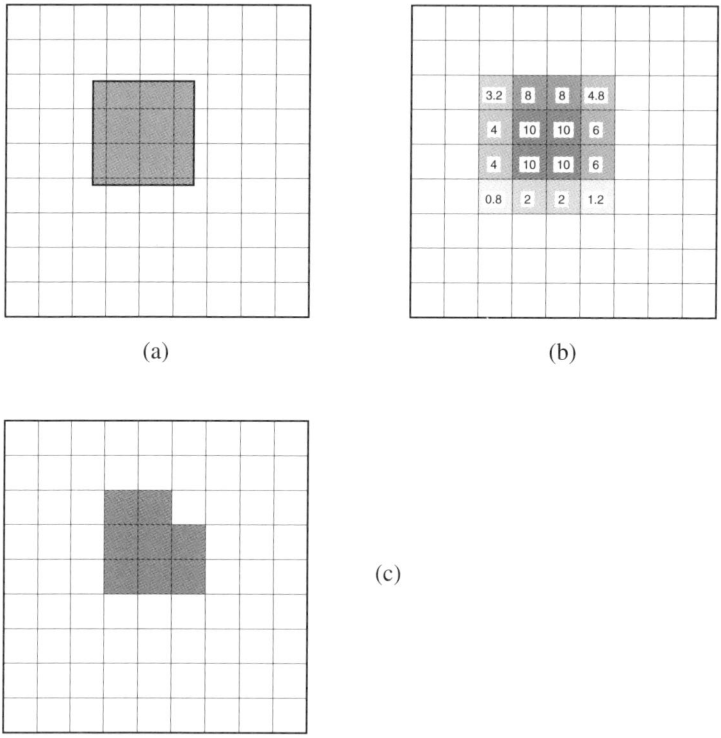

Figure 13.12: A square-shaped fiducial center, size 60 μm, is ideally imaged on the film (a). The film is digitized with 20 μm pixel size. The resulting gray level image is shown in (b). The numbers indicate the fractional area of each pixel that is "covered" by the fiducial (ten is full coverage). Thresholding the gray level image leads to the binary image in (c). Note that the binary image is different from the original in shape, size and even in location. Sampling and thresholding cause dislocalization.

thresholding effects. Different thresholds may lead to different shapes and consequently to different locations.

Another approach for defining the subpixel location is based on cross-correlation (see Sec. 10.4.1). Here, we generate an ideal image of the fiducial mark (template) and correlate it with the thresholded image. For every correlation position, the correlation factor is determined. A 3×3 area that contains the maximum correlation coefficient in its center is used to fit a quadratic surface whose coefficients can be conveniently determined by least-squares. We have as a linear observation equation

$$r_i = a \cdot x_i^2 + b \cdot y_i^2 + c \cdot x_i y_i + d \cdot x_i + e \cdot y_i + f - \rho_i \qquad (13.17)$$

with ρ_i $i = 1, \ldots, 9$ the correlation coefficients in the 3×3 window. The maxima of the surface are found by solving the following two equations for x and y

$$2ax + cy + d = 0 \qquad (13.18)$$

$$cx + 2by + e = 0 \qquad (13.19)$$

To assess the accuracy of the subpixel location, we analyze the quadratic function. For example, a significant change in the tangent around the maximum signals higher accuracy. This change is manifest in the second derivatives. Hence, coefficients a and b render a measure for the accuracy of the subpixel location.

By now you may have concluded that binarizing the image is probably not the best approach. Regardless of how sophisticated the thresholding procedure, binarization is somewhat arbitrary. A better approach is to leave the gray level image of the fiducial mark as it is and correlate it with a template that is computed at various subpixel locations. Fig. 13.13 illustrates the procedure.

First, the image of an ideal fiducial mark is computed by taking into account the line spread function. Now, this image is sampled at different subpixel locations, for example in intervals of one fifth of the pixel size. Sampling effects cause all of these templates to be slightly different. They are correlated with the image of the fiducial mark. The template that yields the maximum correlation coefficient is considered to be most similar to the image. The subpixel accuracy is now determined in the same fashion as before (Eq. 13.17).

Example 13.1:

An ideal fiducial mark, size 3×3 *pixels, is imaged at the subpixel location* $\Delta r = 0.2$ *and* $\Delta c = 0.6$. *The first entry in Table 13.1 contains the result of computing the centroid of the binarized image. The centroid is off by* 0.09 *and* 0.29 *pixels. The next entry refers to cross-correlating the ideal template with the binarized image. Entries 3—7 represent the results obtained by subsampling the template at pixel locations indicated in the comments column. The last line represents entry 4, except that the right upper pixel of the fiducial mark image was set to zero for simulating a blemish.*

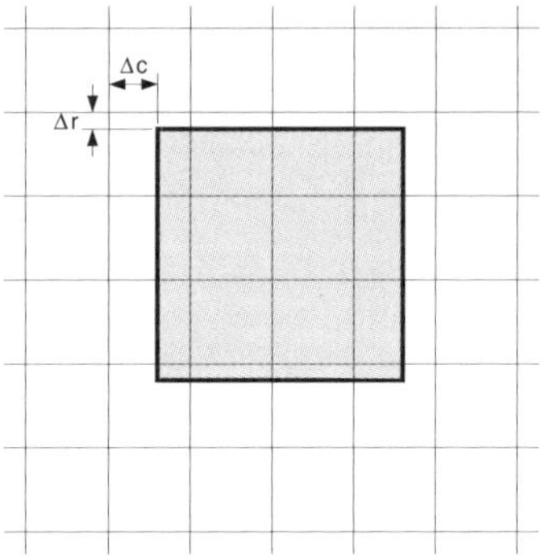

Figure 13.13: The template (ideal fiducial mark) is sampled at different subpixel locations $\Delta r, \Delta c$. The gray values are functions of the fractional part of the pixel that is "covered" by the mark. For example, the left upper pixel is covered by $(1 - \Delta r)(1 - \Delta c)$.

These experimental results confirm that the fiducial mark images should not be binarized. In *Trinder et al.* (1995) and related papers, the authors come to similar conclusions.

13.6 Feature-Based Approach

Fiducial marks are artificial objects that are projected onto the film during exposure time. Almost every fiducial mark has a simple, regular shape. This invites us to represent the fiducials as structural descriptions and to detect the structural elements in the image.

For the following discussion we use the sample fiducial mark shown in Fig. 13.7(c). It is representative of most fiducial marks. Moreover, the blemish—a film scratch right through the fiducial mark—presents one of the many problems that a robust system must cope with in an operational environment.

There are three structural elements, or shape primitives if you wish, that are likely to be present in any fiducial mark: straight line segments, circles, and discs. From the camera manufacturer's design data, exact dimensions are known, such as radius, length of lines, and their spatial relationships (see Fig. 13.9 on page 335).

Table 13.1: Results of subpixel target detection.

Entry	*Corr.* Coeff.	*Location* *row*	*col*	*Error* *row*	*col*	*Comments*
1		0.111	0.889	−0.089	0.289	binary image: centroid
2	0.9258	0.078	0.921	−0.122	0.321	binary image: x-corr.
3	0.9776	0.221	0.600	0.021	0.000	sampled at 0.4/0.6
4	0.9762	0.200	0.608	0.000	0.008	sampled at 0.2/0.4
5	0.9728	0.200	0.621	0.000	0.021	sampled at 0.2/0.8
6	0.9556	0.218	0.618	0.018	0.018	sampled at 0.4/0.8
7	0.9637	0.210	0.474	0.010	−0.126	blemish

With the shape primitives and the relationships we build higher structures. For example, two concentric circles of different radii constitute an annulus, two parallel line segments build a line pair, and line pairs with different orientations make up a cross. Continuing with this line of thought we realize that with a few simple shape primitives and the many relationships among the primitives we can build useful structural descriptions of fiducial marks. Such descriptions are quite general and can be used for many different fiducial mark types, as long as they are made up of circles, straight lines, and perhaps discs.

Fig. 13.14 shows the shape primitives and derived structures, including spatial relationships. With these simple constructs one can devise a powerful strategy for detecting and localizing fiducial marks.

1. Detect edge pixels as the atomic part of shape primitives.
2. Group edge pixels together that belong to one shape primitive, for example straight line segment or circle.
3. Check the relationships among the shape primitives and build higher order structures, for example annuli, line pairs, and crosses.
4. Compute the fiducial mark center from all shape primitives.

You may consider this approach as structural matching; although the matching of shape primitives and the spatial relationships among them are not performed simultaneously. After each grouping process, the relationships are checked. If satisfactory, the confidence related to the recognition increases. One of the most effective checks concerns the centers of the annulus and the cross for they must be identical.

structure	relationships
line segment	graphic primitive
circles	graphic primitive
line pair	2 parallel line segments
line pair with gap	2 parallel line pairs
cross	2 parallel line pairs with gap symmetric; perpendicular
annulus	2 concentric circles

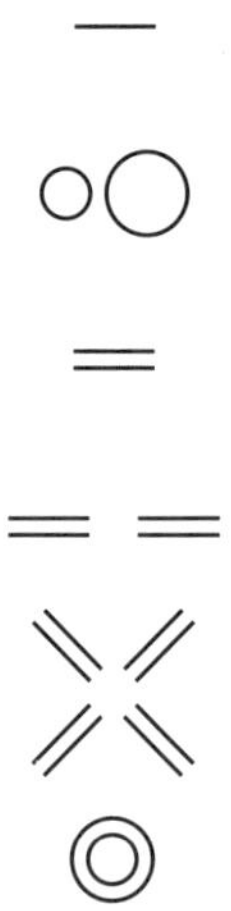

Figure 13.14: Illustration of the structures one can build from shape primitives for fiducial marks. Line segments and circles are the shape primitives. Based on these, annuli and crosses can be constructed that are important parts of the structural description.

The following sections address the problem of detecting the shape primitives (circles and straight lines), of checking the relationships, and of precisely locating the fiducial mark center.

13.6.1 Detecting Circles

We use the Hough transform technique, introduced in Chapter 5, to find circles. A circle can be represented in the spatial domain by

$$(x - x_o)^2 + (y - y_o)^2 = r^2 \tag{13.20}$$

with x_o, y_o the center, r the radius and x, y the variables.

Let x_i, y_i be a point on the circle and let x_o, y_o, r be variables. Then, the representation in the parameter space is again a circle

$$(x_o - x_i)^2 + (y_o - y_i)^2 = r^2 \tag{13.21}$$

We have established a simple relationship between spatial domain and parameter space. A point on a circle in the spatial domain transforms into a circle in the parameter space, where the center is given by the coordinates of the point.

Fig. 13.15 illustrates the relationship. Points 1 to 5 lie on a circle in the spatial domain. They all generate circles in the Hough space that intersect in a single point—the center of the circle in the spatial domain.

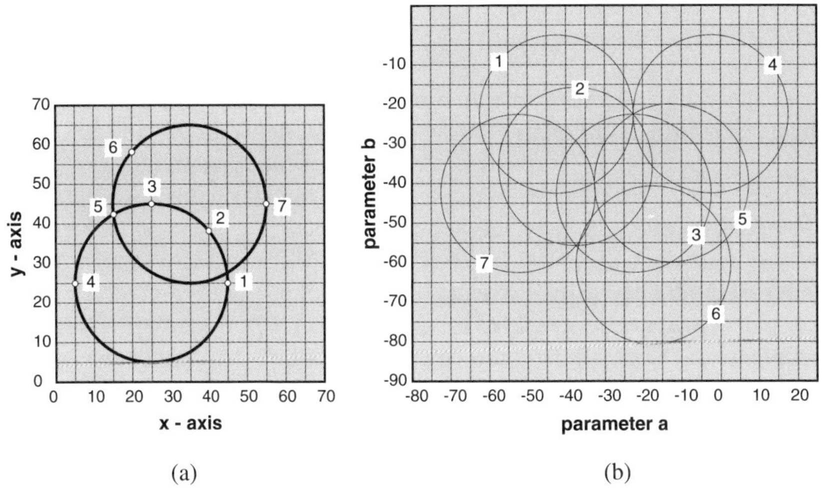

Figure 13.15: Illustration of the relationship between points on a circle in the spatial domain (a) and their representation in the parameter space (b). Points become circles, e.g., point 1 transforms into circle 1. Points on a circle in the spatial domain, e.g. points 1 to 5, generate circles in the parameter space that intersect in a single point—the circle's center in the spatial domain.

Let us apply this procedure to find the circles that are part of a fiducial mark. The radii of these circles are accurately known from design data. Hence we can reduce the 3-D parameter space to two dimensions. Fig. 13.16(a) is a subimage, size 512×512 pixels, pixel size about 40 μm. This fairly large subimage is selected to make sure that it contains the fiducial mark. The first step entails detecting edges, shown in Fig. 13.16(b). For every edge pixel the respective circles are generated in the parameter space.

Table 13.2: Results of circle detection.

Radius	*Circum ference*	*Pixels detected*	*Row*	*Col*
23	90	80	343	64
24	92	62	343	65
25	96	44	343	63
17	74	41	242	255
18	76	58	343	65
19	80	38	347	62

Consider the parameter space as being represented by a two-dimensional array, called the *accumulator array.* The two dimensions of the array correspond to x_o and y_o, respectively. All cells in the accumulator array are initialized to zero. As each edge pixel in the spatial domain is evaluated, the corresponding circle in the parameter space is generated, and represented by incrementing the affected cells of the accumulator array. After processing all edge pixels we search for peaks in the accumulator array. The cell value corresponding to the peak indicates how many pixels lie on a circle, and the peak's row and column indicate the circle's center in the image.

Results of finding circles in the image of Fig. 13.16(a) are listed in Table 13.2. The nominal radius of the larger circle is 24 pixels, and the smaller circle has a radius of 18 pixels. We find the fact that the two circles are concentric confirmed in Table 13.2. The ratio of detected pixels to maximum number of pixels per circle (circumference) is another criterion to assess the performance of the circle detection. These ratios, together with the confirmation of concentricity, justifies the hypothesis that we have found a fiducial mark. This hypothesis is further tested by finding other fiducial mark structures, such as straight lines. Fig. 13.16(c) shows the pixels that are part of the annulus of the fiducial mark. Note the fourth entry in Table 13.2. The row and column values are not consistent with the other results. Here, the circle detector found 41 edge pixels that happen to lie on a circle with radius 17, far from the fiducial mark!

13.6.2 Detecting Straight Lines

We now turn to the problem of detecting straight lines that are part of the fiducial mark. From the design data we expect two sets of parallel straight

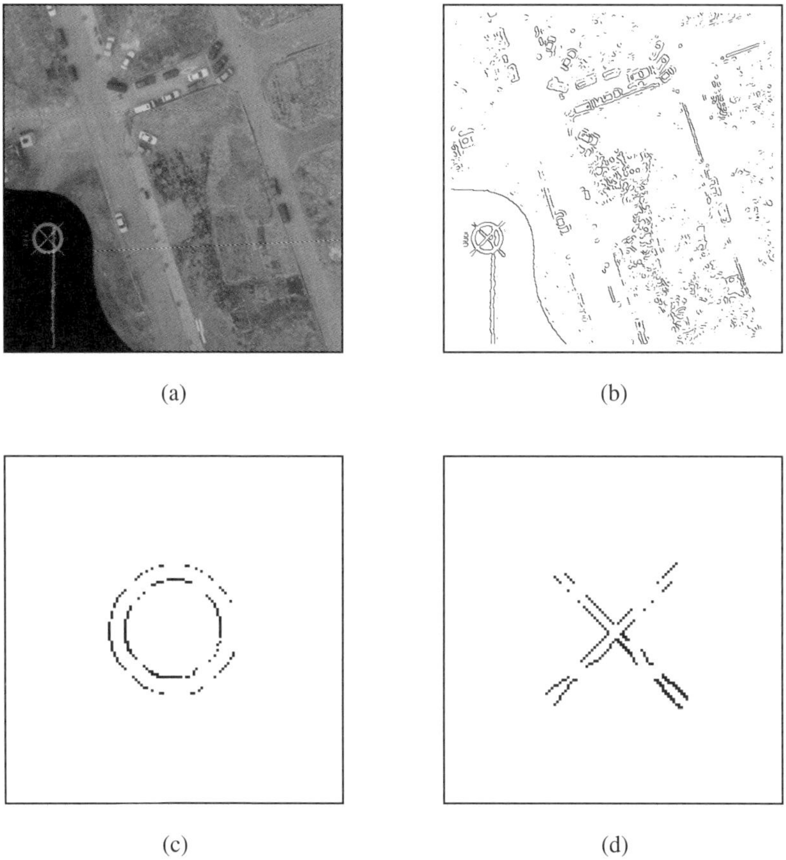

Figure 13.16: Illustration of detecting circles and crosses. The subimage in (a) has 512×512 pixels with a pixel size of 40 μm. A LoG operator with $w = 2$ was used for detecting edges (b). The result of detecting circles is shown in (c). Here, a new subimage, of size 128×128 pixels is chosen such that the hypothesized fiducial mark is centered. Pixels that have been found by the $\pm 45^o$ line detector are shown in (d).

lines at 45° and -45° angles. Moreover, their length is known.

Let $y = m \cdot x + c$ be the representation of a straight line in the spatial domain and let x_i, y_i be point P_i on the line. The parameter space is represented by the slope m and by the intercept c. Then, point P_i transforms into a straight line $c = -x_i \cdot m + y_i$. To generalize: points in the x, y space generate straight lines in the parameter space. At the intersection of these lines we have the parameters m, c of the line in the spatial domain.

The slope/intercept representation of a line is not without problems. Consider a line parallel to the y-axis, for example. However, since we are only interested in lines with slopes of $m = \pm 1$ we do not need to consider another line representation.

Fig. 13.17 depicts the scenario with the fiducial mark (a) and the corresponding accumulator array (b). Its dimensions are chosen such that a slope range of 2 to -2 can be accommodated. The intercept range is designed to detect lines within the area marked by dashed lines in Fig. 13.17(a).

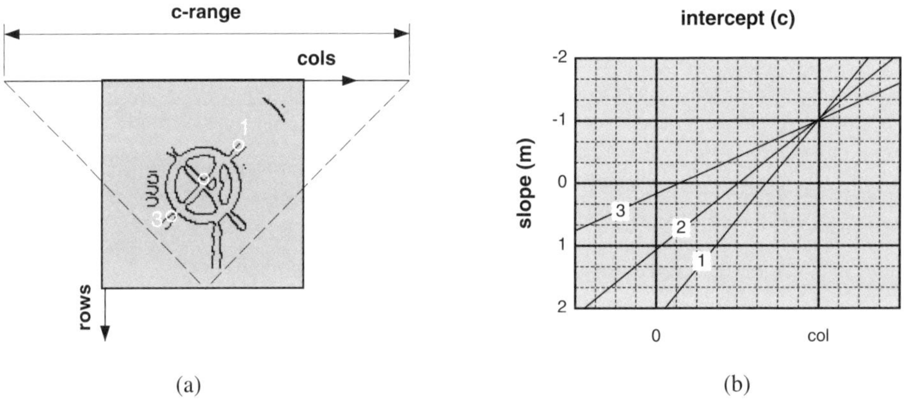

Figure 13.17: Illustration of the relationship between points on a straight line in the spatial domain (a) and their representation in the parameter space (b). Points transform into lines. The intersection of lines determines the parameter of lines in the spatial domain. The range of possible intercepts is shown in (a) by dashed lines. It determines one dimension in the accumulator array (b). The dimension of the slope parameter is application dependent. We need an m-range of at least ± 1 for detecting the crosses of the fiducial mark.

Fig. 13.16(d) shows the pixels that are part of the cross of the fiducial mark. Note that the analysis of the accumulator array lends itself to a robust test of the detected straight lines. For one, we can immediately determine how parallel the line pairs are and how much they are separated. Moreover, from the difference between the two slopes we deduce the angle. Since all these quantities are known from design data, we can determine deviations. If they agree with the expectations, the confidence that we identified a fiducial mark increases.

13.6.3 Precise Localization

The previous section addressed the problem of reliably identifying the fiducial mark. Although we know where it is, we have not made any effort yet to determine its position as accurately as possible.

The success with identifying the major structures of the fiducial mark suggests that we take a different approach to determine its center. That is, we compute the center from the structural elements rather than directly from the center pixels. In our example two concentric circles and four straight line pairs indicate the center.

As a preliminary step we may consider determining the edge pixels more accurately. Fig. 13.18 shows a gray level profile through the fiducial mark center. We clearly recognize the sharp rise in gray levels across the annulus. The fiducial's center also stands out, but not as distinctly as the ring around it. What is the peak between the center and the right cross-section of the annulus? This is the scratch through the fiducial mark that gave us much trouble in the area-based approach. It is conceivable to remove it since we concluded during the recognition process that it is not a structural part of the fiducial. Hence, the edge pixels of the annulus and the cross could be determined to subpixel accuracy.

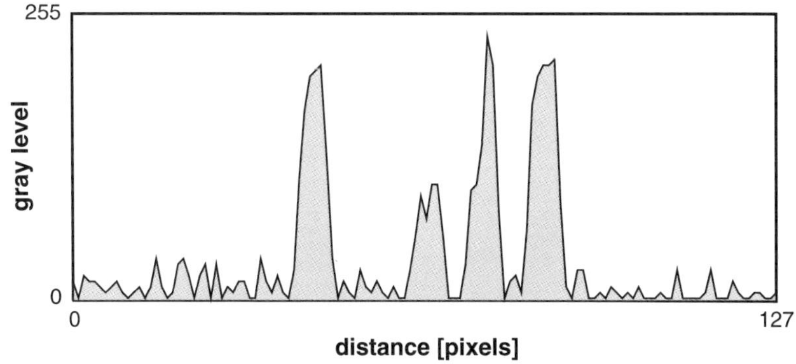

Figure 13.18: This gray level profile cuts through the center of the fiducial mark shown in Fig. 13.16(a). The first and last peak reflect the crossing of the annulus (from left to right). The fiducial center stands out too, but less distinctly than the annulus. The dominant peak is caused by the scratch in the film.

Let us now determine the center by a least-squares adjustment. First, we fit a circle of known radius through the pixels of the outer circle. The same process is repeated for the inner circle. Next, we fit straight lines through the cross pixels and force them to pass through the center of the two circles.

Let the two circles be represented by

$$(x_i - x_o)^2 + (y_i - y_o)^2 - R_1^2 = r_i \tag{13.22}$$
$$(x_j - x_o)^2 + (y_j - y_o)^2 - R_2^2 = r_j \tag{13.23}$$

with x_i, y_i the pixels on the outer circle with known radius R_1 and x_j, y_j the pixels on the inner circle whose radius is R_2. Assuming random errors for the pixel positions, Eqs. 13.22 and 13.23 constitute observation equations where the right-hand side contains the residuals r_i, r_j.

Linearizing 13.22 and 13.23 with respect to the two parameters x_o, y_o leads to

$$\begin{aligned} r_i &= -2(x_i - x_o)\Delta x_o - 2(y_i - y_o)\Delta y_o - (x_i - x_o)^2 + (y_i - y_o)^2 - R_1^2 \\ r_j &= -2(x_j - x_o)\Delta x_o - 2(y_j - y_o)\Delta y_o - (x_j - x_o)^2 + (y_j - y_o)^2 - R_2^2 \end{aligned}$$

Every pixel on a circle gives rise to one observation equation. For example, the circle detection discussed in the previous section leads to approximately 120 circle pixels (see Tab. 13.2 on page 346). We can also merge the cross pixels into the adjustment. Suppose the two narrow parallel lines can be represented by their center line. From the design data of the fiducial mark we know whether or not this centerline passes exactly through the fiducial center. In that case, the centerline can be represented by one point and the slope. We have

$$y_n - y_o - m_1(x_n - x_o) = r_n \tag{13.24}$$
$$y_m - y_o - m_2(x_m - x_o) = r_m \tag{13.25}$$

where x_n, y_n are the pixels of the line with slope m_1 and x_m, y_m the pixels of the line with slope m_2. As before, x_o, y_o is the fiducial center. Assuming the slopes and the center are unknown, Eqs. 13.24 and 13.25 must be linearized. We obtain

$$m_1\Delta x_o - \Delta y_o + (y_n - y_o) - m_1(x_n - x_o) = r_n \tag{13.26}$$
$$m_2\Delta x_o - \Delta y_o + (y_m - y_o) - m_2(x_m - x_o) = r_m \tag{13.27}$$

We now have a total of 4 parameters (x_o, y_o, m_1, m_2) and as many equations as there are detected pixels on the circles and the cross. Due to the large redundancy, blunders can be detected and eliminated more easily. Moreover, the accuracy and reliability increase.

13.7 Summary

The interior orientation of images that are obtained by digital cameras amounts to a simple translation from the pixel system to the image coordinate system. The translation parameters are known from the camera calibration. Although digital cameras are increasingly being used, the majority of imagery used in photogrammetry results from scanning film. Here, the interior orientation involves measuring the fiducial marks and computing the transformation parameters with a similarity or affine transformation. This is indeed very similar to the procedure in traditional photogrammetry, with a subtle difference, however. In digital photogrammetry, the interior orientation needs to be established only once, while in analytical photogrammetry it must be repeated every time the diapositives are measured on the analytical plotter.

We have focused on measuring the fiducial marks automatically. This was motivated by the desire to establish an automation chain, starting with the interior orientation. We examined two approaches. The simple radiometric model of a fiducial mark suggests an area-based approach with an attempt to separate the fiducial mark from the background by thresholding the histogram. We figured that binarization is not entirely satisfactory because changing the threshold value may change the pixel distribution. This, in turn, affects the shape of the fiducial mark and ultimately its position. To make the area-based approach work, the sampling errors should be taken into account, for example by generating gray level samples of the ideal fiducial mark at different subpixel locations, followed by matching them with the gray level image.

The feature-based approach pursues an entirely different strategy. It is driven by the simple structural description that one can build for a fiducial mark. The success of this approach depends on how well the features that make up the structural description can be extracted from images. The predominant structural elements of fiducial marks are straight lines, circles, and discs. They can be extracted quite simply from edge images by the Hough method. The rich set of spatial relationships among extracted features lends itself to a reliable method for identifying fiducial marks.

We also realized that the precise localization of the fiducial mark should consider all of the pixels that have been identified as structural elements. Usually, hundreds of edge pixels are available but only a few that belong to the very center. Hence it stands to reason to determine the center by an adjustment procedure, taking advantage of the rigorous geometric constraints that can be imposed on the edge pixels. For the feature-based approach to work, the design data of the fiducial mark must be available based on which a suitable structural description is generated. Also, this description must be available as part of an extended calibration report.

Problems

1. Discuss the pros and cons of the area-based vs. the feature-based approach for detecting and locating fiducial marks.
2. How is the interior orientation performed when digital cameras are used to obtain imagery?
3. Suppose a diapositive was scanned upside down. What are the consequences for the interior orientation?
4. Binarizing the subimage that contains the fiducial mark may dislocate the mark. Discuss under what circumstances this may happen.
5. We used the Hough technique for finding straight lines and circles. Elaborate on the relationship between the tessellation of the accumulator array and the original image.
6. Suppose the circular structure of a fiducial mark was detected with the Hough technique. A close examination of the accumulator array reveals that there is no distinct peak but rather a blur. How would you interpret this result?
7. Show that neglecting the linearization of the observables in the adjustment problem on page 350 is justified.

References

Haralick, R. and L. Shapiro (1992). *Computer and Robot Vision.* Addison-Wesley Publishing Company, Inc.

Heipke, C. (1997). Automation of interior, relative, and absolute orientation. *ISPRS Journal of Photogrammetry & Remote Sensing*, **52**(1), 1–19.

Kersten, T. and S. Haering (1997). Automatic Interior Orientation of Digital Aerial Images. *Photogrammetric Engineering and Remote Sensing*, **63**(8), 1007–1011.

Lue, Y. (1997). One step to a higher level of automation for softcopy photogrammetry Automatic Interior Orientation. *ISPRS Journal of Photogrammetry & Remote Sensing*, **52**(1), 103–109.

Otsu, N. (1979). A threshold selection method from gray-level histograms. *IEEE Transactions on Systems, Man, and Cybernetics*, **9**, 62–66.

Sahoo, P. (1988). A survey of thresholding techniques. *Computer Vision, Graphics, and Image Processing*, **41**, 233–260.

Schickler, W. (1995). Ein operationelles Verfahren zur automatischen inneren Orientierung von Luftbildern. *Zeitschrift für Photogrammetrie und Fernerkundung*, **63**(3), 115–122.

Schickler, W. and Z. Poth (1996). The automatic interior orientation and its daily use. In *International Archives of Photogrammetry and Remote Sensing*, **31**(B3), 746–751.

Trinder, J.C., J. Jansa, and Y. Huang (1995). An assessment of the precision and accuracy of methods of digital target detection. *ISPRS Journal of Photogrammetry and Remote Sensing*, **50**(2), 12–20.

Chapter 14

Automatic Relative Orientation

Relative orientation is a fundamental and well established photogrammetric procedure. Its purpose is to determine the orientation parameters of a stereomodel so that subsequent processes, such as measuring DEMs and compiling maps can be performed stereoscopically.

Recent advances in platform orientation expedite the orientation of stereopairs but they do not supersede it. For one, the relative orientation also renders information about the surface (model points). Moreover, the orientation parameters compensate model deficiencies to a certain degree, that is, small differences between the mathematical model and the physical reality.

Although all examples in this chapter refer to the standard aerial case, the solutions discussed are general and can be adapted accordingly, for example for terrestrial applications. The next section elaborates on the major differences between classical relative orientation and automatic approaches and provides a summary of common mathematical models.

The section about interactive relative orientation on digital photogrammetric workstations is kept rather short because we do not want to get too deeply involved in describing workstation dependent aspects. Additionally, the interactive relative orientation is remarkably similar to the traditional orientation procedure on the analytical plotter. The main thrust of this chapter is on automatic and autonomous orientation procedures.

Research in automatic relative orientation (ARO) has primarily focused on image matching, that is, on finding conjugate points—a subject treated in the two image matching chapters (Chapters 10 and 11). Of great interest is the question of how to deal effectively with features rather than with conjugate points for determining the orientation parameters. Sections 14.5 and 14.6 address important issues and describe solutions. The chapter ends with a brief summary, a selection of problems, and a bibliography.

The motivation for developing ARO systems is the desire to build automation chains, beginning with the autonomous interior orientation (AIO), followed by ARO and systems for determining the exterior orientation (see Fig. 13.6 on page 330).

This chapter builds on material presented in previous chapters. I particularly suggest reviewing the two image matching chapters (Chapters 10, 11) as well as Chapter 12.

14.1 Background

14.1.1 Traditional vs. Automatic Orientation

In a traditional setting, a human operator performs the relative orientation. Analytical plotters offer various support functions, leading to a very efficient procedure that is accomplished in 15 minutes or less for standard aerial stereopairs. The crucial task for the operator is to identify and to measure suitable conjugate points. To verify the result, more than the absolute minimum of five points should be measured. Economic reasons put a limit to the number of measured points, however. Simple, but effective adjustment procedures have been developed to determine the orientation parameters and to assess their quality. The algorithms are point based and are implementations of the collinearity or the coplanarity model (see Sec. 14.1.2).

In contrast to a human, a computer operator is far inferior in terms of selecting (extracting) suitable points and measuring (matching) them. It is easier to extract features instead. Strictly speaking, a point in a mathematical sense does not exist in an image. In fact, what the human operator measures as a "point" is the result of an elaborate image interpretation process—an ability that is far beyond the reach of a computer operator.

Extracted features, for example edges, consist of a sequence of edge pixels. Suppose now that edges are matched. However, conjugate edges may not have conjugate pixels. Hence, existing point-based orientation models must be applied with care. Subsequent sections address this issue.

Table 14.1 summarizes the most relevant differences between traditional and automatic orientation. The concerned photogrammetrist will raise the accuracy question. Points measured by a human operator are as accurate as a few microns. Contrast this with features. Taking into account that individual pixels of features may not even be conjugate, the accuracy

Table 14.1: Relevant differences between traditional and automatic orientation methods.

Traditional Methods	*Automatic Methods*
human operator • selects and measures points • well selected • few points • stereoscopically measured • very accurate	computer operator • extracts and matches features • randomly selected • many features • matches features • less accurate
adjustment procedures are point-based, using collinearity or coplanarity model	same point-based adjustment procedures used although matched entities are features and NOT points
precision of orientation parameters $\sigma_p \sim \sigma_o \sqrt{c/r}$	
σ_o small, redundancy r small, cofactor c depends on point distribution	σ_o larger, $c \approx$ same, r large $\Rightarrow$ compensates for the larger σ_o!

is probably not better than a few pixels. Assuming 15 μm pixel size for a moment suggests an accuracy of “conjugate” pixels of 50 μm or so—ten times worse than in classical approaches. By and large, the accuracy of the adjusted orientation parameters depends on the measuring accuracy and redundancy. With a redundancy of 100 you can expect an accuracy of the parameters that is roughly ten times better than the measuring accuracy, or comparable to the classical result! The compelling conclusion is that the poorer quality of points is compensated by the greater redundancy—quantity for quality!

14.1.2 Mathematical Models for Relative Orientation

This section provides a short summary of common mathematical models used for the solution of the orientation parameters and model points. The traditional model used in photogrammetry for orienting aerial stereopairs is the coplanarity model. We begin, however, with the more general collinearity model and show the connection between the two.

Collinearity Model

The classical collinearity equations define the condition that the perspective center, a point in object space, and its corresponding image point are on a straight line. With the notation sketched in Fig. 14.1 we express this relationship as follows

$$\mathbf{p} = \mathbf{c}' + \lambda' \mathbf{R}' \mathbf{p}' \tag{14.1}$$

$$\mathbf{p}' = 1/\lambda' \mathbf{R}'^{\mathrm{T}} (\mathbf{p} - \mathbf{c}') \tag{14.2}$$

The first equation defines an object point $\mathbf{p}$ as a function of the exterior orientation $\mathbf{c}'$ (position of perspective center of camera) and $\mathbf{R}'$, a 3-D rotation matrix that contains the three independent angles of the camera's attitude. The scalar quantity λ' is a scale factor. Rearranging this equation leads to the classical collinearity equation (14.2). Here, the observed quantity $\mathbf{p}'$ is a function of the orientation parameters ($\mathbf{c}'$, $\mathbf{R}'$) and points to $\mathbf{p}$ in object space. The popularity of Eq. 14.2 stems from the fact that it lends itself to a simple adjustment procedure where observations and unknowns are neatly separated.

The image vector $\mathbf{p}'$ has the three components $[x'_p, y'_p, -f]^T$. Therefore, we obtain three component equations from Eq. 14.2. To eliminate the scale factor λ', the first and second component equations are divided by the third, leading to the more familiar collinearity equations

$$\begin{aligned} x'_p &= -f \frac{(X_P - X_C)r_{11} + (Y_P - Y_C)r_{12} + (Z_P - Z_C)r_{13}}{(X_P - X_C)r_{31} + (Y_P - Y_C)r_{32} + (Z_P - Z_C)r_{33}} \\ y'_p &= -f \frac{(X_P - X_C)r_{21} + (Y_P - Y_C)r_{22} + (Z_P - Z_C)r_{23}}{(X_P - X_C)r_{31} + (Y_P - Y_C)r_{32} + (Z_P - Z_C)r_{33}} \end{aligned} \tag{14.3}$$

where $r_{11}, \ldots, r_{33}$ are the elements of the 3-D orthogonal rotation matrix $\mathbf{R}'$, containing the three rotation angles ω, ϕ and κ:

$$\mathbf{R} = \begin{vmatrix} \cos\phi\cos\kappa & -\cos\phi\sin\kappa & \sin\phi \\ \cos\omega\sin\kappa + \sin\omega\sin\phi\cos\kappa & \cos\omega\cos\kappa - \sin\omega\sin\phi\sin\kappa & -\sin\omega\cos\phi \\ \sin\omega\sin\kappa - \cos\omega\sin\phi\cos\kappa & \sin\omega\cos\kappa + \cos\omega\sin\phi\sin\kappa & \cos\omega\cos\phi \end{vmatrix}$$

Conceptually, the collinearity model establishes a relationship between object and image space by way of a 3-D similarity transformation. Such transformations have an inherent *datum defect* of seven. For solving the seven transformation parameters, common information between the two coordinate systems must be available, for example, control points, or the knowledge of some of the transformation parameters.

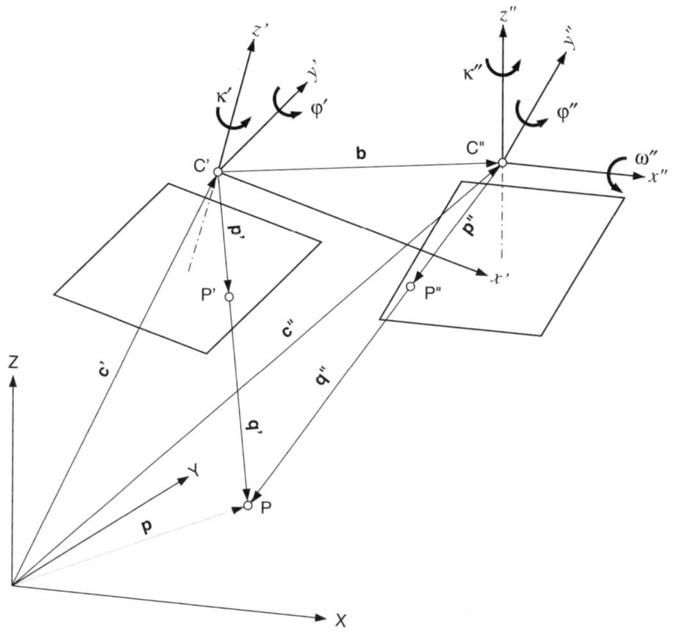

Figure 14.1: Stereopair with photo coordinate systems located at the perspective centers C' and C''. The collinearity condition requires the vectors $\mathbf{p}'$ and $\mathbf{q}'$ to be collinear, while the coplanarity condition states that the base $\mathbf{b}$ and the two vectors $\mathbf{q}' = \lambda'\mathbf{p}'$ and $\mathbf{q}'' = \lambda''\mathbf{p}''$ lie in one plane.

To solve the relative orientation with the collinearity model, we eliminate the datum defect by declaring seven of the twelve orientation parameters as constants. Of the many solutions, two have practical importance. Table 14.2 summarizes the assignment of constants, indicated by an underlined symbol. The dependent relative orientation is important in aerial triangulation for it permits us to form strips from consecutive models.

Conjugate points P_i', P_i'' render four collinearity equations. In contrast to $4n$ observation equations, we have a total of $5 + 3 \cdot n$ unknowns, where n is the number of conjugate points. The advantage of the collinearity model for relative orientation is its generality and simultaneous computation of orientation parameters and object points. An analysis of the structure of the normal equation system reveals that the unknown object points can be easily eliminated before solving for the parameters. Hence, the number of points—usually hundreds in digital photogrammetry—does not increase the computational cost significantly.

Table 14.2: Fixing the datum defect by declaring certain orientation parameters as constants (underlined symbol).

	Independent RO		Dependent RO	
perspective centers	$\mathbf{c}' = \begin{bmatrix} \underline{0} \\ \underline{0} \\ \underline{0} \end{bmatrix}$	$\mathbf{c}'' = \begin{bmatrix} \underline{\mathrm{bx}} \\ \underline{0} \\ \underline{0} \end{bmatrix}$	$\mathbf{c}' = \begin{bmatrix} \underline{0} \\ \underline{0} \\ \underline{0} \end{bmatrix}$	$\mathbf{c}'' = \begin{bmatrix} \underline{\mathrm{bx}} \\ by \\ bz \end{bmatrix}$
orientation angles	$\mathbf{a}' = \begin{bmatrix} \underline{0} \\ \varphi' \\ \kappa' \end{bmatrix}$	$\mathbf{a}'' = \begin{bmatrix} \omega'' \\ \varphi'' \\ \kappa'' \end{bmatrix}$	$\mathbf{a}' = \begin{bmatrix} \underline{0} \\ \underline{0} \\ \underline{0} \end{bmatrix}$	$\mathbf{a}'' = \begin{bmatrix} \omega'' \\ \varphi'' \\ \kappa'' \end{bmatrix}$

Coplanarity Model

Referring to Fig. 14.1 we note that point vector $\mathbf{p}$ can be derived independently from the left or right image. Thus, we obtain the following equation

$$\mathbf{c}' + \lambda'\mathbf{R}'\mathbf{p}' = \mathbf{c}'' + \lambda''\mathbf{R}''\mathbf{p}'' \tag{14.4}$$

Substituting $\mathbf{q}' = \mathbf{R}'\mathbf{p}'$ and $\mathbf{q}'' = \mathbf{R}''\mathbf{p}''$ leads to

$$\mathbf{c}' + \lambda'\mathbf{q}' - \mathbf{c}'' - \lambda''\mathbf{q}'' = 0 \tag{14.5}$$

With $\mathbf{b} = \mathbf{c}'' - \mathbf{c}'$, the base vector, the above equation reads

$$\lambda'\mathbf{q}' - \lambda''\mathbf{q}'' = \mathbf{b} \tag{14.6}$$

Let us dotmultiply both sides of Eq. 14.6 by the cross product $\mathbf{q}' \times \mathbf{q}''$. Then, the left side becomes zero and the scale factors are successfully eliminated.

$$\mathbf{b}(\mathbf{q}' \times \mathbf{q}'') = 0 \tag{14.7}$$

Eq. 14.7 is usually the starting point for explaining the coplanarity model. This scalar vector product, of course, expresses that the three vectors $\mathbf{b}$, $\mathbf{q}'$, $\mathbf{q}''$ are confined to one plane—the epipolar plane, since it contains the base. Eq. 14.7 can be written in the form of a determinant

$$\begin{vmatrix} BX & BY & BZ \\ u' & v' & w' \\ u'' & v'' & w'' \end{vmatrix} \tag{14.8}$$

Again, $\mathbf{q}' = \mathbf{R}'\mathbf{p}' = [u', v', w']^T$ and $\mathbf{q}'' = \mathbf{R}''\mathbf{p}'' = [u'', v'', w'']^T$ are the image vectors rotated with the orientation angles such that they become coplanar with the base vector **b**.

The coplanarity equation 14.8 contains 8 unknowns of which only 5 are independent. Like in the previous section, we declare three of them as constants. This leaves the parameters listed in Table 14.2.

Since there are many more equations than parameters we find the solution by a least squares adjustment with the general model of condition equations with parameters.

$$\mathbf{B}\mathbf{v} = \mathbf{A}\mathbf{x} - \mathbf{f} \tag{14.9}$$

with $\mathbf{A} = \frac{\partial F}{\partial \mathbf{x}}$ and $\mathbf{B} = [\frac{\partial F}{\partial x'}, \frac{\partial F}{\partial y'}, \frac{\partial F}{\partial x''}, \frac{\partial F}{\partial y''}]^T$.

Considering near vertical photography, $\mathbf{B}\mathbf{v} \approx py$. With this, the problem reduces to adjustment with observations where only the $y-$parallaxes $py = y'' - y'$ enter as observations.

The next step involves the linearization of Eq. 14.8, requiring initial values for the five relative orientation parameters. We skip the details of deriving the partial derivatives and present the result for the case of the independent relative orientation

$$p_y = -x'\Delta\kappa' + x''\Delta\kappa'' + \frac{x'y'}{f}\Delta\varphi' - \frac{x''y''}{f}\Delta\varphi'' + (f + \frac{y''y''}{f})\Delta\omega'' \tag{14.10}$$

with x', y' and x'', y'' the photo-coordinates of conjugate points, and f the focal length.

The parallax equation 14.10 is an observation equation. The parallax p_y tells how much the current estimation of the parameters and the observations are off the epipolar plane. By the way, you can derive the same equation from the collinearity model.

With the parallax equations of all conjugate points we form the normal equation system **N**. The coefficients are functions of the product of coordinates of the conjugate points, except the fifth column. The Gruber points[1] are selected at locations where at least one of the photo coordinates is zero. Consequently, some of the normal equation coefficients will become zero, leading to the following structure with x denoting a non-zero coefficient

[1]Special point pattern in a stereomodel (see Fig. 14.12 on page 379 for illustration).

$$\mathbf{N} = \begin{bmatrix} x & 0 & 0 & 0 & x \\ \cdot & x & 0 & 0 & x \\ \cdot & \cdot & x & 0 & x \\ \cdot & \cdot & \cdot & x & x \\ \cdot & \cdot & \cdot & \cdot & x \end{bmatrix} \tag{14.11}$$

This structure is particularly suited for an iterative solution of the orientation parameters as they are independent of each other with the exception of the last parameter. The analog orientation of stereoplotters is entirely based on this principle. You may wonder what this has to do with digital photogrammetry, where points are automatically matched (measured) and an abundance of computing power is available to solve for **N** directly. Sec. 14.6 describes a method for determining the orientation parameters without explicitly matching conjugate points that is based on the principle discussed here.

Linear Models

Both the collinearity and the coplanarity model are not linear in the unknowns. Linearization requires approximations, and to ensure convergence, good approximations for the parameters to be estimated must be known. Determining good approximations may pose quite a challenge for applications in terrestrial photogrammetry or convergent aerial imagery. Hence, models that do not require approximations have been proposed. The treatment of this interesting subject is beyond the scope of this book and the reader is referred to the special literature, to which a few references are given below.

The non-linearity of the traditional models is related to the 3-D orthogonal rotation matrices. *Thompson* (1968) suggested giving up the orthogonality and considering the rotation elements r_{ij} as unknowns. *Horn* (1990) proposes a procedure whereby solutions for the base components and the rotation elements are independently determined in an iterative procedure that does not require approximations. ?), and *Wang* (1994) present direct solutions of the relative orientation, requiring at least 8 conjugate points.

14.2 Interactive Relative Orientation

Interactive relative orientation is performed on Digital Photogrammetric Workstations (DPWs). These stations offer various degrees of support for measuring conjugate points. As a minimum you would expect the same support analytical plotters offer. For example, the drive to feature drives analytical plotters to the Gruber locations (see Fig. 14.12 on page 379) where the operator selects a suitable point and removes the y−parallax. After six points are measured, the system computes orientation parameters and switches from comparator mode to model mode if the result is acceptable (e.g. standard deviation of y−parallax below a preset value).

The same procedure is performed on a DPW, except that the drive to feature is replaced by positioning the cursors accordingly. Since no mechanical parts move, the operation of positioning the cursor at specific locations is considerably quicker, except if loading of new image data into the graphics system is too slow. Yet another advantage is the shape and color of the cursor that can easily be changed for optimal positioning. Fig. 14.2 shows a typical display of LH Systems' SOCET SET during the relative orientation. The two images of the stereomodel are displayed separately on the monitor. Once the cursor is positioned on a feature in one of the images, it is locked and the cursor can be moved to the same feature in the other image.

Most DPWs offer an option to measure conjugate points automatically. A typical scenario would be that you pick a feature in the left image by positioning the cursor on it. The cursor in the right image should be close to the corresponding position. The exact position of the conjugate point would then be computed by matching, using correlation, for example. You can accept or reject the result based on visual inspection of the display.

14.3 Automatic Orientation with Interest Points

Extracting and matching interest points is a straightforward process and quite easy to implement. Not surprisingly then, it is the most popular approach for orienting stereopairs. Furthermore, it offers the advantage that the existing point-based algorithms developed in analytical photogrammetry for determining the orientation parameters can be used. Representative examples of this approach include *Tang and Heipke* (1996), and *Hahn and Kiefner* (1994).

14.3.1 Extracting Interest Points

The basic idea of extracting distinct points in an image or image patch is to identify areas with high variance—areas which are potentially useful for matching. An interest operator performs the extraction process; points with distinct features are called interest points. Compare Sec. 11.1.1 for a summary on detecting interest points.

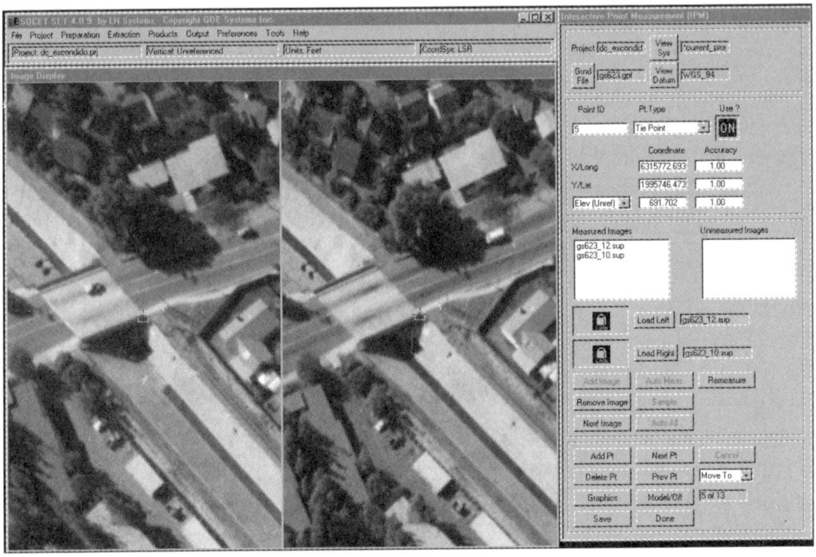

Figure 14.2: Relative orientation performed on LH Systems' SOCET SET softcopy workstation. To measure conjugate points, the cursor is positioned in the left image on a feature. This image is then locked and once the cursor in the right image is moved onto the same feature, the point is measured. This elementary sequence is assigned to mouse buttons for an efficient way of measuring conjugate points interactively.

Luhmann and Altrogge (1986) compare several interest operators. The authors conclude that the operators developed by Moravec (*Moravec*, 1976) and Förstner (*Förstner* , 1986) perform best for aerial imagery. These two operators are most widely used in automatic relative orientation.

The Förstner interest operator is rotation invariant and renders subpixel accuracy. Different point features can be detected, including corner points and circular features (blobs). The Moravec operator is simpler, easier to implement, and executes faster.

14.3.2 Matching Interest Points

Sec. 11.1.2 describes the method of matching interest points with area-based methods. Fig. 11.1 on page 270 demonstrates the concept of matching interest points. The stereopair shows interest points extracted by the

Förstner operator. The template is centered on one of the interest points, say in the right image. Next, the search window in the other image is computed. Most likely, the search window will contain several interest points. The matching window is centered on all of the candidates and correlation coefficients are computed. The point yielding the highest correlation factor is usually considered the conjugate point to the template's center. *Tang and Heipke* (1996) also consider the y−component for selecting the conjugate point.

As discussed in Sec. 10.3.1, the location of the search window depends on the exterior orientation and the surface. The uncertainties of these parameters determine the size of the search window. Since the orientation of the stereopair is unknown, the procedure begins with initial approximations, for example by assuming true vertical imagery and a flat, horizontal terrain surface. This greatly simplifies the estimation of the center of the search window. We have $x_s = x - bx$, $y_s = y$ where x, y are the coordinates of an interest point in one image and x_s, y_s the coordinates of the search window center. The size of the window in the x−direction is primarily determined by the uncertainty in the model surface. The uncertainty in the orientation parameters impacts the y − direction. To keep the number of matching candidates within reasonable limits, a hierarchical approach is chosen.

Tang and Heipke (1996) limit the hierarchical approach to points that have been accepted as conjugate on a previous level of the image pyramid. Once an interest point is found, the authors argue, it is very likely that several interest points will be found around that location on the next lower pyramid level. This leads to an ever increasing number of extracted interest points. The authors present examples that typically begin with a few hundred points on the top level and end with several thousand points on the high-resolution levels. Fig. 14.3 on the next page shows the location and distribution of matched interest points following this approach.

In conclusion, the method of determining the relative orientation parameters by matching interest points is simple and effective. The choice of suitable threshold values and window sizes deserves special attention. The decision-making series begins with selecting the top level of the pyramid where the hierarchical approach begins. Next is the size of the window being used to extract interest points, followed by determining threshold values for accepting interest points. Yet another threshold to be determined refers to the acceptance criterion for matches and conjugate points. Most of these values depend on the imagery. Even more problematic are those values that depend on the image location (content) and the resolution (level in image pyramid). No clear rules exist for how to determine and adapt the values. This adds to the difficulties of developing autonomous

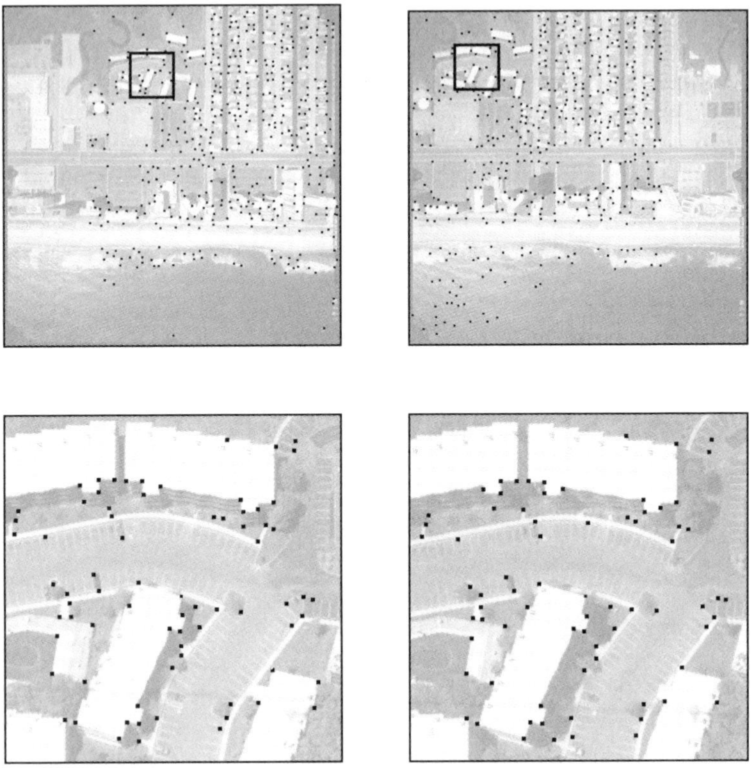

Figure 14.3: Example of matched interest points found by the approach described in *Tang and Heipke* (1996). The top row shows the stereopair at the top level of the image pyramid with extracted interest points superimposed. The bottom row depicts the highlighted area at a higher resolution level. The centers of the two windows were selected from matched interest points on the coarse level. Note that the windows contains many more interest points than the same area on the coarse level.

systems that are based on this approach.

Section 14.6 introduces another approach to matching interest points, based on computing the orientation parameters first. This is accomplished in an iterative fashion, following the principle of manually orienting stereopairs on stereoplotters. The matching of interest points follows after the orientation parameters are established.

14.4 Automatic Orientation with Edge Pixels

Edge matching gained popularity in computer vision in the late seventies when it was realized that the astonishing stereovision ability of humans is based on finding conjugate edges rather than conjugate points. Edges are more robust features than points. Moreover, they are more directly related to object space events, such as object boundaries, markings, or abrupt surface changes. The automatic orientation of images is a prerequisite for applications such as surface reconstruction and object recognition. Since edges play an important role in the solution of these tasks, it makes eminent sense to use them as entities for orienting images.

We have already covered essential parts of edge matching in previous chapters. For example, Sec. 5.3 in the computer vision chapter addresses the issue of detecting edge pixels and of linking them to edge contours. Several different approaches to matching edges are discussed in Chapter 11.

14.4.1 Matching Edges

The major part of Chapter 11 is devoted to matching edges. More general methods are based on comparing the shape of edges, that is, entire edges, or edge segments are compared instead of individual edge pixels. Such is the case with the $\psi - s$ approach and the Generalized Hough Transform, both introduced in Sec. 11.1.4.

Matching edges based on shape and location criteria can be improved further by considering relationships between edges. Relational matching, also referred to as symbolic or structural matching, provides a mechanism to introduce relationships (Sec. 11.2). Sec. 14.6 describes a new approach for matching edges.

14.4.2 Computing Orientation Parameters with Characteristic Edge Points

After edges are successfully matched we face a new problem: how can conjugate edges be used to determine the orientation parameters? This is not a trivial task because the individual pixels of matched edges are, in general, not conjugate. Not much has been reported about this problem. Evidently, researchers focused entirely on the matching problem and have neglected the orientation task.

Schenk et al. (1990) use characteristic points of matched edges as approximate locations for determining conjugate points. Matching is performed with the $\psi - s$ method. The vertices of the segmented $\psi - s$ curve correspond to maximum changes in curvature of the edge contour. These points define centers of windows within which interest points are extracted and matched.

The procedure requires that the windows refer to the same surface patch in object space. Although edges may be dislocalized—depending on the edge operator used—the authors found that most characteristic points are conjugate to a few pixels. The reason is that edges are usually displaced by a similar magnitude and orientation, due to the fact that the image function at the conjugate location is similar.

Fig. 14.4 depicts a stereopair with matched edges. The characteristic points can either be used directly for the computation of the relative orientation parameters, or serve as centers of windows within which interest points are extracted and matched.

To improve the accuracy of characteristic edge points in terms of their correspondence, interest points can be determined within windows centered on characteristic point pairs. As described by *Schenk et al.* (1991), matched interest points are then used for the computation of the orientation parameters.

14.5 Automatic Orientation with Edge Entities

Having matched edges suggests using them as entities for the computation of the relative orientation parameters. Using specific points, for example characteristic points, as discussed in the previous section, utilizes only a small subset of the available information. Moreover, characteristic edge points are only approximately conjugate.

Theoretically, it is possible to use the matched edges as observed entities for the orientation. In that case, the collinearity model would have to be extended to cope with lines instead of points. A bundle ray, for example, would correspond to a surface defined by the projection center and the edge. Solutions are known for the case where edges are approximated by straight lines.

Several interesting problems arise when straight lines are used. First, straight lines must be extracted from the images. Since edges are hardly ever perfectly straight, we first need to fit a straight line through edge pixels. A more fundamental question is related to the 3-D representation of straight lines and their perspective relationship between image and object space. Four parameters suffice for a unique representation of straight 3-D lines. Several models have been proposed. *Mulawa and Mikhail* (1988) use one point and the line's direction vector. Since this representation is not unique, the authors introduce two constraints. The closest point to the origin must be chosen and the norm of the direction vector is unity.

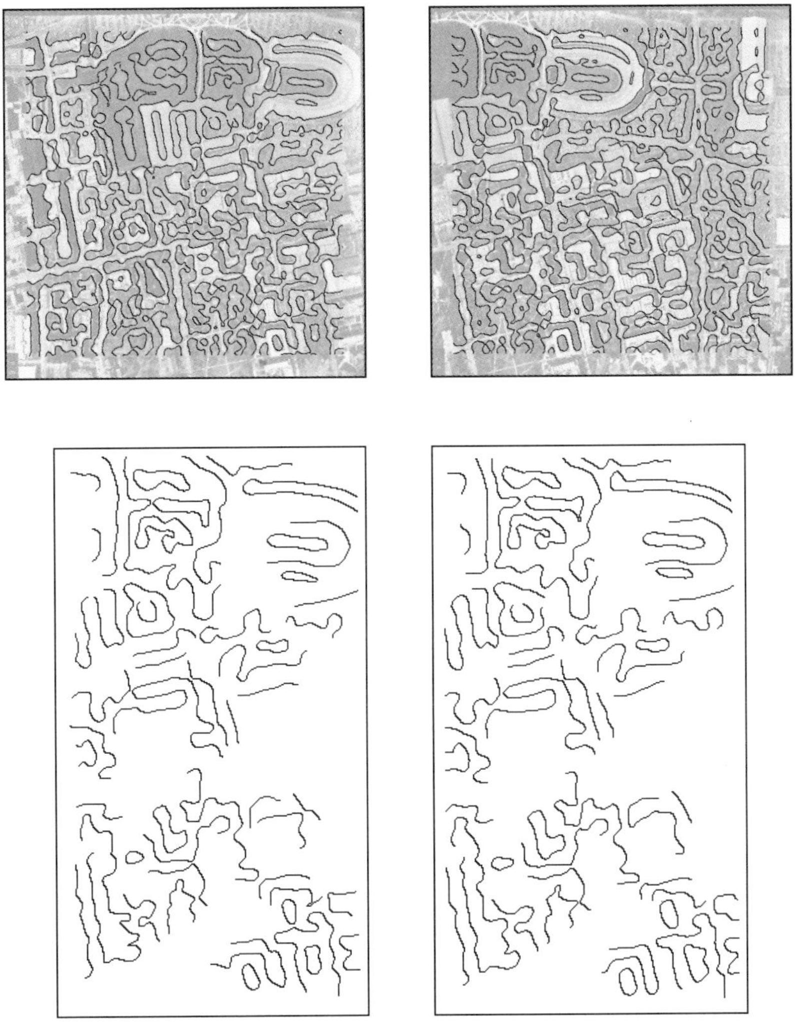

Figure 14.4: Stereopair with matched edges. The characteristic points are the vertices of the segmented $\psi - s$ representation of edges and correspond to maximum changes in curvature. On coarser pyramid levels, these points are used directly to determine improved orientation parameters. On the fine resolution level, the points serve as centers of windows within which interest points are extracted and matched.

and Faugeras (1989) represent 3-D lines as the intersection of two planes that are parallel to the x−axis and y−axis, respectively. This four parameter representation has a singularity for lines that are parallel to the first principal plane, however. To avoid singularities, *Habib* (1999) suggests a mathematical model for the back and forth perspective transformation of straight lines between image and object space.

This section discusses a new approach of using matched edges as entities, without approximating them by straight lines or by curves.

14.5.1 Relevant Edge Characteristics

Let us first identify some edge properties that are relevant for the computation of the relative orientation parameters.

Edge Pixels Are Not Necessarily Conjugate

There are several reasons why conjugate edges may have non-conjugate edge pixels. Consider Fig. 14.5, for example. The vertical edge of a building is imaged at very different lengths (highlighted edge) in both images. Therefore, not all pixels can be conjugate.

Figure 14.5: Example of a foreshortened edge. The vertical building edge is imaged at very different lengths. Since the pixel size is the same in both images, the edge in the right image is represented by twice as many pixels as the one in the left image. Hence, not all pixels can be conjugate.

Ends of Edges are Ill-Defined

Quite often we observe conjugate edges that are strong but do not have well defined end points. Fig. 14.6 exemplifies the problem. Here, the thresholding of edges leads to two edge segments in the left image, separated by a gap, while the right edge passed thresholding unaffected. Thresholding edge attributes, such as gradients, is a common procedure to select strong edges for the matching and orientation.

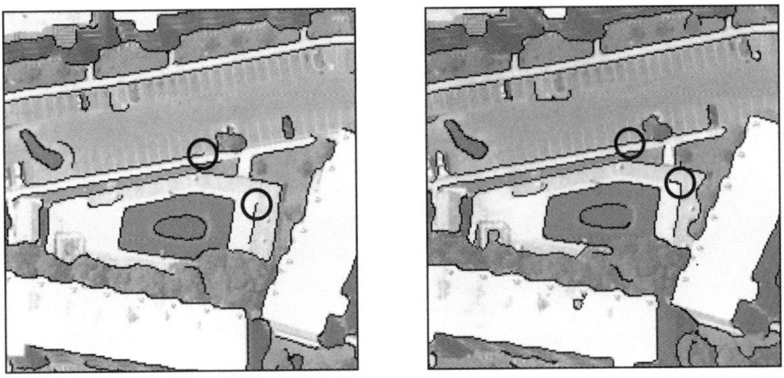

Figure 14.6: Here, edges have been thresholded in both subimages, using the same parameters. The two highlighted areas show that the edges responded differently to the thresholding. In the subsequent edge matching process, most likely, the left edge segments would be matched to the unaffected edges in the right image. In this case the terminal points are not well defined.

Accuracy of Edge Pixels

The above examples confirm our intuitive feeling that edges are more accurate across and less accurate along the edge. As a first approximation, we can stochastically model an edge by assigning error ellipses to the edge pixels. Fig. 14.7 shows a few examples. Pixels of a straight edge have an elongated error ellipse whose major axis lies in the direction of the tangent at the pixel's location.

We argued that the ends of edges are usually less well-defined than other edge pixels—hence, one could assign an error ellipse with a larger major axis. The magnitude of the minor axis of the error ellipse—with its direction across the edge—is largely defined by the distinctness of the edge. Edge operators usually offer a measure for the strength, for example, the

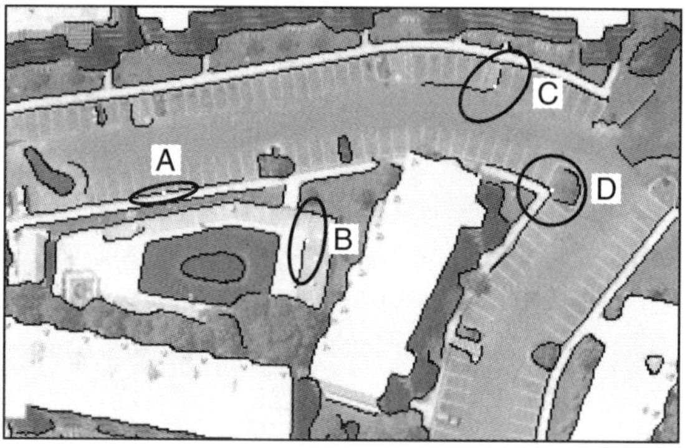

Figure 14.7: The accuracy of edges can be expressed by assigning error ellipses to edge pixels. Straight edges have elongated ellipses whose minor axes depend on the strength of the edge (A). Because ends of edges are usually not well defined, the major axis would become longer towards the end(B). Edges around corners have more rounded error ellipses (C) and (D).

gradient of the convolution values. Some edge operators, such as the LoG operator, exhibit typical dislocalization patterns (see Fig. 5.6 on page 114). Consequently, the error ellipse would assume a more circular shape for pixels at expected locations of displacements, such as corners.

14.5.2 Computation of the Orientation Parameters

Suppose we use the coplanarity model for determining the five angles of the independent relative orientation. In this case, the coplanarity plane is identical with the epipolar plane, as pointed out in Sec. 14.1. It is important to realize that errors in the observations, e.g. conjugate points, have no impact on the orientation parameters as long as they are confined in the epipolar plane. Fig. 14.8 depicts a stereopair, a surface patch with point S, and the epipolar plane, defined by the two projection centers and S. The imaged points S' and S'' are conjugate.

Let T and R be two points in the epipolar plane but off of the surface. Are T' and T'' and R' and R'' still conjugate? Since they are in the epipolar plane you may still consider them as being conjugate. If you take conjugacy in a more restrictive sense, however, the points are not conjugate because they represent different points on the surface. For example, R' is the image of surface point U while its conjugate point R'' refers to surface point S.

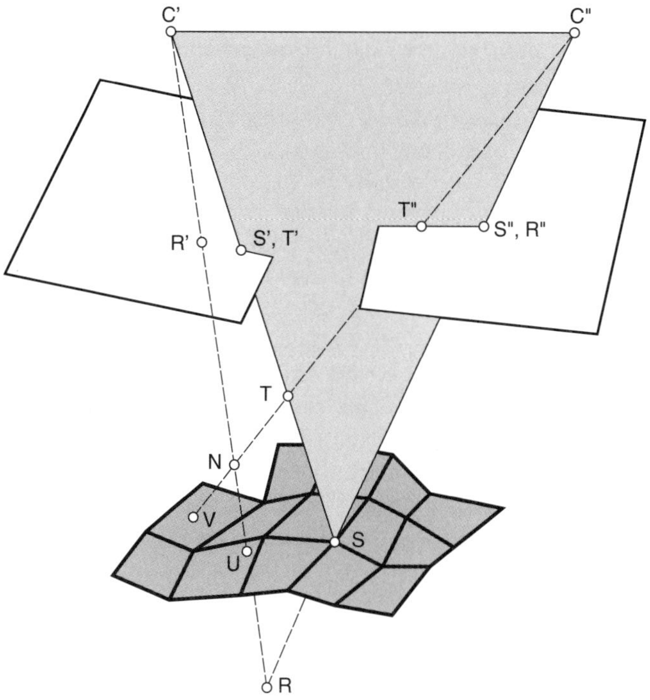

Figure 14.8: Surface point S and the projection centers C', C'' define the epipolar plane. The images of S are the conjugate points S', S''. Points T and R lie in the epipolar plane but not on the surface. In a broader sense, their images are also conjugate as far as the coplanarity condition is concerned. However, they are not conjugate with respect to the visible surface. For the computation of the orientation parameters, all points in the epipolar plane can be used, even though they may not be on the visible surface. What we see at the location of points off the surface is the intersection of the bundle ray through that point with the surface. For example, the image of point R' refers to the surface point U.

Let us perform a mental experiment with the situation depicted in Fig. 14.8 repeated at the Gruber locations. In a traditional setting we compute the five orientation parameters with the conjugate point pairs (S', S''). Now, let us construct new pairs, such as (S', T'') and (R', T''). The latter pair is more interesting because it involves two image points that correspond to a new point in object space, N, found by intersecting the respective bundle rays. Note that N is not on the visible surface. In fact, what is imaged at the positions of this artificial pair (R', T'') is the object points V and U. Now, regardless what point pairs we use for the orientation, the parameters will always be the same.

The compelling conclusion of this exercise is that for determining the orientation parameters it does not matter what point pairings we use as long as the points are in the epipolar plane. We just found a way of performing the relative orientation without conjugate points, but still using the coplanarity model. Suppose we have a matched edge that happens to lie in the epipolar plane. Then, the error along the edge does not influence the orientation, only errors perpendicular to the plane will impact the result.

Edges in the epipolar plane have error ellipses with the major axis parallel to the epipolar lines. In this case, only the minor axis, perpendicular to the epipolar line, determines the uncertainty of the orientation parameters—the magnitude of the major axis does not matter. Now, if we were to select matched edges that are in the epipolar plane, then the problem that edge pixels are not conjugate is irrelevant for we can use all independent pairings of edge pixels. Comparing again the error of edges as illustrated in Fig. 14.7 on page 372, we realize that only the error across edges matters. This fact forms the cornerstone of an ARO approach that uses original edges for orienting stereopairs.

Of course, only a few conjugate edges, or perhaps none at all, will lie in epipolar planes. Moreover, the epipolar plane is only approximately known since we do not have the orientation parameters that would be needed to determine it. Let us now examine the error that will result from non-conjugate pixels and edges that are at an angle with the epipolar line. Fig. 14.9(a) shows the conjugate edges g', g'' and the epipolar lines e', e''. Select pixel P' on edge g' and let P'' be its true conjugate point. We argued that we will make an error in estimating P'' (point Q instead of P''), because edge pixels are not conjugate. We quantified this error by the major axis of the error ellipse at pixel P''. Remembering that only error components perpendicular to the epipolar plane will affect the orientation, we find the simple relationship

$$v = u \sin \alpha \tag{14.12}$$

where v is the effective error, u is the error we make in estimating the conjugate pixel to P', and α is the angle between conjugate edges and epipolar lines.

Curved edges will cause an additional error. Fig. 14.9(b) depicts curved conjugate edges g', g'' and the effective error in the orientation. The max-

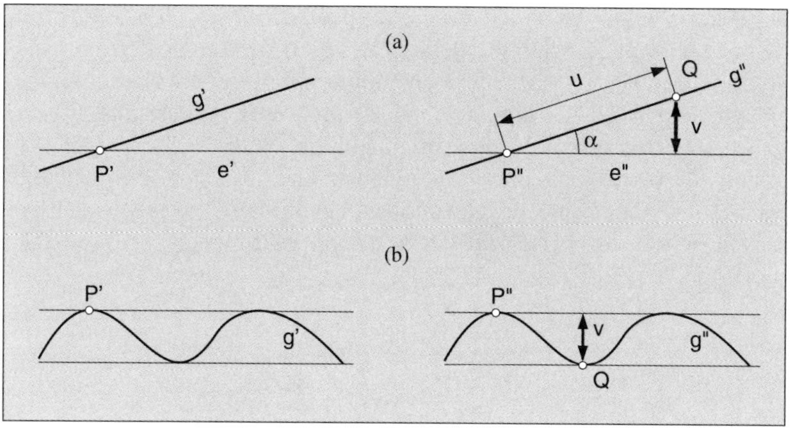

Figure 14.9: In (a), an angle α between epipolar lines e', e'' and the respective conjugate edges g', g'' causes an effective error of $v = u \sin \alpha$. The error u results from estimating the conjugate location to pixel P' at point Q instead of the true conjugate location P''. Curved edges may cause an additional error as illustrated in (b). The maximum error is related to the waviness that may be expressed by the minimum diameter of a tube within which the edge fits.

imum error can be estimated by determining the minimum diameter of a tube within which the edge fits.

The error u that we make in estimating the conjugate point to P' is related to the matching accuracy along edges. It is also related to the surface, however. Suppose we knew the shape of the surface around point P. In this case the uncertainty of predicting the position of P'' would only depend on the uncertainty of the orientation parameters—in other words, u would be smaller.

To be precise, there are two different errors that will affect the accuracy of the orientation parameters. First we have to take the error of an edge pixel, expressed by its error ellipse, into account. In addition, we will make an error u in associating conjugate pixels (matching error), but only its y−component will affect the result (Eq. 14.12).

14.5.3 Computation of the Surface

Theoretically, with the determination of the orientation parameters, the relative orientation is complete. For subsequent applications, it is quite likely that the surface is needed, however. Moreover, if we imbed the ARO procedure in a hierarchical scheme, surface elevations need to be determined to limit error u in Fig. 14.9(a). For the computation of the orientation parameters, edges parallel to epipolar lines are preferred. Unfortunately, such edges are less than ideal for determining points on the surface as Fig. 14.8 on page 373 reveals. Errors along epipolar lines translate into elevation errors. Hence, for computing elevations we should use edges whose minor error ellipse axes are perpendicular to the epipolar lines.

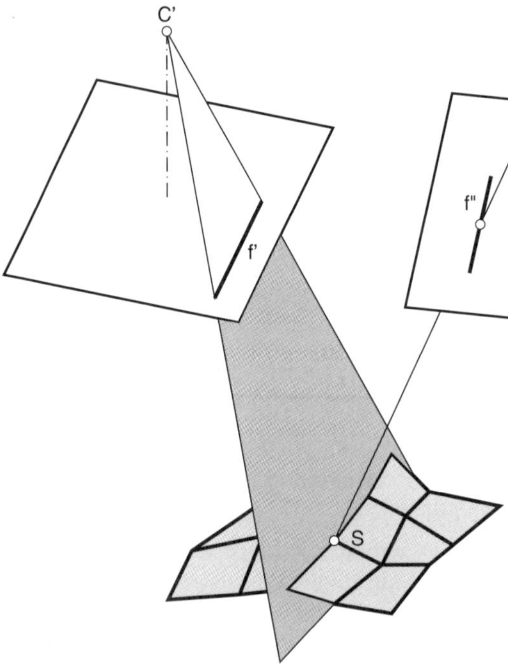

Figure 14.10: Elevations of non-conjugate edge pixels can be found by defining a surface through the perspective center C' and edge f', and intersecting it with a ray from C'' to an edge pixel of edge f''.

Fig. 14.10 depicts the situation. Elevations are obtained from conjugate edges that are oriented perpendicular to the base. Imagine a surface defined by the perspective center, say C', and edge f'. Now, any ray from C'' through a pixel of edge f'' will intersect on the surface. Again, pixels need not be conjugate.

14.5.4 Strategy

We are now in a position to formulate a strategy for the computation of the orientation parameters and elevations that is based on matched edges and non-conjugate edge pixels.

1. Extract edges in both images and match them with any of the methods discussed in Chapter 11.
2. Select elongated edges that are more or less parallel with epipolar lines.
3. Compute orientation parameters with all independent edge pixel pairs that can be formed with the subset of conjugate edges found in step 2.
4. Select elongated edges that are more or less perpendicular to epipolar lines.
5. Compute normalized images with the orientation parameters found in step 3. This will render approximate conjugate pixels that can now be used to compute approximate elevations for the edges selected in step 4.
6. Update the parameters and repeat the procedure through the pyramid.

Fig. 14.11 on the next page illustrates the procedure. The stereopair is shown on the top level of the image pyramid with the matched edges superimposed. Only those edges that are suitable for the computation of the orientation parameters are shown.

14.6 Computing Orientation Parameters Without Prior Matching

14.6.1 Principle

The ARO methods described in the previous sections are based on the traditional approach of first matching conjugate entities followed by the computation of the orientation parameters by an adjustment procedure. In this section we abandon this traditional principle and describe a method that computes the orientation parameters without a set of matched features. The motivation is rooted in the observation that matching—however sophisticated the approach—is still the Achilles' heel of automated orientation methods.

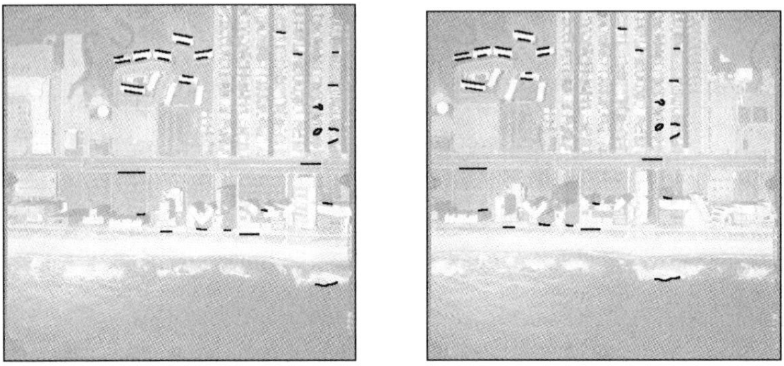

Figure 14.11: Example of edges suitable for computing the orientation parameters. The edges have been selected from the set of matched edges according to the criterion that they should be straight and parallel to epipolar lines (see text for detailed explanations). A total of 535 edge pixels were used in the coplanarity model.

Instead of an adjustment procedure one can determine the most likely orientation parameters by trial and error. Imagine a 5-D space where the coordinate axes correspond to the five relative orientation parameters. This space is finite—it is bounded by the plausible range of the orientation parameters. For example, the rotation angles of aerial stereopairs are within $\pm 5^o$ and the base components hardly exceed 10% of the base. Now consider the situation with extracted interest points. Instead of matching them we can build all possible pairs, compute a solution for every combination of five points, and keep track of all of these solutions in the parameter space, e.g. by an accumulator array. All combinations of points that are conjugate render the same solution. Hence, a peak in the 5-D parameter space identifies the correct solution. You may consider this procedure a trial and error approach. The success depends on the number of independent pairs of five points that can be formed, and on the assumption that there are, in fact, many conjugate points. The method is statistical—the more correct pairings, the higher the peak in the accumulator array.

For a realistic implementation, several obstacles must be circumvented. The most serious problem is the combinatorial explosion. For example, one thousand points per image would lead to 5.7×10^{15} possible combinations of 5 points in a stereopair. Another problem is related to the size of the parameter space. Suppose we employ an independent relative orientation (five independent rotation angles) and discretize the search space in units of $1'$. Given the range of $\pm 5^o$, one coordinate axis would have 600 units, requiring 7.8×10^{13} memory cells, each cell large enough to hold the number of solutions! Yet another challenge is the analysis of the accumulator array for detecting clusters.

An obvious solution to the combinatorial explosion is to reduce the number of points, either globally, or by dividing the image into subregions. Possible solutions for decreasing the size of the parameter space include a coarser tessellation or a reduction of the number of parameters.

Habib et al. (2000) describe an elegant approach for avoiding the combinatorial explosion, the memory problem, and the cluster analysis all together. We skip the detailed derivation here but summarize the most salient features of the approach. First, the stereopair is partitioned into 6 regions, each centered at one of the Gruber locations (see Fig. 14.12). From analyzing Eq. 14.10 on page 361 we conclude that these locations have the interesting property that y−parallaxes are related to one orientation parameter only—each region to a different one. This suggests an iterative solution to the orientation problem, quite like the way you would orient a model on an analog stereoplotter. Consequently, the 5-parameter space reduces to one dimension. Hence, working with the Gruber locations allows the necessary reduction of point combinations to a manageable quantity. At the same time, only a one-dimensional accumulator array is necessary.

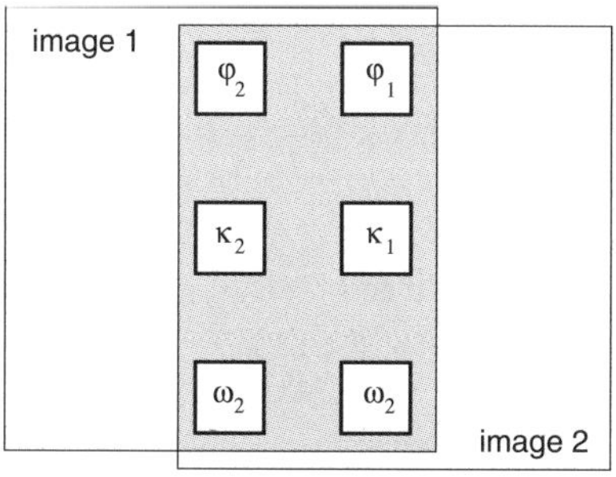

Figure 14.12: Schematic diagram of the Gruber locations and the orientation parameters that are sequentially solved in these locations.

The method follows the following strategy.

1. Determine the location and size of subimages, centered at the Gruber locations. The subimages must be in the stereo overlap region. General knowledge about the forward overlap and the surface elevations is used at this stage.

2. Within overlapping subimages generate all possible pairs. For each pairing, determine the relevant orientation parameter with Eq. 14.10 on page 361 and update the accumulator array.

3. Analyze the accumulator array for a peak. All point pairs that rendered a solution in the peak are considered conjugate.

4. Repeat step 2-4 at the remaining Gruber locations.

5. Perform a relative orientation adjustment with all conjugate point pairs, determine outliers and eliminate ambiguous pairs.

Fig. 14.13 exemplifies the procedure at the first Gruber location. An interest point in the left subimage is paired with every interest point in the right subimage and one κ_2 value is computed and entered into the accumulator array. This is repeated for all interest points in the left subimage. Fig. 14.13(c) depicts the accumulator array.

14.6.2 Matching Edges by Trial and Error

The method described in the previous section lends itself to a new method for matching edges. The role of interest points is now assumed by edge pixels. Let $\{g'\}$ and $\{g''\}$ be sets of edges in a stereopair. Theoretically, we could form all independent pairings of 5 edge pixels, compute a solution, and keep track of all solutions in the parameter space that is defined by the unknown orientation parameters. As before, a single peak in the parameter space would indicate a unique solution. All edge pixel pairings that lead to this solution can be considered conjugate. Finally, from conjugate edge pixels we can infer conjugate edges. If you think that we put the cart before the horse then you are right—edge matching is performed only after having determined the orientation parameters!

To circumvent the combinatorial explosion of forming pairs of 5 edge pixels[2], we resort to the iterative approach described earlier. That is, we deal with only one parameter at any Gruber location. Every possible pair of an edge pixel in the two images of a stereopair renders an equation of the form Eq. 14.10 on page 361, where four of the five parameters are entered with their current values. The value of the fifth parameter is determined from this equation and entered into a histogram (one-dimensional accumulator array). The peak in the histogram corresponds to the solution of the orientation parameter under consideration. The shape of the histogram reveals how unique and distinct the solution is.

[2]This situation exists if all five parameters were to be computed simultaneously.

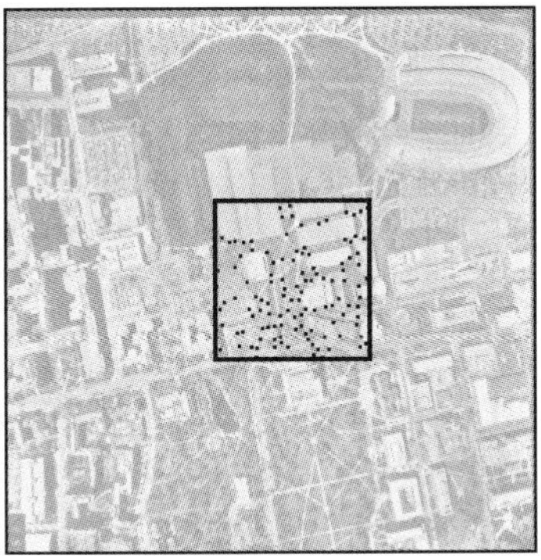

(a)

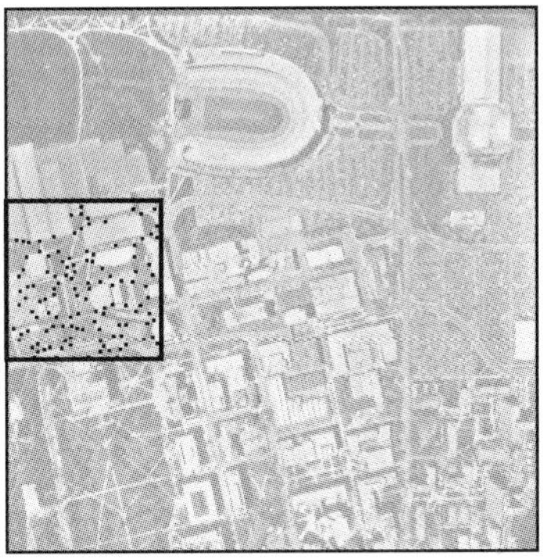

(b)

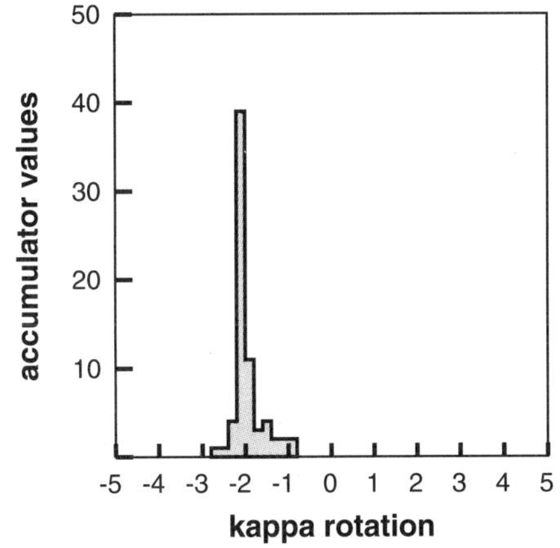

(c)

Figure 14.13: The two subimages in (a) and (b) show the first Gruber locations and the interest points. All possible point pairs are formed and used to compute κ_2 parameters whose values are entered into the accumulator array. The peak in (c) denotes the solution. All point pairs that contributed to this solution are considered conjugate. Multiple conjugate points along epipolar lines will be eliminated at a later step.

After the orientation parameters stabilize in this iterative scheme, conjugate edge pixels are identified. It is important to realize that we are now no longer restricted to the subimages at the Gruber locations. With the orientation parameters available, we can either normalize the images (see Chapter 12), or determine the epipolar plane for finding conjugate edges.

Let us first examine the situation with normalized edge images. Here, epipolar lines are identical with the rows of the stereopair, implying that conjugate edge pixels are in the same row. Hence, edge matching methods such as proposed by *Grimson* (1985), or *Barnard and Thompson* (1982), can be employed. The procedure is relatively simple. Pick an edge pixel, say in the left image. Determine the center of the search window with the current model elevation at that point and find all edge pixels within the window. Repeat this procedure for all edge pixels. Analyze all candidates found, eliminate unlikely candidates based on such principles as continuity and piece-wise smooth surface patches.

The second approach works with the original edge images. Let g'_{ij} be edge pixel j of edge i. Let w''_r, w''_c be the center of the search window corresponding to g'_{ij}, and projected with the orientation parameters and the current model surface. The size of the search window is related to the uncertainty of the orientation parameters and the surface, the latter being the dominant contributor. Now, for all combinations of g'_{ij} with the edge pixels within the search window, the parallax is computed with Eq. 14.10. Pairings with parallaxes below a threshold are stored as candidates. After all pixels of edge g_i have been processed in this fashion, the pool of candidate pixels is examined. The voting scheme for identifying corresponding pixels is governed by the principle of continuity. More details can be found in *Habib et al.* (2000).

14.7 Summary

In traditional photogrammetry, a human operator performs the relative orientation. The crucial task is to identify and to measure suitable conjugate points. For economic reasons, only a few points are measured (less than a dozen). In contrast to a human operator, the computer is far inferior in selecting and measuring suitable points. Instead, it is easier to extract features, such as edges and regions. As a consequence, algorithms for computing the orientation parameters should cope with features and take the lower quality but greater redundancy into account.

We have discussed several ARO approaches. The most popular method is based on interest points, presumably because this is still closely related to the point-based principles of traditional photogrammetry. Although these methods work well, one may argue that interest points are rather abstract quantities, carrying little explicit information about the object space. In the light of reconstructing the object space, including the recognition of objects, it is desirable to deal with linear and areal features.

Research in edge-based ARO has primarily focused on finding conjugate edges. Relational matching is a fairly general approach and is preferred because some useful topological relationships are invariant under projection. This chapter introduced a new matching method. It is sort of a trial and error method where various combinations of points (interest points or edge pixels) in a stereopair are used to compute solutions. Those combinations that render the same solution are considered correct matches. This approach takes advantage of the special relationship between orientation parameters and the location of conjugate points.

A closer examination of matched edges revealed that individual edge pixels may not be conjugate at all. This error is confined to the direction along the edge. An edge is usually much better defined perpendicular to its tangent. We can formally express this interesting property by an error ellipse. We concluded that the error along the edge does not impact the orientation parameters if the conjugate edges are contained in the epipolar plane. This suggests a new approach for using edges in ARO. Edges that are more or less parallel to the epipolar lines are suitable for computing the orientation parameters; edges that are perpendicular to the base are used for determining the model surface.

Problems

1. How does the number of matched features influence the accuracy of the orientation parameters? How does it affect the model surface?
2. Comment on the pros and cons of using interest points or edges for solving the relative orientation.
3. How does an uneven distribution of matched features influence the relative orientation? For example, imagine a high concentration of features in the left upper corner of the model and very light distribution elsewhere.
4. Suppose an ARO system employs a hierarchical approach. What exactly determines the uncertainty in predicting conjugate locations when moving through the pyramid?
5. Assume that edges have been successfully matched in a stereopair. Elaborate on the problem of determining the orientation parameters with matched edges.
6. Explain why edges that are parallel to the base are well suited for relative orientation, even if individual edge pixels are not conjugate.

7. Suppose you have extracted interest points and you also know their error ellipses. Following the discussion in Sec. 14.5, which interest points would you select for computing the orientation parameters?

8. In the chapter about computer vision we made a point that edges can be dislocalized (see Sec. 5.3). How does dislocalization affect the relative orientation? Hint: think of its impact on conjugacy.

9. What is the major advantage of computing the orientation iteratively, as described in Sec. 14.6?

References

Ayache, N. and O. Faugeras (1989). Maintaining representations of the environment of a mobile robot. *IEEE Transactions on Robotics Automation*, **5**(6), 804–819.

Barnard, S. and W. Thompson (1982). Disparity analysis of images. *IEEE Transactions on Pattern Analysis and Machine Intelligence*, **2**(4), 333-340.

Brandstätter, G. (1996). On critical configurations of projective stereo correlation. In *International Archives of Photogrammetry and Remote Sensing*, **31**(B3), 77–81.

Förstner, W. (1986). A feature based correspondence algorithm for image matching. In *International Archives of Photogrammetry and Remote Sensing*, **26**(B3/3), 13–19.

Grimson, W.E.L. (1985). Computational experiments with a feature based stereo algorithm. *IEEE Transactions on Pattern Analysis and Machine Intelligence*, **7**(1), 17–43.

Habib, A. (1999). Aerial triangulation using point and linear features. In *International Archives of Photogrammetry and Remote Sensing*, **32**(3-2W5), 137–142.

Habib, A., A. Asmamaw, D. Kelley, and M. May (2000). *Linear Features in Photogrammetry.* Report No. 450, Department of Civil and Environmental Engineering and Geodetic Science, The Ohio State University, Columbus, OH 43210.

Hahn, M. and M. Kiefner (1994). Relative Orientierung durch digitale Bildzuordnung. *Zeitschrift für Photogrammetrie und Fernerkundung*,**62**(6), 223-228.

Horn, B.K.P. (1990). Relative Orientation. *International Journal of Computer Vision*, **4**(1), 59-78. 826-837.

Luhmann, T. and G. Altrogge (1986). Interest-Operator for Image Matching. In *International Archives of Photogrammetry and Remote Sensing*, **26**(B3/2), 459-474.

Mikhail, E. (1993). Linear features for photogrammetric restitution and object completion. In *Proc. of the International Society for Optical Engineering (SPIE)—Integrating Photogrammetric Techniques with Scene Analysis and Machine Vision*, **1944**, 16-30.

Moravec, H.P. (1976). Towards automatic visual obstacle avoidance. In *Proceedings 5th International Joint Conference on Artificial Intelligence*, 584.

Mulawa, D. and E. Mikhail (1988). Photogrammetric Treatment of Linear Features. In *International Archives of Photogrammetry and Remote Sensing*, **27**(B3).

Schenk, T., J.C. Li and C. Toth (1991). Towards an Autonomous System for Orienting Digital Stereopairs. *Photogrammetric Engineering and Remote Sensing*, **57**(8), 1057-1064.

Schenk, T., C. Toth and J.C. Li (1990). Zur automatischen Orientierung von digitalen Bildpaaren. *Zeitschrift für Photogrammetrie und Fernerkundung*, **58**(6), 182-189.

Tan, Z., G. Brandstätter and X. Xu (1996). A Method for Solving the Inverse Problem of Photogrammetry. In *International Archives of Photogrammetry and Remote Sensing*, **30**(3/1), 806-811.

Tang, L. and C. Heipke (1996). Automatic relative orientation of aerial images. *Photogrammetric Engineering and Remote Sensing*, **62**(1),47-55.

Thompson, E.H. (1968). The projective theory of relative orientation. *Photogrammetria*, **23**(2), 67-75.

Tommaselli, A. and C. Tozzi (1996). A recursive approach to space resection using straight lines. *Photogrammetric Engineering and Remote Sensing*, **62**(1), 57-66.

Wang, Y. (1995). Ein neues Verfahren zur automatischen gegenseitigen Orientierung der digitalen Bilder. *Zeitschrift für Photogrammetrie und Fernerkundung*, **63**(3), 122-130.

Wang, Y. (1994). *Strukturzuordnung zur automatischen Oberflächenzuordnung*. Dissertation. Wissenschaftl. Arbeiten der Fachrichtung Vermessungswesen der Universität Hannover. Nr. 207, 103 pages.

Chapter 15

Automatic Exterior Orientation

In this chapter, we are concerned with determining the exterior orientation parameters of single images (single photo resection) and of digital stereopairs—better known as absolute orientation. The fundamental problem is to establish a relationship between image or model space and the object space by way of a suitable transformation. To compute the transformation parameters, information common to the coordinate systems must be known. Traditionally, this is accomplished through control points. That is, points known in the object space coordinate system are identified and measured in the image or model coordinate system.

The approach of using control points in the exterior orientation works very well in traditional photogrammetry and also in interactive digital photogrammetry where an operator identifies and measures the control points. As alluded to in previous chapters, the concept of "points" poses a real challenge in digital photogrammetry, however. In a strict sense, points do not physically exist in an image. Hence, it is difficult to extract them. Control points are, in fact, targets and the coordinates refer to the centroid, for example. As you realize, the difficult part is in the automatic identification of these targets. Section 15.3 addresses relevant issues and summarizes research related to finding control points in images.

Since it is easier to extract features than points, this chapter focuses on the problem of how to deal with linear and areal features. Section 15.4 presents two solutions that use linear features for the exterior orientation of single images. The first method is based on the *trial and error approach* described in the previous chapter. The exterior orientation parameters are determined iteratively in a parameter space by checking all possible pairs of points in image and object space. In this fashion, the correspondence between the features in image and object space is found, resulting in a new matching method. The second method approximates the features by analytical curves and extends the collinearity model to include the curve parameters.

Section 15.5 introduces the notion of control surfaces and their use in establishing the 3-D transformation between stereo models and object space. Known surfaces in object space may be available from existing DEMs or from laser altimetry, for example. This is yet another example of using features instead of points for orienting models.

The motivation for automating the basic orientation procedures is the desire to build an automation chain for single images or stereomodels as depicted in Fig. 15.1. Photogrammetric projects consist of many images obtained in strips and blocks; and the orientation parameters are usually obtained from a block adjustment. Hence, establishing exterior orientation parameters of single images or stereopairs is not of great practical importance. However, the problems encountered in automatic exterior orientation (AEO) and automatic absolute orientation (AAO) will occur again in aerial triangulation and, to a certain extent, in DEM generation. Thus, the solutions presented in this chapter are of more general use.

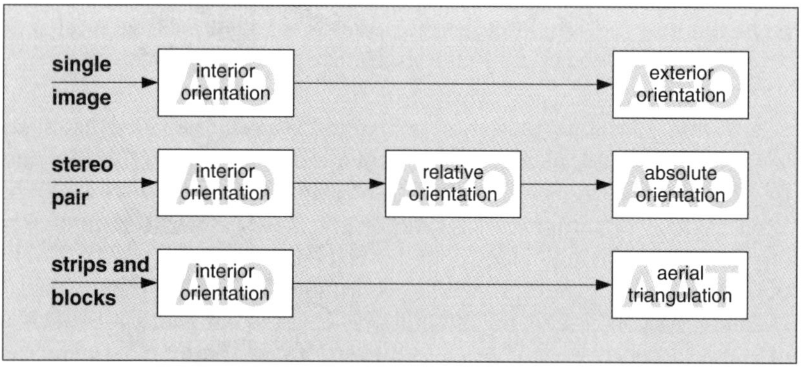

Figure 15.1: Automation chain for single images, stereo models, or strips and blocks. Chapter 13 deals with AIO and Chapter 14 with ARO. Automatic aerial triangulation is treated in the second volume of this book.

15.1 Direct vs. Indirect Orientation

Advanced direct orientation systems[1] are capable of measuring the position and attitude of the perspective center of an aerial camera (exterior orientation). The accuracy achieved allows direct georeferencing of aerial photographs that is acceptable for some mapping applications. The systems are based on a combination of GPS and inertial measurements. Direct orientation systems have been employed successfully for a variety of sensor systems, such as laser altimetry, remote sensing, and pushbroom scanners. In fact, these systems rely entirely on direct orientation, for there is no alterative way to determine the exterior orientation of the sensor. This is different in photogrammetry, where the exterior orientation of frame cameras is obtained indirectly by spatial resection. Recent advances of direct orientation systems achieve accuracies in the orientation parameters that are comparable to those obtained by indirect orientation. Some photogrammetric organizations report successful applications of direct orientation. Pundits of platform orientation systems predict the end of traditional orientation, including aerial triangulation.

The discussions about direct vs. indirect orientation are sometimes more emotional than technical. The fact that the orientation parameters can be determined with comparable accuracy is only part of the story. This section briefly summarizes a few points one must take into account when comparing direct with indirect orientation methods. Part IV in the second volume offers a more detailed treatment of the subject.

I have occasionally made the point in this book that photogrammetric processes should be viewed in context with other processes. Orientation is a good example. It is an absolute prerequisite for all subsequent processes, such as surface reconstruction and object recognition. The reconstruction of points requires the interior and exterior orientation. With reconstruction, I mean the computation of points and features in the object space from their measurements in the image space.

To illustrate the difference between the two orientation methods with respect to reconstruction, I will use the simple example of an aerial stereopair. Fig. 15.2 illustrates the reconstruction of point P in object space from its conjugate image points P' and P''. The exterior orientation defines the position of the perspective centers C' and C'', and the direction of the camera axes. In order to obtain the spatial direction of the conjugate bundle rays, the interior orientation of the camera is required. The interior orientation is essentially defined by the principal point PP, the calibrated focal length c, the radial distortion curve, and the origin of the radial dis-

[1] Also called *direct georeferencing.*

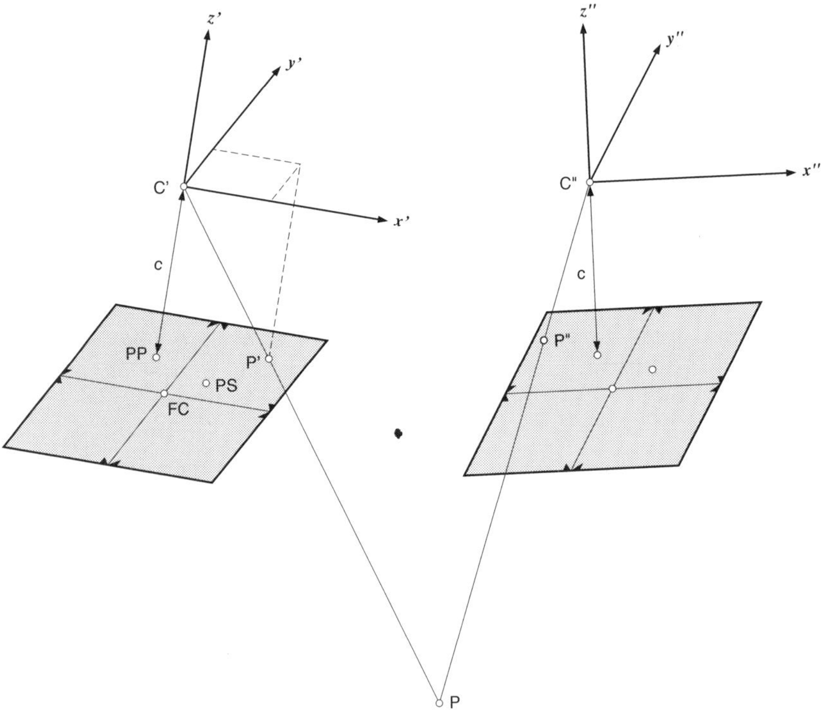

Figure 15.2: Principle of reconstruction. For reconstructing point P, its two conjugate image points are necessary, together with the interior and exterior orientation parameters of the images. The interior orientation is defined by the principal point PP, the calibrated focal length c, the radial distortion curve, and the point of origin of the radial distortion, PS.

tortion, PS (point of best symmetry).

We now examine the reconstruction of point P for the cases of direct and indirect orientation, considering different scenarios. Clearly, if the interior orientation is correct (ideal scenario), then the two methods are identical. Let us now be more realistic and consider errors in the interior orientation of frame cameras that are difficult to determine[2]. It is well

[2]An elegant solution is to include the interior orientation as additional unknowns in a block adjustment. In this case, interior and exterior orientation parameters are determined simultaneously and render an optimal reconstruction of the object space. In direct orientation, the exterior orientation parameters are observed directly and are completely independent of the interior orientation.

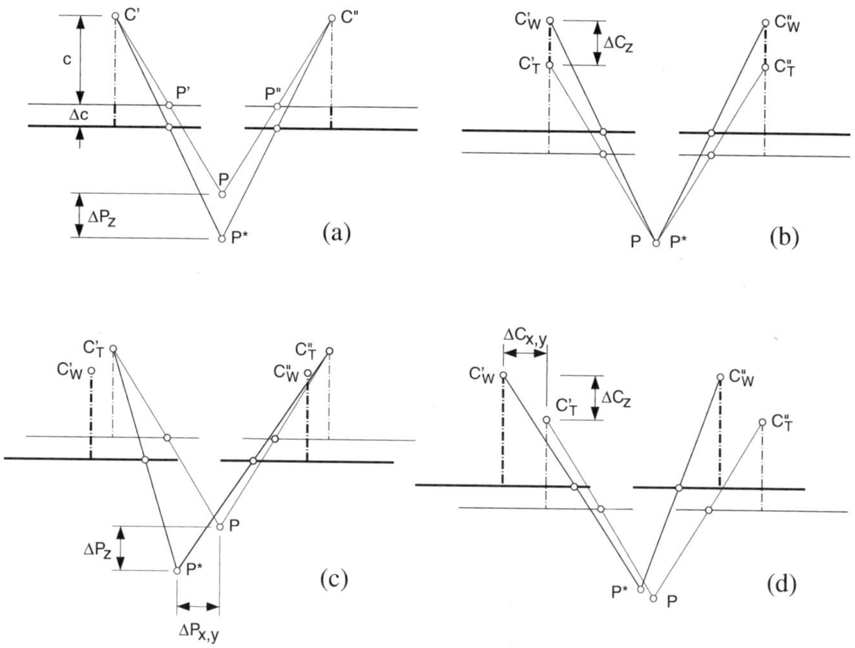

Figure 15.3: Illustration of reconstruction errors. Figs. (a) and (c) refer to reconstruction in the case of direct orientation and (b) and (d) are reconstruction examples for indirect orientation. In (a) and (b), an error, Δc, in the focal length is assumed. This error propagates into object space, ΔP_z, where it is magnified by the photo scale. In the indirect orientation, the error manifests itself as a $Z-$error of the perspective centers. This, in turn, leads to a correct reconstruction. The lower part of the figure demonstrates the effect of a shift of the principal point. Again, in indirect orientation, most of this error is absorbed by the exterior orientation (d).

known that the actual focal length frequently differs from the calibrated value. Let us for the moment call the actual focal length during the time of exposure the *true* focal length (c_T), and the calibrated focal length the *wrong* focal length (c_W). Differences in temperature and pressure during the flight mission may cause such deviations. Also, the principal point may change for the same reasons. In direct orientation systems, time synchronization problems between the GPS clock and the exact time of exposure cause translational errors that are equivalent to a shift of the principal point.

First, we examine the impact of an error, Δc, in the focal length on the reconstruction ($\Delta c = c_W - c_T$). Fig. 15.3(a) depicts this situation in the case of direct orientation. The exterior orientation is known (measured). With the true interior orientation, the reconstructed point P is, of course, correct. With the wrong interior orientation, however, the reconstruction is wrong. Since the true focal length is not known, we now obtain point $P^\star$, and a reconstruction error of $\Delta P_Z \approx \Delta c \cdot s$ results, with s the photo scale.

In contrast to the direct orientation method, in the indirect method the exterior orientation is not known. It is computed based on control points and the wrong interior orientation. As a consequence, the exterior orientation will be wrong. This is illustrated in Fig. 15.3(b) by C'_W and C''_W. An error in the focal length causes essentially an error in the Z−component of the perspective center. The reconstruction, on the other hand, is correct—the error is absorbed in the exterior orientation!

Fig. 15.3(c) and (d) illustrate a combined error in focal length and a shift of the principal point. It should be clear that the reconstruction in the direct orientation is now even more problematic. The reconstructed point $P^\star$ has error components in all coordinate components, indicated by $\Delta P_{x,y}$ and ΔP_Z in Fig. 15.3(c). In the case of indirect orientation, the reconstruction error is considerably smaller, because the exterior orientation parameters absorb a large part of the errors. The compelling conclusion is that in the presence of remaining errors and modeling errors, the indirect orientation renders a better reconstruction.

Example 15.1:

Assume a flying height of 4,500 m and a calibrated focal length of 150 mm. The real focal length at the time of exposure was 150.02 mm. We obtain a reconstruction error of −0.67 m in elevation in the case of direct orientation. The reconstruction error in indirect orientation is $[-0.02, -0.02, 0.01]^T$ *m. Adding a shift of* $[-0.01, 0.015]^T$ *mm of the principal point, the reconstruction errors are* $[-0.33, 0.33, -0.92]^T$ *m in direct orientation, and* $[-0.03, -0.17, -0.22]^T$ *in indirect orientation. These results, obtained from a simulation with a true vertical stereopair, refer to a point near the corner of the model.*

15.2 Background

15.2.1 Single Image Resection

The determination of the six exterior orientation parameters of single frame images is known as single photo resection. It is accomplished by modeling the image formation process. For frame imagery, the collinearity equations constitute a suitable mathematical model (see Sec. 14.1.2 for details). Eq. 14.3 on page 358 are the classical collinearity equations that are repeated here for convenience.

$$
\begin{aligned}
x_p &= -f\frac{(X_P - X_C)r_{11} + (Y_P - Y_C)r_{12} + (Z_P - Z_C)r_{13}}{(X_P - X_C)r_{31} + (Y_P - Y_C)r_{32} + (Z_P - Z_C)r_{33}} \\
y_p &= -f\frac{(X_P - X_C)r_{21} + (Y_P - Y_C)r_{22} + (Z_P - Z_C)r_{23}}{(X_P - X_C)r_{31} + (Y_P - Y_C)r_{32} + (Z_P - Z_C)r_{33}}
\end{aligned}
\qquad (15.1)
$$

These equations express the observed photo coordinates x_p, y_p of an object point X_P, Y_P, Z_P as a function of the exterior orientation parameters X_C, Y_C, Z_C—the location of the perspective center—and the rotation matrix **R**—the camera's attitude.

Every measured point x_p, y_p renders two equations. Assuming control points are measured in image space, the right hand side of the collinearity equations only contains the six exterior orientation parameters, because the coordinates X_P, Y_P, Z_P are known. Hence, three control points suffice to determine the exterior orientation. For n points, $n > 3$, the redundancy is $r = 2 \cdot n - 6$ and the parameters are found by an adjustment. Eqs. 15.1 are non-linear observation equations. Any standard photogrammetry textbook contains the linearization and the formulation of the adjustment procedure.

As always with non-linear adjustment problems, a critical question is how to obtain suitable approximations for the parameters to be estimated. Approximate values for the perspective centers may be available from the flight mission. A crude approach for aerial imagery is to determine the vector between the centroid of extracted features in image space and control features in object space. This would render X_C^o, Y_C^o. The third component, Z_C^o, is either found by multiplying the focal length by the estimated photo scale, or simply by calculating the flying height above ground from the flight altitude as recorded on the images, for example. Obviously, for near-vertical imagery, the two angles ω, ϕ are approximated by zero, while suitable values for κ must be derived from information from the flight mission (e.g. navigational data, flight map).

15.2.2 Absolute Orientation

With the notation introduced in Fig. 15.4, we have the following equation that describes the relationship between the model coordinate system and the object space coordinate system.

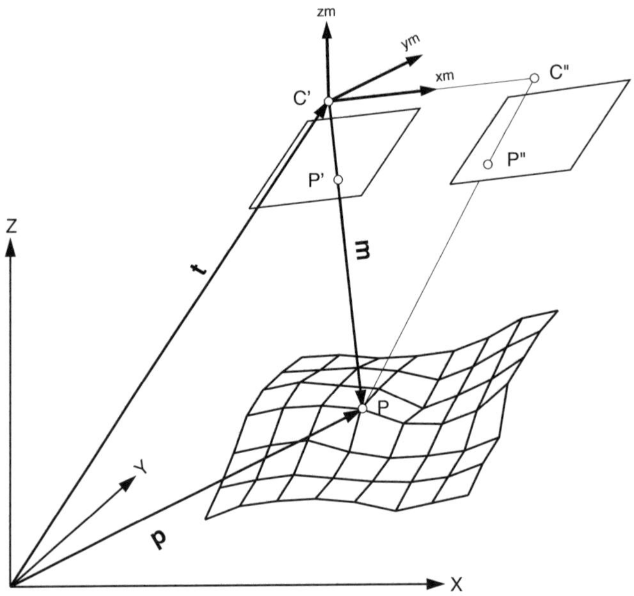

Figure 15.4: Absolute orientation of a stereomodel in the model coordinate system x_m, y_m, z_m, with respect to the object coordinate system X, Y, Z. A 3-D similarity transformation is usually employed as the mathematical model.

$$\mathbf{p} = \mathbf{S}\,\mathbf{R}\,\mathbf{m} + \mathbf{t} \tag{15.2}$$

where **S** is a scale matrix, **R** a rotation matrix, and **t** the offset between the two coordinate systems. As is normal in photogrammetry, a uniform scale is assumed, that is, $\mathbf{S} = s \cdot \mathbf{I}$ with s the scale factor. Thus, we obtain the familiar similarity (7-parameter) transformation

$$\boxed{\begin{aligned} X &= s(r_{11}x_m + r_{12}y_m + r_{13}z_m) + X_t \\ Y &= s(r_{21}x_m + r_{22}y_m + r_{23}z_m) + Y_t \\ Z &= s(r_{31}x_m + r_{32}y_m + r_{33}z_m) + Z_t \end{aligned}} \tag{15.3}$$

with $r_{11}, \ldots, r_{33}$ the elements of the 3-D rotation matrix, X_t, Y_t, Z_t the translation parameters, and s the scale factor.

The seven parameters are usually determined through an adjustment procedure, whereby the differences between transformed values and given

Table 15.1: Computing exterior orientation parameters from relative and absolute orientation.

orientation	orientation parameters	
	position	*attitude*
relative	$\mathbf{c}'_\mathrm{r}$ $\mathbf{c}''_\mathrm{r}$	$\mathbf{R}'_\mathrm{r}$ $\mathbf{R}''_\mathrm{r}$
absolute	$\mathbf{t}$	$\mathbf{R}_\mathrm{a}$
exterior	$\mathbf{c}'_\mathrm{e} = \mathbf{t}$ $\mathbf{c}''_\mathrm{e} = \mathbf{t} + \mathbf{b}$	$\mathbf{R}'_\mathrm{e} = \mathbf{R}_\mathrm{a}\mathbf{R}'_\mathrm{r}$ $\mathbf{R}''_\mathrm{e} = \mathbf{R}_\mathrm{a}\mathbf{R}''_\mathrm{r}$

values are minimized. Again, standard photogrammetric literature contains details about the linearization or alternative approaches to cast the non-linear 3-D transformation as a sequence of independent transformations for planimetry and elevation, both of which are linear.

For aerial stereopairs, a straightforward way to obtain suitable approximations is to first perform a 2-D transformation that yields X_t^o, Y_t^o, s^o, κ^o. Then, $Z_t^o \approx s^o \cdot f$ and $\omega^o \approx \phi^o \approx 0$.

15.2.3 Exterior Orientation Parameters from Relative and Absolute Orientation

The parameters from the relative and absolute orientation can be combined to obtain the 6 exterior orientation parameters. Table 15.1 summarizes the computation. As usual, the model coordinate system in relative orientation is assumed to be defined by $\mathbf{c}'_\mathrm{r} = \mathbf{0}$ and $\mathbf{c}''_\mathrm{r} = \mathbf{b}$, where the base $\mathbf{b}$ is either $[bx, 0, 0]^T$ (independent relative orientation) or $[bx, by, bz]^T$ (dependent case). Note that in the latter case $\mathbf{R}'_\mathrm{r} = \mathbf{I}$.

15.3 Automatic Orientation with Control Points

The preferred procedure for identifying control points is to target them appropriately. An alternative approach is to use topographic control points, for example road intersections. The following sections describe methods for detecting the control points automatically.

15.3.1 *Automatic Detection of Targets*

To aid the identification of control points, targets are laid out. Common target shapes include three- and four-legged crosses, occasionally "T" and "V" shapes, or circular targets. To provide a distinct image that can be reliably identified, targets must be of appropriate size and shape. Moreover, high foreground/background contrast is essential. For example, white plastic sheets work satisfactorily on dark soil and grasses, while dark felt building paper would give greater contrast on light reflective surfaces. The size of targets is a function of the photo scale. A general rule for the minimum size of the center mark is 0.01 to 0.1 mm (side of square or diameter of circle) in the image. The legs should be separated from the center mark twice this distance. Finally, the lengths of the legs should be about five times the width of the center mark.

A straightforward approach to identify and precisely measure targeted control points is by way of matching. Here, an image of the target is generated and sought in the image. Of course, you realize that this is the template matching method discussed in Chapter 11. Also, we already came across a similar problem in the interior orientation, where the fiducial mark constitutes the target that has to be found in the image.

Several problems need to be solved before we can employ, say, cross-correlation or least-squares matching to find the target in the image. For one, we ought to model the target; that is, we need suitable images of the target. To be consistent with terminology, let us call these models templates or masks. To obtain as similar a template as possible we should generate it with the actual image formation process that lead to the image in the first place, including the illumination, the reflectance properties, the topography of the surface, and the sensor. This is not quite possible, however, and generating templates is generally achieved by neglecting several aspects of image formation. The common approach generates ideal images of the target, based on an assumed image scale. In case of rotation variant matching methods, such as cross-correlation, a series of templates with different orientations must be used. Modeling targets is quite restricted, however. Note that templates do not consider geometric distortions caused by tilted surfaces, nor do they take varying foreground/background contrasts into account. As a consequence, the differences between the templates and the actual data may be considerable, resulting in low similarity measures and high failure rates.

Fig. 15.5 shows targets under various conditions. The targets in (a) and (b) contrast well with the dark background. Fig. 15.5(c) is an example of a feature that has been selected and surveyed after the flight mission. Parking lots are not recommended because it is easy to misidentify the parking space, except corner spaces as shown here. Fig. 15.5(d) and (e) are less satisfactory examples with decreased contrast and resolution. An ideal image of a target is shown in (f). It serves as the template in an area-based matching approach to determine the exact location of the control point. Note that there are considerable differences between the template and the images. This will result in a low correlation coefficient.

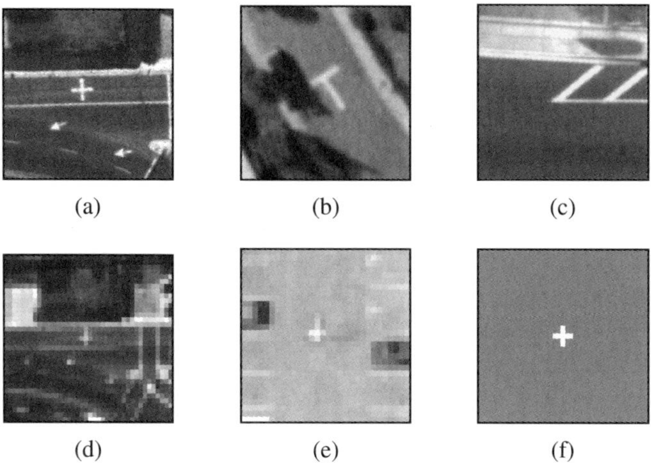

Figure 15.5: Examples of targeted ground control points. The white target in (a) contrasts well with the dark background. In (b) we have a "T" type target, also with good contrast to the background. Occasionally, ground control points are only determined after the flight mission. In (c), a corner parking space of a parking lot was selected and measured. The target in (d) is more difficult to identify because of lower resolution. Not all targets contrast well with the background as (e) demonstrates. A synthetic target is generated in (f). It serves as the template for matching with the targets shown in (a), (d) and (e). The differences between the template and the true image will result in low correlation coefficients.

To overcome some of the problems related to locating the control points by matching the image with a template, *Gülch* (1995) proposes a feature-based approach that is based on region segmentation, contour extraction, and shape analysis. The author assumes that the target is rather homogeneous and can be extracted from the image with region segmentation, followed by detecting its boundary by snakes (*Gülch* (1996)). This procedure is essentially a foreground/background separation. It allows analyzing the shape of the target which not only increases the reliability of the target identification but also its precise localization. To appreciate this point, suppose we have a cross-shaped target where one of the legs is only partially visible. As demonstrated in Chapter 13 with fiducial marks, the centroid of a corrupted instance of a template is not identical with the center of the control point. *Gülch* (1994) presents interesting experimental results, including comparisons using different methods. The experiments were conducted with image patches of 30×30 pixels. This requires very good approximations for the locations of the control points, however.

15.3.2 Automatic Detection of Topographic Control Points

Control points are not always targeted. Quite often, imagery must be oriented using points that are defined by features, such as road intersections, bridges, or any other characteristic points of linear features. This approach is widely used in remote sensing for correcting imagery, and for registering the image to the object space. The task entails two steps. First, the control feature must be identified. This is followed by determining the precise location.

Of course, the identification is the crux of the problem. Picture the scenario of Fig. 15.6(a) and (b). We have a small scale image and a map. The problem is to identify and measure suitable topographic control points, say, road intersections. You compare the map with the image, identify road intersections by way of analyzing the roads, for example, how many roads intersect and at what angle. How do you proceed if there are several candidate intersections, e.g. two road intersections at right angles? Now, you take other clues into consideration, for example, road width, paving, or even features that do not belong to the road. We just realize that the identification requires a fair amount of image interpretation. Of paramount interest now is the question of how to automate this process.

Let's pretend that the control point descriptions are available from a GIS, perhaps as a result of a query that requested intersections or what have you. Ideally, the result should also contain all relevant geometric and topological information. In analogy, we now query the image for the same information and compare the two descriptions. Querying the image is a lot more cumbersome because the raw image does not contain direct information about roads, or objects in general. All we can do at this stage is to extract elongated features that—hopefully—will correspond to roads, build road network descriptions and compare them with similar descriptions obtained from the GIS.

There is a fair amount of uncertainty involved in extracting road networks. First, we definitely will miss some roads or road segments. Moreover, we surely will extract elongated features that do not belong to roads.

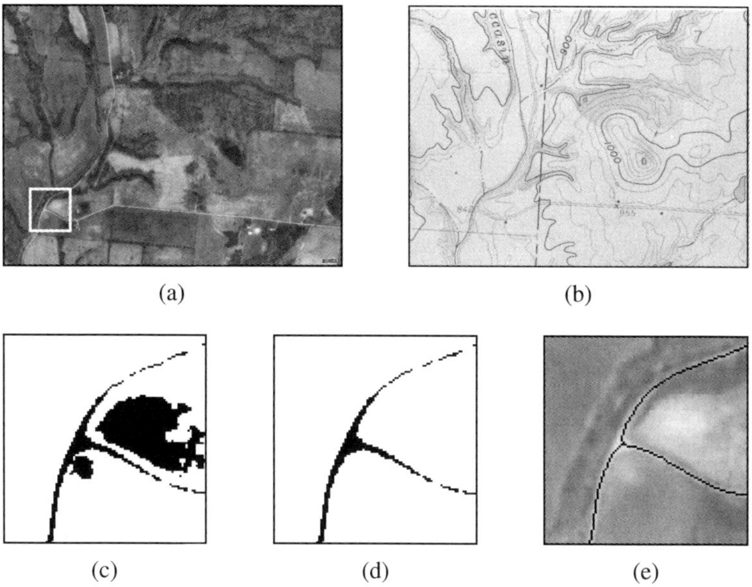

Figure 15.6: Example of orienting an image with topographic control points. The highlighted area shows a road intersection which is extracted from the image by adaptive histogram thresholding (c,d), followed by thinning (e). This is the starting point for building a structural description of the road intersection in the image, to be matched with the corresponding description of the map.

The matching method must take this into account. Relational matching has the potential for solving this intricate problem (see Sec. 11.2). *Vosselman and Haala* (1992) investigated the potential of relational matching for identifying topographic control points. Relational descriptions of features extracted from color imagery are compared with relational descriptions of the topographic control points in object space. The comparison (matching) is performed with the classical tree approach described in Sec. 11.2. Color and IR imagery help in extracting features that are known in object space.

15.4 Automatic Orientation with Control Features

The previous sections dealt with the traditional notion of a control point as an abstract point, defined either by a target or by topographic features. In this section, we abandon the point concept and use features instead.

We describe two solutions for orienting single images with linear features. The first method is based on the *trial and error approach* introduced in the previous chapter. Although this method yields the orientation parameters, its chief advantage lies in the matching of extracted features with known control features[3]. The second method describes the computation of orientation parameters with linear features as entities.

The automatic orientation of single images with features entails three distinct problems.

1. Extract linear features in the image. Hopefully, features will correspond to the control features. Hence, the extraction process should be guided by domain knowledge. For example, if the control features are trajectories of a GPS vehicle, then roads should be extracted.
2. Establish the correspondence between extracted and control features. This is a typical matching problem that must take into account shape differences arising from the projection of 3-D control features into images.
3. Determine the exterior orientation parameters of the image.

We skip the extraction problem here (see, e.g., Sec. 5.3) and explore a solution for matching extracted features with control features in the next section. This is followed by the computation of the exterior orientation parameters using a parametric representation of the control features.

15.4.1 Matching Extracted Features with Control Features

Fig. 15.6 depicts an aerial image with extracted features and a set of control features, represented in the object space coordinate system. Which portions of the extracted features correspond to the control features? Several methods that we have discussed in Chapter 11, such as relational matching or the Generalized Hough Transform may be suitable to solve the problem. In the following, we employ the trial and error method that was introduced in Sec. 14.6

We assume that the extracted features are represented by a sequence of pixels in image space. In a similar fashion, the control features are also defined by a sequence of 3-D points in object space. The trial and error approach assumes that both representations have approximately the same

[3]Control features may be obtained by tracing the trajectory of a vehicle. Such vehicles are equipped with GPS and additional sensors, such as navigation systems and cameras, to obtain positional and pictorial information along roads (mobile mapping systems). Other likely sources of control features are GIS or digitized maps.

point density. This is crucial because the matching is accomplished by examining correspondences between points in image and object space. If the control features are defined as polylines, they must be rasterized to satisfy the equal density constraint.

The collinearity equations 15.1 on page 393 define the relationship between image and object space. Three pairs of these equations suffice to determine the six exterior orientation parameters. Now imagine now a 6-D parameter space that contains all plausible solutions obtained from all possible combinations of three points in image and object space. All correct combinations would yield the true solution, represented by a single point in the parameter space. Suppose for a moment that we have just three points. We get nine independent combinations of image and object points and six different solutions, of which only one is correct. More than three points will result in a greater number of combinations and in more than one correct solution. If we were to keep track of all the solutions in the parameter space, for example by accumulator arrays, the correct solution would then emerge as a peak.

Example 15.2:

Suppose we have n edge pixels in the image and m control pixels from control features in the object space. Then, the maximum number of combinations of edge pixels with control pixels is $n \cdot m$. We need three independent pairs for obtaining a solution. Within these groups, no repetition of image or ground pixels is allowed. Thus we obtain for the total number of solutions, s

$$s = \frac{n!m!}{k!(n-k)!(m-k)!}$$

With a modest number of $n = m = 100$ and $k = 3$ we get 1.510^{11} solutions!

Example 15.2 is indeed a vivid demonstration of the combinatorial explosion. What can be done? *Habib et al.* (2000) suggest forming pairs of one image point with one object point only. Now we have $p = n \cdot m$ possible combinations, but at the cost of giving up the possibility to compute all six exterior orientation parameters simultaneously for with one pairing (2 collinearity equations) we can only determine two parameters at a time. A possible iterative sequence involves the following parameter groups: $\{X_C, Y_C\}$; $\{Z_C\}$; $\{\omega, \phi\}$; $\{\kappa\}$.

Fig. 15.7 on the next page shows the accumulator array for the first parameter group $\{X_C, Y_C\}$. The accumulator array is dimensioned according to the anticipated change of the parameters from their initial estimates.

The position of the peak indicates the correction to be applied to the approximate values X_c^o, Y_c^o. The height of the peak is a direct measure for the uniqueness of the solution.

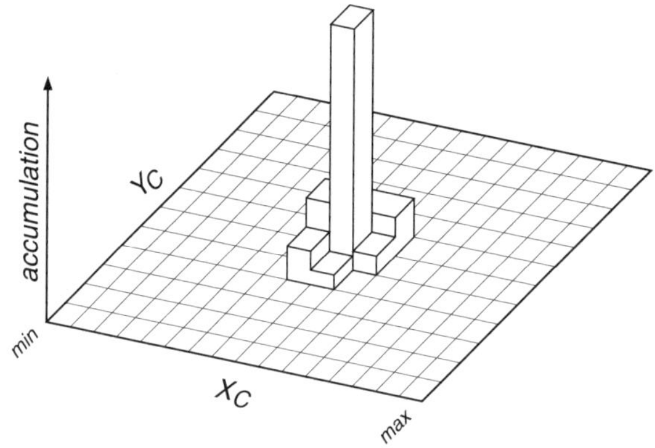

Figure 15.7: Example of a 2-D accumulator array. The position of the peak indicates the solution for the parameters $\Delta X_c, \Delta Y_c$. The array is dimensioned according to the anticipated changes of the parameters from their initial approximations X_c^o, Y_c^o. Note that a distinct peak means a unique solution.

We still have not explicitly matched image features with control features in object space. Two possibilities exist. Once a stable solution for the orientation parameters is found with the procedure just described, we repeat the entire process one more time. Since the correct solution is now known, we label every correct pair as a match. In this fashion, the list of edge pixels is augmented with the corresponding pixels of the control features.

Let us now briefly sketch an alternative approach to explicitly match image features with control features. With the known orientation parameters we can image the control features, using the collinearity equations directly. Now, the matching is performed in image space and it is based on proximity and continuity. Image features and control features should nearly overlap. In rare cases where proximity leads to ambiguous matches, the continuity principle comes to the rescue because one control feature corresponds generally to one, or occasionally to a few, image features.

15.4.2 *Computing Orientation Parameters with Control Lines*

The matching method described in the previous section also yields the exterior orientation parameters. However, the solution ultimately depends on identical points in image and object space. It is assumed that the pixels delineating a feature in image space correspond to 3-D points describing the control feature in object space. We now relax this constraint and describe a more general solution that is based on a parametric representation of the control lines, as proposed by *Zalmanson* (2000).

The parametric form of a space curve[4] can be expressed by

$$L(t) = \begin{bmatrix} X(t) \\ Y(t) \\ Z(t) \end{bmatrix} \qquad a \le t \le b$$

As illustrated in Fig. 15.8 on the following page, the space curve is described by the loci of vectors $L(t)$ as the curve parameter t assumes values in the interval $< a, b >$.

In order to express that a measured point x_p, y_p in image space corresponds to the control curve $L(t)$, the collinearity equations 15.1 on page 393 must be modified as follows:

$$\begin{aligned} x_p &= -f\frac{(X(t)-X_C)r_{11}+(Y(t)-Y_C)r_{12}+(Z(t)-Z_C)r_{13}}{(X(t)-X_C)r_{31}+(Y(t)-Y_C)r_{32}+(Z(t)-Z_C)r_{33}} \\ y_p &= -f\frac{(X(t)-X_C)r_{21}+(Y(t)-Y_C)r_{22}+(Z(t)-Z_C)r_{23}}{(X(t)-X_C)r_{31}+(Y(t)-Y_C)r_{32}+(Z(t)-Z_C)r_{33}} \end{aligned} \qquad (15.4)$$

Note that the traditional point to point relationship (image point $P' = [x_p, y_p, -f]^T$ corresponds to object point $P = [X_P, Y_P, Z_P]^T$) is abandoned. Instead, the image point corresponds to a curve. Its corresponding location on the curve is defined by curve parameter $t = t_p$. In the traditional case of exterior orientation, point $P = [X_P, Y_P, Z_P]^T$ is known (control point) and the collinearity model contains only the six exterior orientation elements as unknowns. Using control curves adds an additional unknown for every measured point—its curve parameter t.

The modified collinearity equations must be linearized for use in a linear adjustment scheme. We skip the details of the derivation and refer the

[4]The terms *lines* and *curves* are used interchangeably. In this section we use curves or *space curves* to emphasize that the procedure derived here is not restricted to straight lines. Also, we refer to the parameter used in the parametric representation of a 3-D line as the *curve parameter*, to distinguish it from the general use of the term parameter to express unknowns.

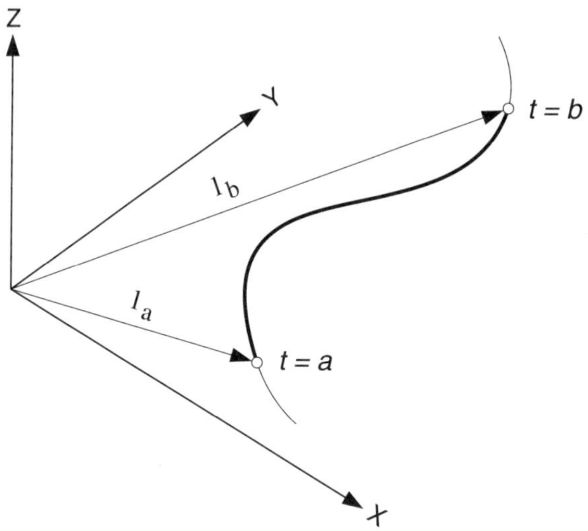

Figure 15.8: Parametric representation of a 3-D line. Curve parameter t assumes values in the interval $< a, b >$. The curve begins at $t = a$, defined by the vector $\mathbf{l}_a$ and ends at $t = b$, defined by the vector $\mathbf{l}_b$.

interested reader to *Zalmanson* (2000). Linearization requires initial values of the parameter vector. Apart from approximations for the six exterior orientation parameters, we also need initial estimates of curve parameter t^o. One way to obtain a reasonable estimate is to compute the point on the control curve that is closest to the ray defined by the perspective center X_C^o, Y_C^o, Z_C^o and point x_p, y_p (image vector) in image space. With the linearized collinearity equations, a Gauss-Markov model is established which expresses measured image points x_p, y_p as a function of the incremental parameter vector $[\Delta X_C, \Delta Y_C, \Delta Z_C, \Delta\omega, \Delta\varphi, \Delta\kappa, \Delta t]^T$.

So far we have used the condition that the image vectors must intersect the control curves. Imagine we were to measure points along the image of a control curve. Every point gives rise to two equations and increases the number of unknowns by one (curve parameter t). Every new point we add is independent—the only constraint introduced in this scheme is that its scaled image vector must lie on the control curve. Aren't we missing some additional information here? Of course, because an analytical curve does not just consist of a sequence of unrelated points—there is more independent information we can exploit for solving the orientation problem. An example of additional and independent information is the direction of the tangent[5] in object space and image space.

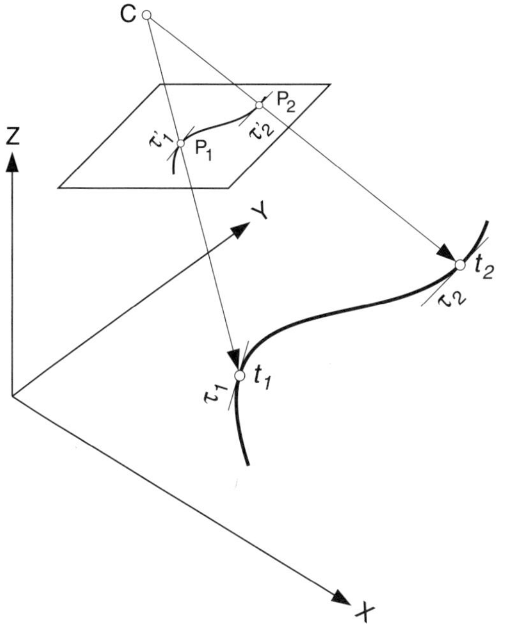

Figure 15.9: Points P_1 and P_2 are measured on the image of a control curve. The curve parameters t_1 and t_2 define the location of the measured points in object space. Additionally, the tangent directions τ_1' and τ_2' are observed. They correspond to the tangent directions of the control curve. This can be used as additional, independent information (see text for details).

The direction of the tangent at a point S of the parameterized curve is defined by $[X'(t_s), Y'(t_s), Z'(t_s)]^T$. Point S can be imaged with the current estimates of the exterior orientation parameters. The image coordinates of the projected point S depend on curve parameter $t = t_s$ as follows:

$$\begin{bmatrix} u(t) \\ v(t) \\ w(t) \end{bmatrix} = \mathbf{R} \begin{bmatrix} X(t) - X_C \\ Y(t) - Y_C \\ Z(t) - Z_C \end{bmatrix} \tag{15.5}$$

[5]Higher order derivatives of the control line contain additional independent information.

where $u(t), v(t)$ correspond to the numerators of the collinearity equations; $w(t)$ is the denominator. **R** is the rotation matrix of the exterior orientation. That is, $x(t)/f = -u(t)/w(t)$ and $y(t)/f = -v(t)/w(t)$, with f the focal length. Taking the derivatives with respect to parameter t yields

$$\begin{aligned} x'(t) &= -f\frac{u'(t)w(t) - w'(t)u(t)}{w^2(t)} \\ y'(t) &= -f\frac{v'(t)w(t) - w'(t)v(t)}{w^2(t)} \end{aligned}$$

By dividing these two equations we obtain the tangent direction in image space, expressed as a function of the tangent direction in object space and of the exterior orientation parameters:

$$\tan(\tau) = \frac{y'(t)}{x'(t)} = \frac{v'(t)w(t) - w'(t)v(t)}{u'(t)w(t) - w'(t)u(t)} \tag{15.6}$$

Eq. 15.6 must be linearized with respect to the parameter vector. Then, we obtain an observation equation that expresses the observed tangent direction as a function of the exterior orientation parameters and the curve parameter t. Suppose the image of the control curve is given as an edge, that is, as a sequence of edge pixels. With this representation, the tangent direction is readily obtained from the chain-code representation. The chain-code is the discrete version of the first derivative (first difference)—hence it gives the tangent direction directly, although only eight directions. Fig. 15.9 shows an example with two points P_1 and P_2. The tangent directions τ_1' and τ_2', respectively, can be derived from averaging a few chain-codes around P_1 and P_2, for example.

The modified collinearity equations express the positional relationship between points in image space and the control line. The tangent equation relates to the shape property. Now, for each point we have two observation equations; every tangent contributes another equation. In this case, the total number of observation equations amounts to $3 \cdot n$, where n is the number of points and tangents measured. On the other hand, there are $6 + n$ unknowns. Thus, it appears that three points with tangents just solve the problem. Let us examine this further for a moment. Suppose we have one known straight line in object space. How many points, including their tangents, are independent—or put differently, what is the information content of a straight line? Only one point and its tangent is independent[6]. This leaves us with three equations and seven unknowns. For

[6]Additional points on a straight line are determined by the parametric definition of the

single photo resection, the information deficit from using one straight line is -4. Now, take two points but no tangents, resulting in four equations and eight unknowns—again, the deficit is -4. Repeating this exercise for three different straight lines reveals that we will have just enough equations for solving the exterior orientation. Moreover, for straight lines one point and the tangent direction is equivalent to two points without directional information.

Higher order curves contain more information. For example, parabolas have a deficit of only -1. Third order polynomials have no deficit. As a consequence, the single photo resection problem can be solved with only one third order polynomial, for example by measuring three points and their tangent direction.

15.5 Automatic Orientation with Control Surfaces

So far, we have used control points and known linear features in object space for orienting single images or stereopairs. We now turn to the problem of using known surfaces—called control surfaces, in analogy to control points—to solve the absolute orientation of a relatively oriented model. Originally, the idea of using surfaces in the form of DEMs for orienting models was suggested by *Ebner and Strunz* (1988). The approach is based on minimizing the Z−differences between model points—subject to an unknown transformation to object space—and points in object space found by DEM interpolation. Subsequent studies demonstrated the applicability of the method (*Ebner et al.* (1991), *Ebner and Ohlhof* (1994)).

The method described in this section is based on minimizing distances along surface normals, instead of minimizing Z−differences. This is quite important because, for a robust solution, surfaces with different slopes are required. The Z−minimization is not suitable for differently tilted surfaces. Take the extreme example of a vertical surface where Z−differences are meaningless.

In the following we summarize the most salient features of the method. More details are provided in *Schenk* (1999) or *Jaw* (1999).

15.5.1 Problem Statement

Let $S_1 = \{\mathbf{p}_1, \mathbf{p}_2, \ldots, \mathbf{p}_n\}$ be a surface described by n discrete points $\mathbf{p}$ that are randomly distributed. Let $S_2 = \{\mathbf{q}_1, \mathbf{q}_2, \ldots, \mathbf{q}_m\}$ be a second surface described by m randomly distributed points $\mathbf{q}$. Suppose that the two sets, in

line. Hence, they are not independent. Regardless of how many points and tangents we measure on a straight line, the rank deficiency remains -4.

fact, are describing the same surface but in different reference systems. In our absolute orientation problem, set S_2 is the model system and set S_1 is referenced in an object space system. After proper transformation we have $S_1 = S_2$, except for differences due to random errors of the observed points **p** and **q**. Yet another difference may arise from the discrete representation of the surfaces, for example, $n \neq m$. Even in cases where $n = m$, the different distribution may cause a differently interpolated surface. Suppose further that no points in the two sets are known to be identical (same surface point). The problem is now to establish a transformation between the two sets such that the two surfaces S_1 and S_2 become as similar as possible in terms of closeness and shape.

15.5.2 Solution

The problem described is cast as an adjustment problem where the second set of points **q** is transformed to the first set such that the differences between the two surfaces are minimized. Additionally, the orientation of surface normals between S_1 and S_2 can also be minimized. Minimizing the distances assures the best positional fit while minimizing differences in surface normals assures the best shape fit.

Since the problem is non-linear, approximations are needed. For the following discussion it is assumed that reasonable approximations exist.

Geometric Considerations

The following considerations help us to understand what conditions must exist to solve the transformation problem. Let us assume that the absolute orientation is solved by a 3-D similarity transformation involving three translation parameters x_t, y_t, z_t, three rotation angles ω, φ, κ, and one scale factor s.

Suppose S_1 is a horizontal plane. With one point of set **q** we can solve for z_t. With two additional points, the two rotation angles ω and φ are determined. Obviously, the farther apart the points, the better the solution for the rotation angles. Regardless of how many more points we use, the remaining four parameters cannot be determined because we can shift, rotate and scale the two planes with respect to each other without changing their closeness. To solve for the two translation parameters x_t, y_t, two planes parallel to the coordinate planes are needed. To determine the scale, a fourth plane is needed, ideally parallel to one of the other planes. Note that points in vertical planes also solve for the κ-rotation.

In conclusion, the ideal configuration is a surface that consists of one horizontal plane and three vertical planes. Moreover, one pair of vertical planes should be perpendicular while the other pair should be parallel. The distances to be minimized between the two planes should be taken along surface normals. From this it is clear that minimizing elevation differences does not work satisfactorily.

Topographic surfaces hardly ever have vertical planes. Consequently, solving the problem for topographic surfaces will not result in an ideal solution. To assure a reasonable determination of all parameters, as many sloping surfaces as possible must be used. Moreover, the surface normals should be oriented in as many directions as possible.

From the above considerations it also follows that only small surface patches are needed for computing the distances. For example, the three points needed in the horizontal plane may be on three small horizontal surface patches that are not related to each other.

Mathematical Model

Let the model points $\mathbf{q}$ be transformed into the object space by a 3-D similarity transformation

$$\mathbf{q}' = s\mathbf{R}\mathbf{q} - \mathbf{t} \tag{15.7}$$

The observation equations are defined by the shortest distance from $\mathbf{q}'$ to the surface S_1. Two scenarios are feasible for expressing surface S_1. Let us first approximate S_1 in the neighborhood of $\mathbf{q}'$ by a plane, for example by fitting a plane through points $\mathbf{p}$ confined to a small spatial extent, called surface patch SP_q thereafter. Then, the shortest distance d from $\mathbf{q}'$ to the surface patch, expressed in Hessian normal form, using the three directional cosines and the distance p from the origin, is

$$d = \mathbf{q}' \cdot \mathbf{h} - p \tag{15.8}$$

with $\mathbf{h} = [\cos\alpha, \cos\beta, \cos\gamma]^T$

Thus, Eq. 15.9 expresses the observation equation for point $\mathbf{q}$

$$r = (s\mathbf{R}\mathbf{q} - \mathbf{t}) \cdot \mathbf{h} - p - d \tag{15.9}$$

This equation must be linearized with respect to the seven transformation parameters. Since approximations are needed, relatively small rotation angles can be anticipated. Thus, the correct rotation matrix can be approximated by differential angles

$$\mathbf{R} \approx \begin{vmatrix} 1 & -d\kappa & d\varphi \\ d\kappa & 1 & -d\omega \\ -d\varphi & d\omega & 1 \end{vmatrix} \tag{15.10}$$

With the partial derivatives of **R**

$$\frac{\partial \mathbf{R}}{\partial d\omega} = \begin{vmatrix} 0 & 0 & 0 \\ 0 & 0 & -1 \\ 0 & 1 & 0 \end{vmatrix} \quad \frac{\partial \mathbf{R}}{\partial d\varphi} = \begin{vmatrix} 0 & 0 & 1 \\ 0 & 0 & 0 \\ -1 & 0 & 0 \end{vmatrix} \quad \frac{\partial \mathbf{R}}{\partial d\kappa} = \begin{vmatrix} 0 & -1 & 0 \\ 1 & 0 & 0 \\ 0 & 0 & 0 \end{vmatrix}$$

the linearized observation equations are

$$\begin{aligned} \frac{\partial}{\partial \omega} &= -z\cos\beta + y\cos\gamma \\ \frac{\partial}{\partial \varphi} &= z\cos\alpha - x\cos\gamma \\ \frac{\partial}{\partial \kappa} &= -y\cos\alpha + x\cos\beta \\ \frac{\partial}{\partial s} &= (x - y \cdot d\kappa + z \cdot d\varphi)\cos\alpha + \\ & \quad (x \cdot d\kappa + y - z \cdot d\omega)\cos\beta + \\ & \quad (-x \cdot d\varphi + y \cdot d\omega + z)\cos\gamma \\ \frac{\partial}{\partial x_t} &= -\cos\alpha \\ \frac{\partial}{\partial y_t} &= -\cos\beta \\ \frac{\partial}{\partial z_t} &= -\cos\gamma \end{aligned} \tag{15.11}$$

If the points **p**, contained in the surface patch SP_q, cannot be satisfactorily approximated by a plane, then a second order surface can be used. Should this also fail, then SP_q is not suitable (not smooth enough) for the proposed procedure and no observation equation is formed for this particular point **q**. If it can be sufficiently approximated, then the situation depicted in Fig. 15.10 applies. The distance from **q** to the surface is measured along the surface normal. Hence, we need point **v** on the interpolated surface SP_q. Since the surface normal in **v** passes through **q**, the distance between these two points is in fact the shortest distance from **q** to SP_q. The distance d is then simply determined by

$$d^2 = (x_v - x_q)^2 + (y_v - y_q)^2 + (z_v - z_q)^2 \tag{15.12}$$

Assuming that minimizing d^2 instead of d leads to the same result for the parameters we have for the observation equation

$$r = (x_v - x_q)^2 + (y_v - y_q)^2 + (z_v - z_q)^2 - d^2 \tag{15.13}$$

Eq. 15.13 must be linearized with respect to the transformation parameters.

Let us now turn to the problem of determining **v**. This can be solved iteratively in the following fashion. First, a few surface normals in surface patch SP_q are determined. The directional derivatives of SP_q in the selected locations are the components of the normals. Since we want smooth surface patches, the direction of the normals changes smoothly. Next, an average surface normal is determined. Its direction and point **q** define a straight line that intersects SP_q at a point near **v**. This location can be used to compute an improved direction and the procedure is repeated. It is not extremely critical to find **v**. Approximate locations lead to a slightly larger distance d but the minimization process still yields good transformation parameters.

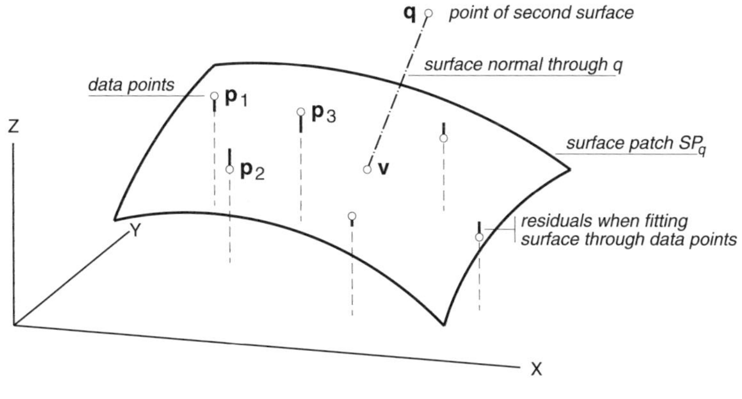

Figure 15.10: Illustration of determining the shortest distance between point **q** and surface patch SP_q.

15.5.3 Procedure

For every point **q** on surface S_2 do the following:

1. Pick the nearest points **p** on surface S_1. These points define surface patch SP_q.

2. Approximate surface patch SP_q either by a plane or by a second order polynomial. If it cannot be approximated well enough (e.g. residuals too large), skip this point and proceed with the next one.

3. With the approximated surface patch compute the nearest distance and form an observation equation (Eq. 15.9 or 15.13).

4. Store the surface normal of surface patch SP_q.

5. After all points have been processed, analyze the surface normals, for example by generating a histogram. A solution should only be attempted if the distribution is somewhat similar to that outlined in Section 15.5.2.

6. Build normal equations and solve for the parameters. The residuals, for example the weighted square sum, give a good measure for the closeness of the two surfaces.

15.6 Summary

The exterior orientation of single images (single photo resection) and stereopairs (absolute orientation) establishes a relationship between image and object space by way of a suitable transformation. To determine the transformation parameters, known points in object space (ground control points) are measured in the image or model. If the control points are determined prior to the flight, targets are laid out for better identification in the images. Another approach is to select points after data acquisition and to identify and measure them in the object space.

In the attempt to automate this traditional procedure we have encountered formidable obstacles. The crux of the problem is the identification because targets are rather small and their images are often corrupted. Even experienced photogrammetrists sometimes have problems uniquely identifying control points—a sure case for disaster with an automated system. The automatic identification and precise localization of targets is performed by area-based matching methods: an ideal image of the target is matched against the image. For this method to succeed, very good approximations must be available.

The preferred way to orient single images in remote sensing is by topographic control points. Here, features such as roads are extracted and matched with their object space representation that might be available from a GIS. Compared to targets, the subimages that contain the control points are much larger, thereby facilitating feature extraction. On the other hand, it is quite unlikely that exactly the same features that are known in object space are extracted. Thus, the matching is faced with another challenge that can probably best be met by relational matching. The transformation is still performed with points. For example, if roads serve as topographic features, then road intersections are used as control points.

Instead of using points for the computation of orientation parameters, we have argued for using features, such as lines and surfaces. We demonstrated the use of natural features for orienting single images, and also presented a new way of matching extracted linear features with their known representation in object space. The trial and error method is based on building all plausible associations of edge pixels with ground feature pixels and storing the respective solutions in an accumulator array. Then the correct solution emerges as a peak in the accumulator array. The combinatorial explosion is avoided by solving the parameters sequentially. This approach yields the solution in an iterative fashion. We have also described a direct solution by extending the collinearity model to accommodate parameterized lines and curves.

Another example of using features for computing orientation parameters was demonstrated in the classical absolute orientation. Here, the common information between model and object space is surfaces. Known surfaces in object space may come from laser altimetry, existing DEMs, or other sources. Since the distance between corresponding surfaces is minimized, no explicit point to point association is necessary. The method is also suitable for comparing surfaces sampled by points at different locations.

"From points to features" is the common theme of the third part of this book. We have emphasized it in this chapter by including linear and areal features for computing exterior orientation parameters. Features are easier to extract and to deal with in an automatic setting than points. Moreover, features carry more meaningful information about the scene. Thus, it makes sense to include features at as early a stage as possible.

Problems

1. The common theme of this part is "from points to features". How are features used for determining the exterior and absolute orientation parameters? What role do points play in the computation?

2. One approach for detecting targets of control points in an image is to match a template with an image. Elaborate on the problems of this approach. What is the advantage of LSM over cross-correlation?

3. In Sec. 15.4 we have examined the problem of using linear features

for orienting images and stereomodels. What is the motivation for the "trial and error" method that attempts to find the orientation parameters without matching?

4. Describe a strategy for determining suitable dimensions and tesselation for the accumulator array in the "trial and error" method of establishing the exterior orientation of a digital image.

5. Another approach to using control features for the orientation of images is based on a parametric description of the control feature. Suppose such a control feature is described by a second order polynomial. How many independent equations do we obtain in this case?

6. Explain why the exterior orientation cannot be solved with the conditions stated in the previous question, regardless of how many points we measure.

7. Suppose we want to solve for the absolute orientation with control surfaces as described in Sec. 15.5. How many different surfaces are necessary for establishing the seven orientation parameters?

8. Why is it important to minimize the distance between a point of one surface parallel to the surface normal of the second surface when using control surfaces?

9. Elaborate on the practical aspects of using control surfaces for establishing the absolute orientation.

References

Ebner, H. and T. Ohlhof (1994). Utilization of Ground Control Points for Image Orientation without Point Identification in Image Space. In *International Archives of Photogrammetry and Remote Sensing*, **30**(3/1), 206–211.

Ebner, H., G. Strunz and I. Colomina (1991). Block Triangulation with Aerial and Space Imagery Using DTM as Control Information. In *ACSM-ASPRS Auto-Carto 10*, **5**, 76–85.

Ebner, H. and G. Strunz (1988). Combined Point Determination Using Digital Terrain Models as Control Information. In *International Archives of Photogrammetry and Remote Sensing*, **27**(B11/3), 578–587.

Gülch, E. (1994). Using feature extraction to prepare the automated measurement of control points in digital aerial triangulation. In *International Archives of Photogrammetry and Remote Sensing*, **30**(3/1), 333–340.

Gülch, E. (1995). From control points to control structures for absolute orientation and aerial triangulation in digital photogrammetry. *Zeitschrift für Photogrammetrie und Fernerkundung*, **63**(3), 130–136.

Gülch, E. (1996). Deformable models as a photogrammetric measurement tool—potential and problems. In *International Archives of Photogrammetry and Remote Sensing*, **31**(B3), 279–284.

Habib, A., A. Asmamaw, D. Kelley, and M. May (2000). *Linear Features in Photogrammetry.* Report No. 450, Department of Civil and Environmental Engineering and Geodetic Science, The Ohio State University, Columbus, OH 43210.

Jaw, J.J. (1999). *Control Surface in Aerial Triangulation.* PhD dissertation, Department of Civil and Environmental Engineering and Geodetic Science, The Ohio State University, Columbus, OH 43210.

Schenk, T. (1999). *Matching Surfaces.* Technical Notes in Photogrammetry, No. 15, Department of Civil and Environmental Engineering and Geodetic Science, The Ohio State University, Columbus, OH 43210, 21 pages.

Vosselman, G. and N. Haala (1992). Erkennung topographischer Passpunkte durch relationale Zuordnung. *Zeitschrift für Photogrammetrie und Fernerkundung*, **60**(6), 170–176.

Zalmanson, G. (2000). *Hierarchical recovery of exterior orientation from parametric and natural 3-D curves.* PhD dissertation, Report No. 452, Department of Civil and Environmental Engineering and Geodetic Science, The Ohio State University, Columbus, OH 43210.

Author Index

P

R

S

T

V

W

X

Y

Z

Subject Index

L

M

N

O

P